A Geography of the Carolinas

D. Gordon Bennett
and Jeffrey C Patton, Editors

2008
Parkway Publishers, Inc.

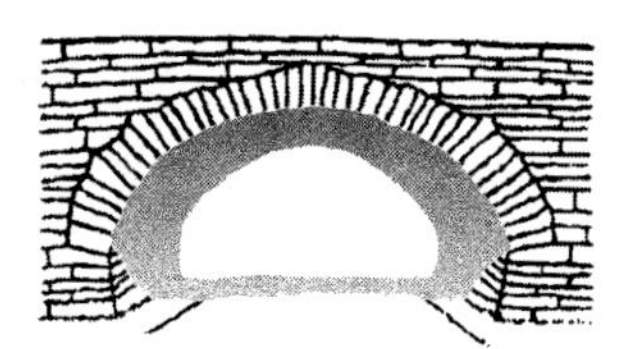

Available from:
Parkway Publishers, Inc.
P. O. Box 3678, Boone, North Carolina 28607
Telephone/Facsimile: (828) 265-3993
www.parkwaypublishers.com

Library of Congress Cataloging-in-Publication Data

A geography of the Carolinas / D. Gordon Bennett and Jeffrey C. Patton, editors.
p. cm.
Includes bibliographical references and index.
ISBN 978-1-933251-43-1
1. North Carolina--Geography. 2. South Carolina--Geography. I. Bennett, D. Gordon (David Gordon), 1941- II. Patton, Jeffrey C.
F254.8.G46 2007
917.56--dc22

2007011760

Book and Cover Design by Aaron Burleson, spokesmedia

Table of Contents

List of Figures, Photos, Tables & Appendices iv

Preface viii

Acknowledgments ix

Chapter 1: Introduction 1
D. Gordon Bennett (University of North Carolina at Greensboro)

Chapter 2: Changing Landforms and Rivers 13
Jeffrey C. Patton (University of North Carolina at Greensboro)

Chapter 3: Climate, Vegetation and Soils 35
Gregory Carbone (University of South Carolina at Columbia)
and John J. Hidore (University of North Carolina at Greensboro)

Chapter 4: Settlement Geography of the Carolinas Before 1900 53
John J. Winberry (University of South Carolina at Columbia)
and Roy S. Stine (University of North Carolina at Greensboro)

Chapter 5: The Evolving Urban and Economic Structure since 1900 95
Ole Gade (Appalachian State University)

Chapter 6: Agriculture and Forestry 141
John Fraser Hart (University of Minnesota)

Chapter 7: The Population of the Carolinas 161
Melinda Meade (University of North Carolina at Chapel Hill)

Chapter 8: Tourism and Recreation in the Carolinas 187
Robert L. Janiskee (University of South Carolina at Columbia)

Chapter 9: Higher Education in the Carolinas 215
Roger A. Winsor (Appalachian State University)

Chapter 10: Planning for the Future 241
D. Gordon Bennett (University of North Carolina at Greensboro)

Index 251

List of Figures

1.1. Movement of Continents and Geologic Plates
1.2. Early Exploration and Settlement in the Carolinas
1.3. Eighteenth Century Settlement and Conflict
1.4. Physiographic Regions of the Carolinas
2.1. A Subduction Zone and Associated Volcanic Activity
2.2. The Collision and Suturing of Continents
2.3. Physiographic Provinces of the Carolinas and Major Features
2.4. Sounds and Outer Banks of North Carolina.
2.5. Distribution of Carolina Bays
2.6. The Sandhills Region
2.7. Development of the Fall Line
2.8. Major Geologic Belts of the Carolina Piedmont
2.9. The Mountains of Western Carolina
2.10. Major Rivers of the Carolinas
2-11. Drainage Pattern of the Carolinas
3.1. July 30-Year Average Temperature (1961-1990)
3.2. Location Of Bermuda High
3.3. 30-year Precipitation Average (1961-1990)
3.4. January 30-year Average Temperature (1961-1990)
3.5. Representative Counties in Five Regions of the Carolinas
4.1. The 4 Major Language Groups of Native Americans
4.2. Early Settlement and Agricultural Regions
4.3. South Carolina Townships
4.4. Crop Regions
4.5. Antebellum Railroads
5.1. Carolina Metropolitan Statistical Areas (as of December, 2006)
5.2. Cotton Spindles, 1925
5.3. Plant Closings in the Carolinas, 2000-2006
5.4a. Textile for Apparel Employment in North Carolina, 1995
5.4b. Textile for Apparel Employment in North Carolina, 2005
5.5a. Manufacturing Employment as a Percent of the Total Civilian Labor Force, 1980
5.5b. Manufacturing Employment as a Percent of the Total Civilian Labor Force, 2004-2006
5.6. Total Manufacturing Employment Change, South Carolina, 2000-2004
5.7. Carolina Main Street and Other Urbanized Carolina Regions, 2006
5.8. Municipal Population Density Change for a Selection of Cities, 2000-2005
5.9. Carolina In-Commuters, by County, 2000
5.10a. Mecklenburg County In-Commuters, 1990
5.10b. Mecklenburg County In-Commuters, 2000
5.11. Unemployment by County, 2005 Annual Average
5.12. Per Cent Population Change, By County, 2000-2005
6.1. Major Physical Subdivisions of the Carolinas
6.2. Cropland Harvested as a Percentage of Total Area, 2002
6.3. Major Land Uses in North Carolina and in South Carolina, 1920-2002
6.4. Average Size of Farm in the Carolinas, 1920-2002
6.5. Acreage of Major Crops in North Carolina and in South Carolina, 1920-2002
6.6. Sales of Major Agricultural Commodities, North Carolina and South Carolina, 1949 and 2002
6.7. Tobacco, 2002
6.8. Soybeans, Corn, and Wheat, 2002
6.9. Poultry Sales, 2002
6.10. Sales of Hogs, 2002
6.11. The Regeography of Cotton
6.12. Peanuts as a Percentage of Harvested Cropland, 2002
6.13. Cropland Harvested per Farm, 2002
6.14. Cropland Harvested in 2002 as a Percentage of Cropland Harvested in 1949
6.15. Forage Crops as a Percentage of Harvested Cropland, 2002
6.16. Sales of Cattle and Calves, 2002
6.17. Sales of Dairy Products, 2002
6.18. Sales of Nursery and Greenhouse Products, 2002
6.19. Percentage of Farm Operators in 2002 Whose Principal Occupation was Not Farming
6.20. Primary Wood-using Mills, 1994
6.21. Forest Land and Forest Industry Land
6.22. Net Cash Farm Income, 2002
7.1. Total Population, 2000

List of Figures

7.2. Population Density, 2000
7.3. Percent Population Growth, 1990-2000
7.4. Metropolitan Statistical Areas, 2000
7.5. Metropolitan Growth; 1980, 1990, 2000
7.6. Percent of Population White, Non-Hispanic, 2000
7.7 Percent of Population Black (African-American), 2000
7.8a. Percent of Population American Indian, 2000
7.8b. Percent of Population Asian, 2000
7.8c. Percent of Population Hispanic, 2000
7.9a. Percent of population 65 and over, 2000
7.9b. Percent of Population under 5, 2000
7.10. Racial Composition of the U.S. and Carolinas, 2000
7.11. Age Structure of the Carolinas and of the U.S., 2000
7.12. Proportionate Population Growth in the 20th Century for Whites and Nonwhites
7.13. Physiographic Regions
7.14. Racial Composition for Sample Counties, 2000
7.15. Age Structure in Sample Counties.
8.1. Select Tourist Destinations
9.1. Diffusion of Existing Four-Year Colleges and Universities
9.2. Four-Year Historically Black Colleges
9.3. Ethnicity at Private Four-Year Colleges and Universities
9.4. Ethnicity at Public Four-Year Colleges and Universities
9.5. Public and Private Higher Education Institutions
9.6. Diffusion of Existing Two-Year Colleges
9.7. Distribution and Total Full Time Enrollment and Grand Total of All Student Enrollment at Public Community, Technical and Two-Year Colleges, Fall 1996
9.8. Research Triangle Park
9.9. UNCC's University Research Park
9.10. South Carolina Research Authority's Four Sites
10.1. Employment in Major Industries in the Carolinas, 1980-2004
10.2. Employment in Major Sectors of the Economy in the Carolinas, 1980-2004

List of Photographs

1.1. A View of Salem, Van Redken, 1787 (Courtesy of Old Salem)
1.2. Hanging Rock in Stokes County, N.C. (Courtesy of Jeff Patton)
1.3. A Scene from the Outdoor Play *Unto These Hills* (Courtesy of the Cherokee Historical Association)
2.1. Drum Inlet. The Pattern of Sand Deposition (Courtesy U.S. Dept. of Agriculture)
2.2. The Pattern of the Sand Deposits as a Result of Storms (Courtesy U.S. Dept. of Agriculture)
2.3. Infrared Landsat Satellite Image of Many Flooded River Valleys of the South Carolina Coast (Courtesy NASA)
2.4. Aerial Photograph of Several Carolina Bays in Bladen County, N.C. (Courtesy U.S. Dept. of Agriculture)
2.5. Pilot Mountain, N.C. (Courtesy Robin Lynch Bennett)
2.6. Stone Mountain, N.C.
3.1. Spartina Grasses in Expansive Salt Marshes
3.2. Sea Oats Dominate Sand Dune Vegetation
3.3. A Typical Sandhills Landscape
3.4. Mixed Forest in the Piedmont
3.5. Mixed Pine and Hardwood Forest of the Lower Blue Ridge Mountains
4.1. The Fort at the Site of the First English Settlement at Manteo
4.2. The Engine House on Upper Hill at Reed Gold Mine (Courtesy of the North Carolina Historic Sites and the North Carolina Division of Archives and History)
4.3. Fort Sumter
4.4a & 4.4b. Tenant Shack and Farm
8.1. The Beach at Myrtle Beach State Park.
8.2. Myrtle Beach's Broadway at the Beach. (Courtesy Myrtle Beach Area Chamber of Commerce)
8.3. Harbour Town Marina at Sea Pines Plantation, Hilton Head, SC. (Courtesy Hilton Head Island Convention and Visitor Bureau)
8.4. Cape Hatteras National Seashore and the Hatteras Lighthouse, the Nation's Tallest. (shown before being moved inland)
8.5. Chimney Rock. (Courtesy Chimney Rock Park)
8.6. Linn Cove Viaduct Carrying the Blue Ridge Parkway Across the Southeastern Flank of Grandfather Mountain. (Courtesy Daniel Stillwell)
8.7. Charlotte's Mint Museum of Art in Former U.S. Mint Building. (Courtesy Mint Museum of Art)
8.8. Congaree River, SC.
8.9. Hunters After Dove Shoot in South Carolina Low Country.
8.10. Commercial Campground at Myrtle Beach, "Seaside Camping Capital of the World." (Courtesy South Carolina Parks, Recreation and Tourism)
8.11. Okra Strut Festival in Irmo, SC.
8.12. Cockaboose Railroad Tailgating Stadium-side at the University of South Carolina in Columbia.
8.13. Carolina Hurricanes Game. (Courtesy Carolina Hurricanes)
10.1. Converting the Rural Fringe of an Urban Area to Commercial Uses (Courtesy of U.S. Department of Agriculture)
10.2. Part of the Redeveloped Waterfront District in Wilmington, N.C.

List of Tables

2.1. The Geologic Time Scale and Associated Events in the Formation of the Carolinas
5.1. Carolina Metropolitan Statistical Areas, 1990-2005 (Dec. 2005 definition)
5.2. Carolina Incorporated Places of 30,000 and Over in 2005 (Estimates)
5.3. Air Passenger Volume at Larger Carolina Airports, 1973-2005
5.4. North Carolina Motor Vehicle and Heavy Equipment Cluster by Industry SIC, 2005
5.5. North Carolina's Largest Private Employer in Order of Employment Size, 2006
5.6. Largest Banks in the United States, December, 2006
5.7. Leading Carolina Financial Institutions by Consolidated Assets, 2006
5.8. Civilian Labor Force and Unemployment, by MSA, 2005 and 2006
7.1. 20th Century Demographic Measures
7.2. Socioeconomic Characteristics and Population in 2000
7.3. Population Characteristics of Wilkes County, N.C.
7.4. Population Characteristics of Tyrell County, N.C. and Charleston County, S.C.
7.5. Population Characteristics of Lenoir County, N.C. and Dillon and Orangeburg Counties, S.C.
7.6. Population Characteristics of Mecklenburg, Durham and Chatham Counties, N.C. and Greeneville County, S.C.
9.1 Resident Credit Headcount Enrollments at Four-Year Colleges and Universities, Fall, 2005
9.2 Largest Resident Credit Headcount Enrollments at Four-Year Colleges and Universities, Fall, 2005
9.3 Total Full Time Undergraduate Enrollment and Grand Total of All Student Enrollment at Public Community, Technical and Two-Year Colleges, Fall, 2005
9.4 Headcount Enrollments at Largest Community, Technical and Two-Year Colleges, Fall, 2005

List of Appendices

9.1. Chronological Establishment of Four-Year Colleges and Universities
9.2. Chronological Establishment of Four-Year Historically Black Colleges
9.3. Chronological Establishment of Two-Year Colleges

PREFACE

This book is the outgrowth of the earlier volume, *Snapshots of the Carolinas: Landscapes and Cultures,* a series of vignettes and field trip guides on the region, which was prepared by numerous geographers at universities mainly in the Carolinas for the 1996 annual meeting of the Association of American Geographers. This book expands on several of the topics covered in the earlier volume and presents a broad updated overview of the major geographical aspects of the region. Geographers and laypersons (both residents and visitors in the Carolinas) are presented with a general background of the region, including the landforms, weather and climate, historical development, changing agricultural, manufacturing and services economy, tourism and recreation, and higher education of these two states.

The contributors to this volume are noted geographers who share a wealth of experience and knowledge of the Carolinas. While working at major universities in the Carolinas and the nation for more than three decades, all of the contributors have spent their careers studying the changing geographical patterns of the region. Much of the material presented here has drawn upon the earlier works of those who contributed to the initial *Snapshots* volume. Hopefully, this new look at the Carolinas in the early twenty-first century, including a glance at how the region evolved historically to its current status, will serve well the citizens, visitors, and prospective migrants of the Carolinas who are seeking a better understanding of this region.

ACKNOWLEDGMENTS

The editors of this volume wish to thank their colleagues who contributed substantial insights into the various geographical aspects of the Carolinas and who generously agreed to donate their recompense to the Southeastern Division of the Association of American Geographers for scholarship money for a graduate student to use to defer the cost of attending the annual meeting of the Division. The editors and contributors are also appreciative of the initial cartographic work by several individuals for various chapters, which is specifically acknowledged in each one. We wish to offer our special thanks to Rebecca Roush Sanderson, who did the early cartographic editorial work for the volume. Many thanks are also expressed to Jim Nelson, our GIS Lab Director, and to Lois Carney for their technical computer assistance. Finally, the editors and contributors are highly appreciative of the helpful comments made by the anonymous reviewer of the manuscript. Many thanks are also expressed to Nancy Ryckman, geography librarian, for her thoughtful suggestions.

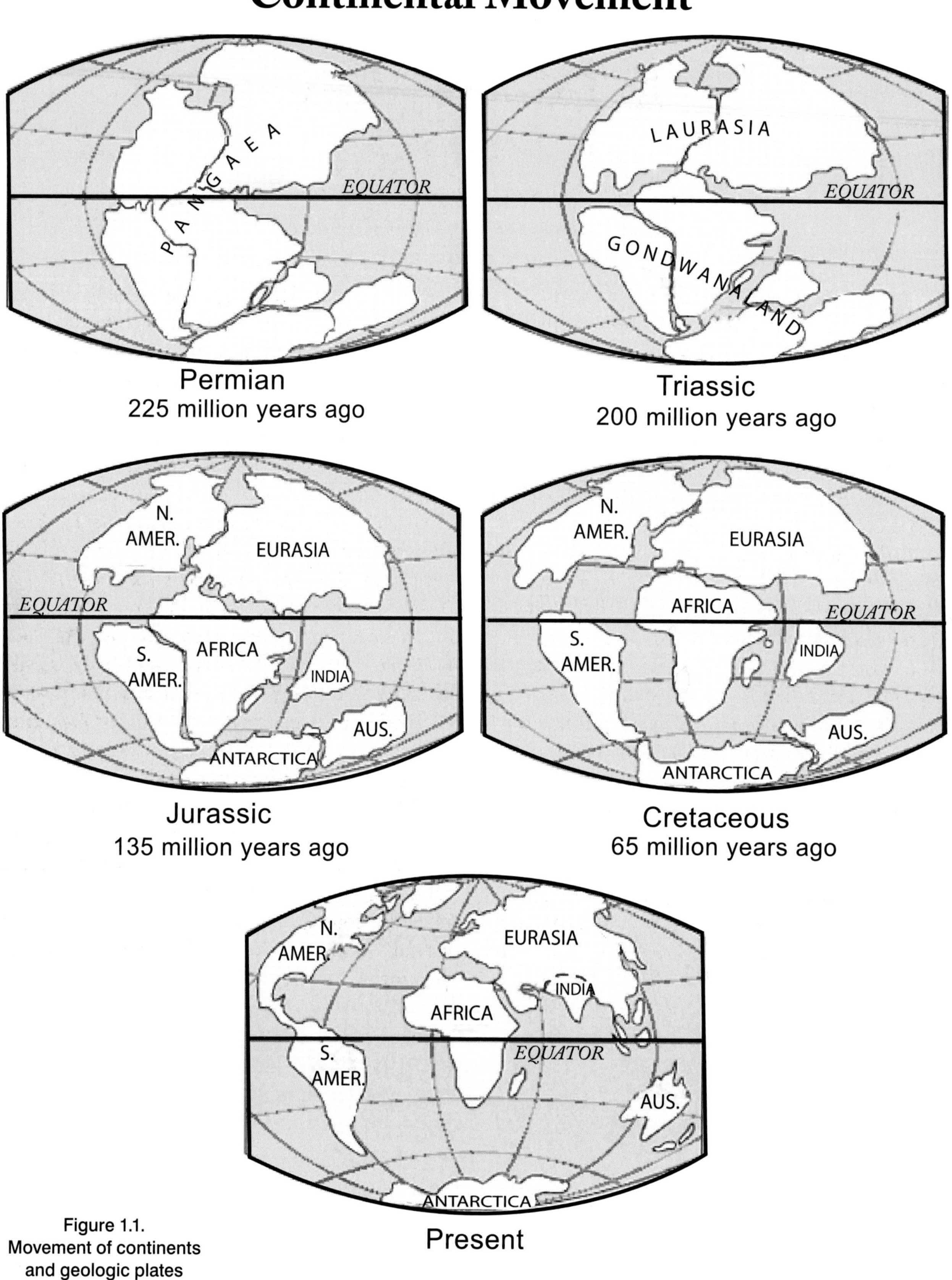

Figure 1.1.
Movement of continents
and geologic plates

Chapter 1

INTRODUCTION

D. Gordon Bennett
University of North Carolina at Greensboro

What must Gordillo, Verrazzano and the other early European explorers have thought as they plied the wind-swept sandy coast of the Carolinas nearly 500 years ago? Or deSoto as he led his men north from Florida into the western part of the Carolinas? What questions must have raced through the minds of those long lost colonists who came ashore on the edge of a vast forested wilderness in northeastern Carolina? And what about the wonder of the Powhatan and Tuscarora natives as they gazed upon the jagged peaks of those white sails slowly moving toward shore like white caps on ocean waves and upon the white-skinned intruders who came ashore?

This pristine and bountiful land, first called *Carolana* by the English, had already been inhabited by earlier newcomers and their descendants for some 10,000 years. The landscape, which had changed relatively little during the occupation by these earliest settlers, had itself been formed over a period of millions of years prior to their arrival. The gradual evolution of the geology and the slow reshaping of the physiography continued almost imperceptibly during the time of the earliest inhabitants.

The fractured geologic plates which shifted about the earth over the eons of time not only led to the formation of majestic mountains that would have rivaled the Himalayas of today, but also resulted in changes in climate that led to the appearance and disappearance of various plants and animals until the landscape took the shape and provided attributes which welcomed the earliest Americans and much later the newcomers from Europe and Africa (Figure 1.1). Of course, by the time people invaded this region, the once towering peaks had been eroded to the much lower, rounded Appalachians. That section, once attached to Africa, had been broken away, with the edge upon which the Europeans stepped ashore covered with tiny broken fragments of the western mountains carried hundreds of miles by the rivers of the Carolinas. Inland from the beaches, strewn with sands from the river mouths by ocean waves and longshore currents, were marsh grasses backed by dense forests. Cyprus, cedar and conifers along and near the coast gradually intermingled with deciduous trees that lost their foliage in the late autumn.

Exploration and Settlement

This still largely pristine physical environment and the natives of the region both welcomed and resisted the early settlers. The Spanish made an unsuccessful attempt to colonize the region in 1526, when de Ayllon and about 500 others tried to settle in the Winyah Bay area (called San Miguel de Gualdape) near present-day Georgetown, S.C. (Figure 1.2). In the early 1560s, the French tried to settle Santa Elena near today's Port Royal, S.C. Soon after Queen Elizabeth I gave permission to colonize America, the geographer Thomas Hariot, artist John White, and over 100 others spent about a year on Roanoke Island (1584-1585), but they so angered the natives that they were forced to leave and return to England. Two years later, John White led 150 settlers to the same area. Here, his granddaughter, Virginia Dare, was the first English child born in America. White soon sailed to England for supplies, but when he returned from his three-year trip, the early colonists, including his family, had vanished without a trace. The mystery surrounding their disappearance has never been solved.

It is estimated that up to half of the early immigrants to the southern colonies were indentured servants, who came both voluntarily and under forced conditions to work off payment of their voyage or criminal penalty. They were used primarily to offset serious labor shortages in agriculture. But when their period of servitude ended, they were free to return to England or begin a new life in America.

When the use of captured natives and indentured servants did not meet farm labor demand, African slaves, who first arrived in Jamestown, Virginia, in 1619, were increasingly imported for this purpose. They were especially useful in the tobacco fields of northern Carolina and the rice plantations of coastal South Carolina. The growing of indigo for use in the preparation of blue dyes also became important in the southern section.

On the plantations, disease was rampant and mortality high among slaves, especially the women and children. Most slaves lived in deplorable conditions, usually with two or three families cooking and sleeping in a one-room hut, often with a dirt floor and no window.

In the Carolinas, as elsewhere in America, the earliest settlers were along the Atlantic coast. Most early English immigrants became farmers, but towns also developed. The only one of any size in the Carolinas was Charleston, initially called Charles Town. Later, some of the Germans, French and Scotch-Irish who had entered the middle colonies made their way through the Great Valley of the Appalachians and passed southeast through the wind and water gaps into the northwestern Piedmont of the Carolinas.

More than 400 years have passed since John White returned to gaze upon the vanished colony at Manteo with its eerie carved message "CROATAN," the name of the local indigenous tribe. For over 60 years, an outdoor drama has told the story of this "Lost Colony." Until recently, most of the nearby area retained much of its pristine nature. But only within the last 20 to 30 years has the area witnessed considerable tourist and retirement development. By the 1980s, Dare had become the fastest growing county in the Carolinas.

Why such a great lapse of time between the earliest attempt at settlement in Sir Walter Raleigh's colony and major development here? Part of the reason is the treacherous shoals of the ocean waters along the Outer Banks, especially at Cape Hatteras and Cape Lookout. About 150 years ago, a lighthouse over 160 feet high (the tallest in the U.S.) was built at Hatteras in order to warn ships passing nearby. The lighthouse had to be rebuilt in 1872, and in the late twentieth century was moved inland before the incessant action of ocean waves and currents could claim it as they have hundreds of ships since the Europeans first arrived here.

Pirates, especially Blackbeard (Edward Teach) and Stede Bonnet, were another deterrent to trade and settlement in the early 1700s, nearly halting

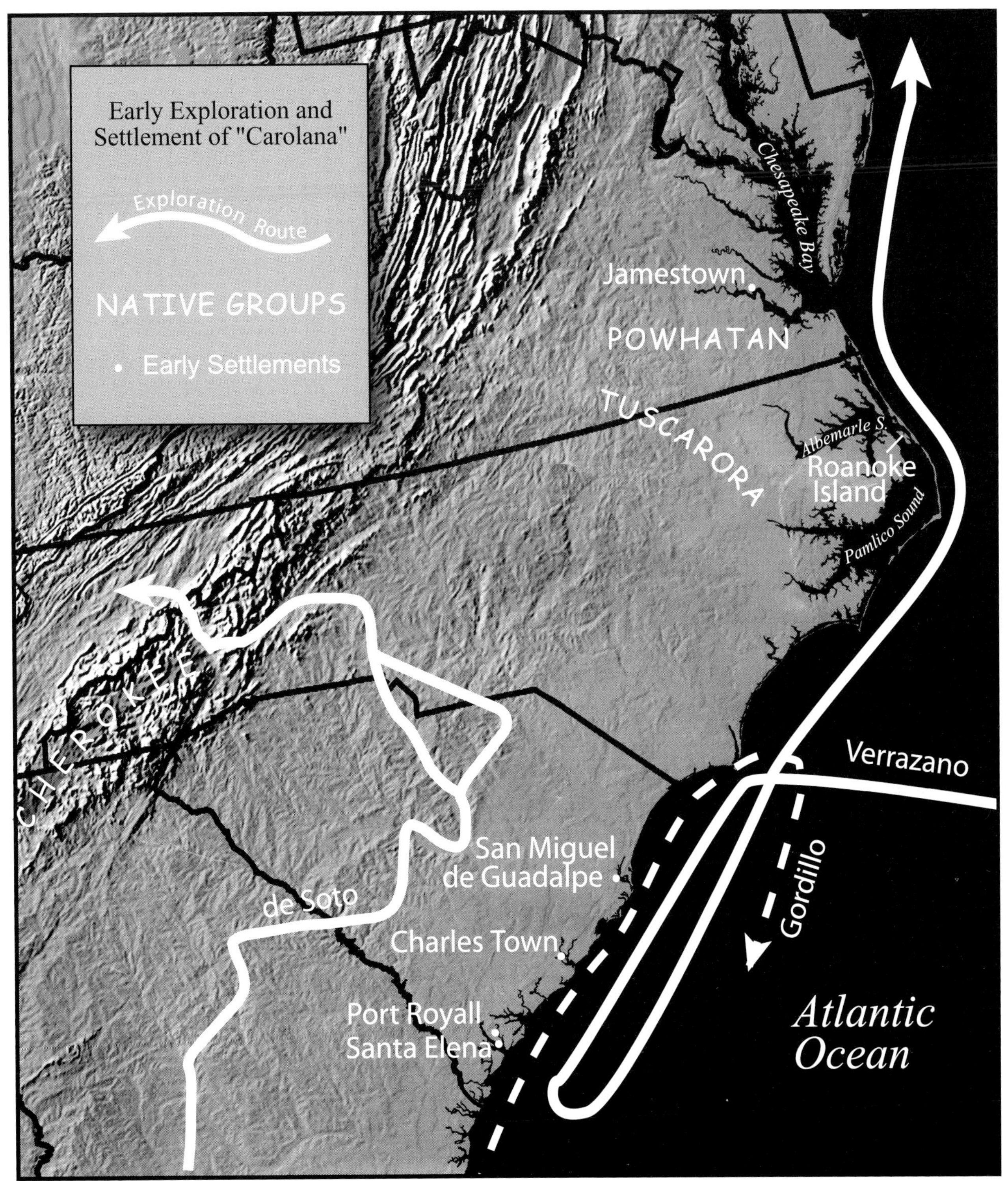

Figure 1.2. Early exploration and settlement in the Carolinas

trade to Carolina before they were captured and killed. Blackbeard, a pirate whose appearance matched his formidable reputation, hid with his men along the Outer Banks between Cape Hatteras and Cape Lookout near the small fishing village of Ocracoke (Figure 1.3). Stede Bonnet was apprehended nearby at Southport.

Earlier interest in settlement had been focused to the north of the Outer Banks near the first permanent English settlement in America at Jamestown, Virginia (1607). In the 1650s, a few settlers from Virginia began moving south

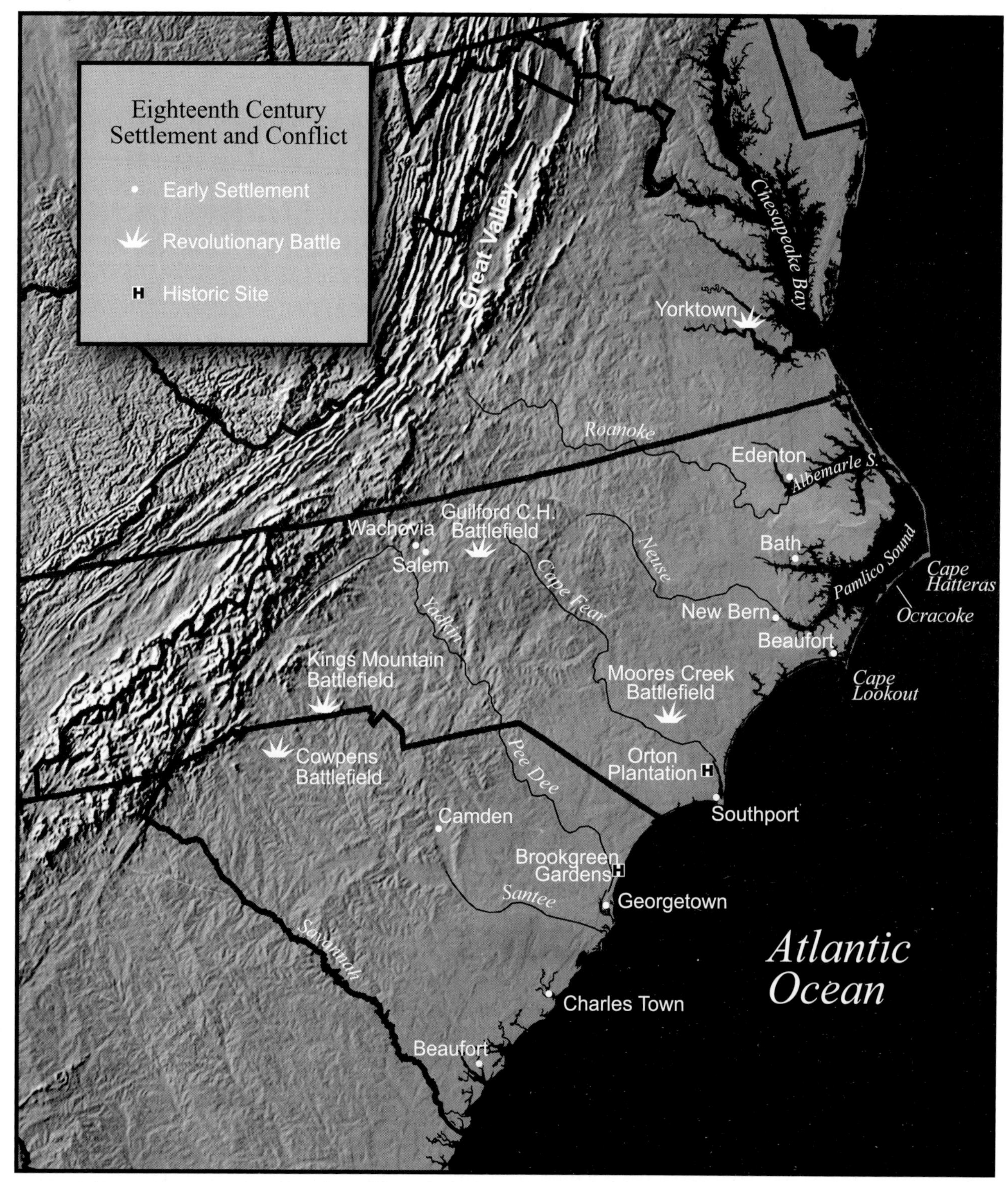

Figure 1.3. Eighteenth Century settlement and conflict

into the Albemarle Sound area, and the colony of Carolina was chartered anew to eight noblemen in the mid-1660s. By 1670, the first permanent settlement was established at Charles Town (Charleston after 1783) to give the Carolina colony an outlet to the sea. In this same year, the Quakers converted many settlers in the Albemarle Basin, thus laying the foundation of this group as the dominant religion here for several decades. But the 1670s and 1680s was a period of much discord and rebellion between rival political groups, including Culpepper's Rebellion, which was the culmination of the conflict between England's representatives who tried to collect tax duty on tobacco destined for New England and local tobacco growers/exporters. In 1695, the young settlement of Charles Town became the center of government for Carolina, and a deputy governor was appointed to oversee the land north of the Cape Fear River, called "North Carolina."

By 1700, settlement was still confined to the northeastern part of the colony and to within 20 to 30 miles surrounding Charles Town in the southeastern part. The first decade of the eighteenth century was a period of conflict between the Quakers and the Anglicans for control of governing North Carolina, until the Quakers were defeated in 1711. During this period, the first town of the northern part of Carolina, Bath, was established. In an attempt to settle this dispute, the English Crown approved a plan by the Lords Proprietors to appoint a separate governor for North Carolina, thus separating it from South Carolina (1712).

Swiss settlers founded New Bern (1710), but European agitation of the native Tuscaroras, whereby various settlers cheated them and took their land, led to attacks on both New Bern and Bath. The uprising was put down in 1712, only after South Carolina twice sent help (see Figure 1.3). Soon after, Beaufort and Edenton in North Carolina were founded.

The Crown bought control of South Carolina in 1719 and North Carolina in 1729 and ruled them until independence. During the 1730s, the port of Wilmington was established, and with the pirates and Indians defeated and the Crown in control, large groups of settlers began to push inland, led by the Highland Scots who moved into the Fayetteville area. A quick parliamentary maneuver by the Cape Fear-Neuse contingent before the Albemarle delegation arrived at a scheduled governing meeting led to New Bern being selected the capital of the colony of North Carolina (1746).

Settlers in the Carolinas spread out along the Coastal Plain and moved up the Savannah, Santee, Pee Dee, Cape Fear, Neuse, and Roanoke Rivers. In addition, Moravians and others either traveled southwest through the Great Valley from the middle colonies before cutting through the Blue Ridge to the Piedmont or moved inland along rivers and early trails from the Coastal Plain. In the early 1750s, the Moravians settled "Wachovia" in present-day Forsyth County, N.C. and established the planned town of Bethabara. In the latter 1760s, Salem (Photo 1.1) and Charlotte were founded. Agriculture, forestry and fishing were the bases of life in the Carolinas. Tobacco and rice provided most of the revenue needed to sustain these colonies. Blacks, who were mainly slaves on farms, outnumbered whites by two to one.

By 1770, the colonists of the Carolinas were embroiled in a struggle for control over the region. In 1780, they were engaged in widespread armed conflict with the British. Although the first battle of the Revolutionary War in the Carolinas occurred in 1776 at Moore's Creek near Wilmington, the first revolt against the British had taken place five years earlier in the Piedmont area of Alamance. Although Cornwallis won the battles at Cowpens, Kings Mountain, and the Guilford Battlefield (Greensboro, N.C. today), he incurred so many losses that these fights led to his defeat several months later at Yorktown in Virginia.

Around the time of the Revolutionary War, the laborious process of cleaning cotton began to adversely affect the economics of planters, making the employment of slave labor uneconomical. There was a small window of opportunity where slaves could have been headed for freedom. But a chance meeting between Eli Whitney, a northern teacher vacationing near Savannah, Georgia, and a group of cotton planters closed this window. Whitney, intrigued by the problem of cleaning cotton, invented a simple cotton gin, which would greatly increase the speed of cotton cleaning. This improved process, in turn, allowed the planters to profitably grow more cotton and to continue the employment of slave labor. Slavery persisted in the U.S. for another three quarters of a century before a civil war between the states put it to an end.

It was this legacy of slavery which split the young nation and eventually led South Carolina in late 1860 to become the first state to secede from the Union. It was only months later, after ten other states had seceded that North Carolina became the last one to leave the Union. The battle which began at Fort Sumter outside Charleston ended with a region as devastated as much of Europe was after the Second World War.

Would not those who fought in the War of Independence or the Civil War be amazed at the changes which have occurred in the Carolinas since those conflicts, just as the natives and early settlers would have been at the events which unfolded during the previous 200 years? Who could have foretold the journey that the men and women of the Carolinas would undertake during the ensuing period?

The Regions of the Carolinas

What lies before today's Carolinians is a region molded by both the forces of nature and the creativity and conflict of human interaction. The geomorphologic and climatologic evolutions shaped a landscape which was further influenced by the attitudes and abilities of those who played their roles on the Carolina stage for the last 400 years. The way the region is viewed today is the result of the culmination of the interplay of all these forces. Many of the place names of towns, counties and physical features of the region are reminders of the natives and settlers who graced this stage long ago.

Both North and South Carolina are composed of three distinct physiographic regions: the Coastal Plain, the Piedmont and the Mountains (Figure 1.4). The variations in climate, vegetation, soils, and way of life roughly parallel them in North and South Carolina. However, each region emerges in its own unique way by incorporating its own physical and cultural attributes.

The Coastal Plain

The Coastal Plain covers two-thirds of South Carolina but less than half of North Carolina. The marshy coastal section is called the Tidewater in North Carolina and the Low Country in South Carolina. Geologically, this region was formed last, but historically, it was settled first by the Europeans. But the coastline where the Europeans landed has been reshaped by the action of wind, waves and currents on the sandy beaches. The grains of sand carried by the rivers from the lofty western heights have both clogged the river mouths and been spread along the coast by the longshore currents and ocean waves which have created and altered the beautiful beaches. Inlets have continued to open and close along the Outer Banks affecting the access of fishermen and traders to the open sea. Parts of the off-shore sandy island "banks" have continued to shift so that these pieces of land have "migrated" toward and parallel to the mainland. Within short distances, flora and fauna which tolerate salt water and marshy conditions have given way to species which can not.

The wild rice of the Native Americans has disappeared, but the maize (corn) and tobacco remain in a hybrid form in the region. Indigo was introduced but was supplanted after a time by other crops. Cotton, once "king," was decimated by the boll weevil but has made a return. The stately plantation homes built on the successes of these crops and on the backs of slaves have largely deteriorated. Some of the plantation homes, such as Orton Plantation and Brookgreen Gardens, near the eastern end of the artificial dividing line between the Carolinas, have been restored for the tourists.

Today, tobacco and maize are joined by farms producing turkeys, chickens, sweet potatoes, peanuts, soybeans, and hogs. Relatively little manufacturing is found in the Coastal Plain, but military bases brought by influential congressmen and, more recently, tourist and retirement developments have become important sources of economic growth along coastal margins and in some other parts usually associated with rivers and lakes.

Moving inland along one of the many rivers

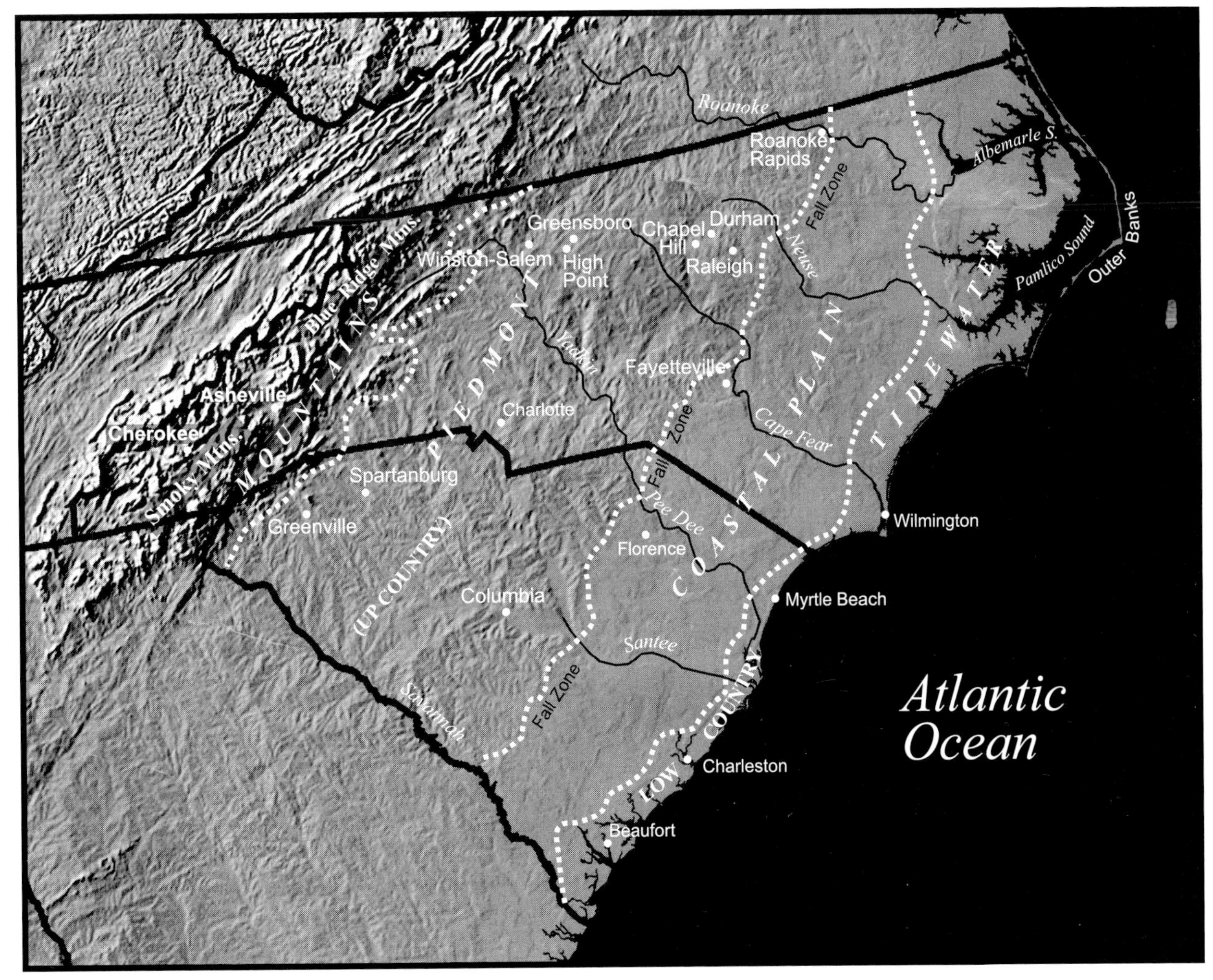

Figure 1.4. Physiographic regions of the Carolinas

that allowed cargo upstream on small boats in the historical past, one eventually reaches a series of rapids where goods and passengers had to be unloaded and placed on other forms of transportation for continued journey into the interior. Just west of Fayetteville, this "fall zone," or "fall line," marks the physiographic border between the younger, loose, unconsolidated material of the mainly flat, sandy Coastal Plain and the older, hard, crystalline rock of the rolling Piedmont.

The Piedmont

The Piedmont of North Carolina, also known as the Up Country in South Carolina, stretches from the fall zone in the east to the foot of the Blue Ridge Mountains in the west. Several fall line cities stretch from Roanoke Rapids near the Virginia line southwest to Fayetteville and Columbia. Inland from the fall zone, a mix of religious groups (Quakers, Moravians, Calvinists, Lutherans, Presbyterians, Methodists, and Baptists) moved into the area from the Coastal Plain or the middle colonies via the Great Valley. Settlement on the Piedmont was based more on small farms than on plantations, though most settlers eventually owned some slaves. Salem (Photo 1.1) was not settled by the Moravians until two decades before conflict broke out between the British and the colonists, and Winston was not established until a century later. Today, agriculture is focused on tobacco, corn, soybeans, and greenhouse plants, plus chickens and turkeys. Manufacturing, which contributes greatly to the economy, is found primarily in the Piedmont, especially in the "Piedmont Urban Industrial Crescent" extending from the Raleigh/Durham area in the northeast through the Greensboro/Win-

Photo 1.1. A view of Salem, Van Redken, 1787 (Courtesy of Old Salem)

ston-Salem/High Point area in the north-central part to Charlotte and Greenville/Spartanburg in the southwest. Textiles, despite an employment decline during the 1980s, 1990s and first part of the 2000s, continues to lead in manufacturing jobs in the Carolinas. However, employment in this industry has been surpassed by jobs in retail and wholesale trade, in finance and real estate, and in the health services sector. The furniture industry ranks second in North Carolina, while the chemical industry, which is linked to textiles, furniture, other industries and even agriculture, ranks second in South Carolina. Tobacco production in the northern Piedmont now ranks only in the lowest quartile of North Carolina industries. Most of the largest cities and universities are found in the Piedmont part of the Carolinas, especially in the "Crescent." The nation's first public university opened in Chapel Hill in 1793. The history of tobacco companies, cities and universities are often intertwined, the Duke and Reynolds fortunes (American Tobacco Company in Durham and R. J. Reynolds Tobacco Company in Winston-Salem) having been responsible for moving Duke and Wake Forest universities to their current locations in Durham and Winston-Salem, respectively. During the 1990s, the introduction of automobile assembly in the South Carolina Piedmont strengthened auto parts manufacturing in both states.

Sports and cultural attractions are also focused on the Piedmont. College and professional basketball and football, plus NASCAR, the National Hockey League, the Professional Golf Association, and minor league baseball and soccer, provide year-round spectator sports. The development in North Carolina of a state-supported art museum in Raleigh (1947) and The School of the Arts in Winston-Salem (1963) are supplemented in both states by strong local arts programs through museums, orchestras, and theaters in all the large cities.

Photo 1.2. Hanging Rock in Stokes County, N.C. (Courtesy of Jeff Patton)

The Mountains

In the western Piedmont, Hanging Rock in Stokes County (Photo 1.2) and other monadnocks of once higher terrain, plus hilly outliers of the northeast-southwest trending Appalachians introduce the old, worn-down mountains that form the western periphery of the Carolinas. Though low and rounded when compared to the younger peaks of the Rockies in the western United States, the Appalachian Mountains long proved to be a great barrier to transportation between the Midwest and the southeast coast, thus slowing the development of the Carolinas and their port cities. They included the highest peak in the eastern United States (Mt. Mitchell, 6,684 feet) and were covered with thick virgin forests and smothered often by clouds, which had already led the natives to call them the "Smoking Mountains."

Those who settled in the Carolina Mountains were largely religious Scotch-Irish and German groups. They moved into numerous small valleys isolated from one another and the outside world by enveloping mountains. The settlers' sustenance was drawn from the tiny fields they cleared and from the nearby forests and rivers. They settled among the rhododendron, mountain laurel, azalea, majestic pine and hemlock, and other greenery that provide a lush rolling carpet over the region. They evolved their own culture of language, music, religion, and politics from a mix of their original European sources, from their personal experiences, and to a lesser degree, from people and ideas filtering in from beyond the hills. Most residents lived a largely subsistence lifestyle, producing most of their own needs and exporting little for trade, except for a few minerals from isolated locations.

The southern part of the Mountains extending into Georgia were already inhabited by the Cherokees, a Native American tribe. The Cherokees tried to prevent the colonists from invading their country and even sided with the British

Photo 1.3. A scene from the outdoor play "Unto These Hills" recounting the removal of the Cherokees from the Carolina mountains (Courtesy of the Cherokee Historical Association)

during the Revolutionary War in hope of driving the settlers away. But after the new nation was formed, they fought with the Americans in the War of 1812 and subsequently were given the right to form their own local government. But the combination of pressure from nearby white farmers, especially in north Georgia, and the finding of gold near present-day Charlotte led to the U. S. government reneging on its promises and force-marching as many of the Cherokee as the army could find westward to Oklahoma along what became known as "The Trail of Tears." That part of the history of the Mountain region is told at Cherokee each summer in the outdoor drama "Unto These Hills" (Photo 1.3).

Even though a few aristocrats from the Coastal Plain/Low Country established summer retreats in selected sites in the Mountains during the 1800s and a few mining sites were developed there during that period, significant contact for most mountain residents with the outside world occurred only since World War II. When the Reynolds, Dukes, Hanes, Cones and other Carolina industrialists were amassing their own fortunes, one of the Vanderbilt heirs of an earlier American fortune had chosen a 125,000-acre tract near the small mountain town of Asheville to build Biltmore House, the largest, most elegant house in America. Since 1930, paying visitors to the Biltmore House have kept it open and in good condition. Indeed, in the Mountains, as on the Coastal Plain, economic activity is often focused on tourism and retirement development more than on manufacturing. In the Mountains, agriculture is much more limited by topography and climate than elsewhere in the Carolinas. Today, much of the Mountain region is plagued by environmental problems, such as erosion, water and air pollution, which beset the rest of the Carolinas.

As a result of a combination of differences in physical environment, source areas and settlement patterns, and resulting economic development, the three major regions of the Carolinas have evolved into rather distinctive geographical units, but the tapestry which unfolds before those who travel throughout the two states is woven together so that without the highway signs to denote the state borders, one would not be able to tell when passing from one political unit into the other.

The remainder of this treatise on the geographical landscapes of the Carolinas provides a thorough foundation for understanding both the physical and human characteristics of what the first English explorers here called the "goodliest land" on the face of the earth. First, the natural setting, both physiographically and climatically, is presented so that the early historical development of the region which follows can be better appreciated. Next, the twentieth century changes in population characteristics, rural-urban patterns, agriculture and manufacturing are presented, as are their present characteristics and problems. Then, the cultural, recreational, and educational attributes of the Carolinas are described. Finally, based on the understandings gained from the geographical assessments of all these topics, a look at some of the challenges and prospects for the future of the region are assessed.

The contributors to this volume are mainly geographer-citizens of the two Carolinas who have lived here for as many as 35 years, and some of them were also reared here. Those few who have not taught in the Carolinas have long studied the region. Thus, those who have offered their insights into *A Geography of the Carolinas* are largely resident-scholars of the region about which they have written. It is the hope of all involved in this undertaking that both the citizens and the visitors in the Carolinas will understand and enjoy the region better after having read these articles.

References

Edgar, W.B. 1998. *South Carolina: A History*. Columbia: University of South Carolina Press.

Gade, O, A.B. Rex, and J.E. Young. 2002. *North Carolina: People and Environments*. Boone, N.C.: Parkway Publishers.

Kovacik, C.F. and J.J. Winberry. 1987. *South Carolina: The Making of a Landscape*. Columbia: University of South Carolina Press.

Powell, W.S. 1989. *North Carolina Through Four Centuries*. Chapel Hill: University of North Carolina Press.

"Unto These Hills." 1998. Cherokee, N.C.: Cherokee Historical Association.

Von Redken. 1787. *A View of Salem*. Winston-Salem, N.C.: Old Salem, Museum of Early Southern Decorative Arts.

Chapter 2

CHANGING LANDFORMS AND RIVERS

Jeffrey C. Patton

University of North Carolina at Greensboro

The following chapter focuses on the processes that formed the physical landscape of the Carolinas. The landforms we see today are the result of repeated cataclysmic collisions and subsequent rupturing of continents. The Carolinas, sitting on the eastern edge of the ancient North American continent, often bore the brunt of these collisions. The forces bent, fractured, thrust upward, and metamorphosed the layers of rock we now walk upon, and let loose unimaginably violent volcanic eruptions and staggering earthquakes. The story of our landforms also involves the persistent gradual forces of erosion and deposition—rivers, wind and waves wearing down and rearranging the landscape over millions of years. It is a complex narrative that earth scientists are only now beginning to understand.

The outer crust of the earth is a solid, brittle rock layer averaging about 100 kilometers in thickness. This thin, shell-like layer, known as the lithosphere, is shattered into dozens of pieces, or plates, each floating on the hot, semimolten rock of the asthenosphere below. Some of these floating fragments carry continents on their backs, while others are covered with a thin film of water, which we call the oceans. As the plates inexorably move, they may collide with one another or slowly drift apart. When collisions occur, one plate may dip beneath the other in a process called submergence. This is most typical when thicker, but less dense continental plates collide with the thinner and denser oceanic plates. The thinner oceanic plate is forced down into the hot asthenosphere where it melts (Figure 2.1). If two continental plates collide, great mountains may form from the impact, and the plates may become sutured together (Figure 2.2). When plates break apart and one part drifts away from the other, molten material from the asthenosphere rises to fill the intervening gap. This molten rock, or lava, cools and becomes a part of the earth's outer shell—and thus a new lithospheric plate material is formed.

Approximately one billion years ago, the vast majority of the plates carrying the continents of the world had sutured together into one supercontinent called Pangaea II. (Table 2.1). The area that would someday become the eastern seaboard of the United States was attached to what

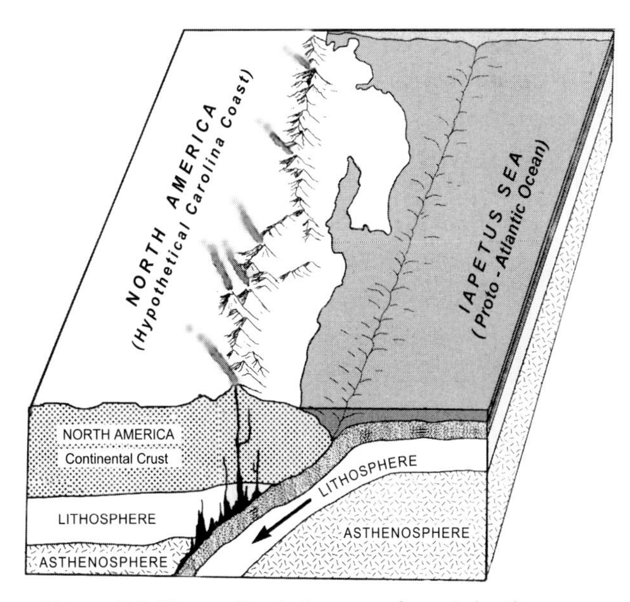

Figure 2.1. Generalized diagram of a subduction zone and associated volcanic activity. During the formation of Pangaea I, II and III, subduction zones formed along the east coast of North America.

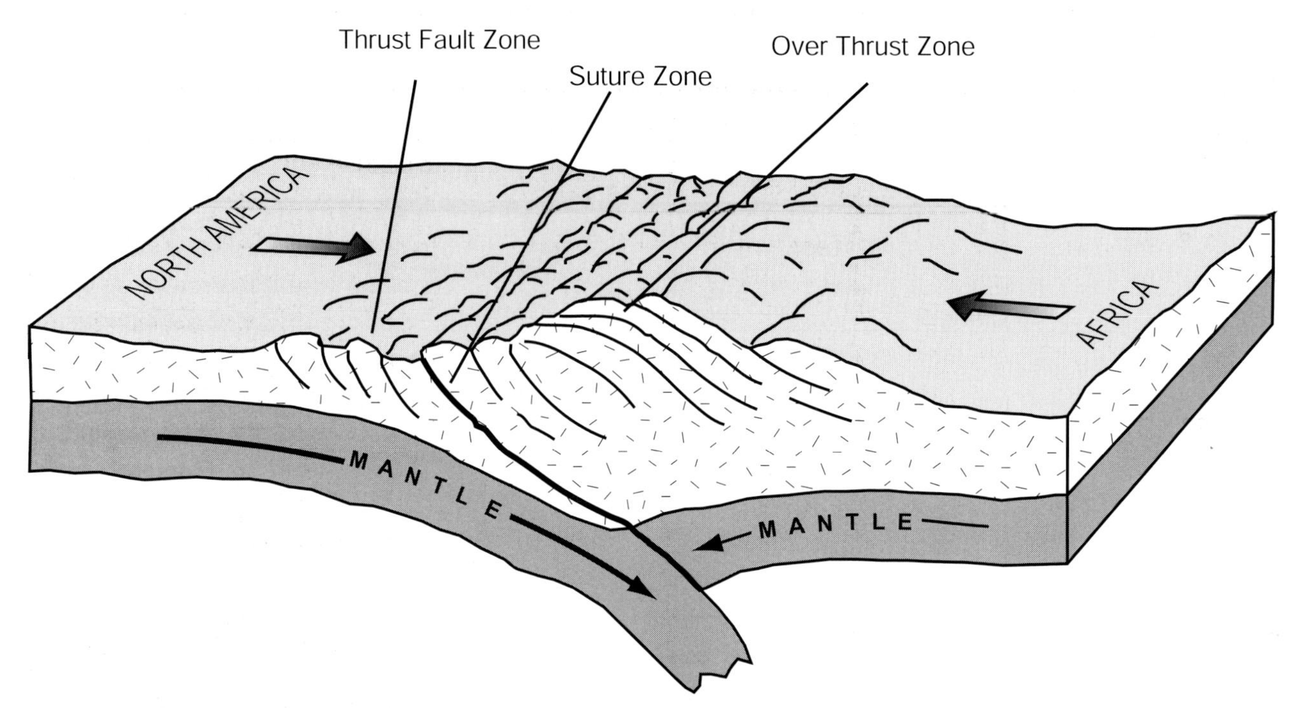

Figure 2.2. Generalized diagram depicting the collision and suturing of continents. The collisions between Africa and North America resulted in the thrust and overthrust belts of the Piedmont and Blue Ridge Provinces of the Carolinas.

is now Europe. At that time, the great supercontinent began to break apart into smaller fragments. One of these great rifts occurred along the east coast of the U.S. Once the rifting was complete, Europe began to drift eastward and North America westward. For over 600 million years, the two continental plates drifted farther and farther apart, the ever-widening gap creating the ancestral Atlantic Ocean. About 480 million years ago, the North American plate stopped its westward migration, and for some unknown reason reversed its course and began to quickly move toward Europe and Africa. Now moving eastward, North America began to override the recently formed lithospheric crust of the Atlantic Ocean, forming a classic subduction zone. The result of the collision with, and the melting of, the oceanic crust was a series of what must have been incredibly violent earthquakes and spectacular volcanic eruptions. In addition, the collision squeezed and compressed the crystalline bedrock and overlying sedimentary rock into great belts of metamorphic rock. As North America continued its eastward course, the ancestral Atlantic was closed and a head-on collision with Europe and Africa began. The force of this collision was so enormous that rock layers folded like a loose carpet pushed against a wall. When the pressure was too great, the bending rock snapped and layers of rock were thrust upward and over rock farther west. In some places, large fragments of the plates along the coast simply broke off and were pushed inland hundreds of miles.

It was this two-step process that created the Appalachian Mountains. First, the subduction of the Atlantic Ocean crust created a series of volcanic peaks. Then, the more violent continent-to-continent collision with Europe and Africa folded, faulted, and thrust the mountains up to heights that undoubtedly would rival the Himalayas of today. As the first dinosaurs were appearing, about 250 million years ago, the suturing of North America to Europe and Africa was complete and Pangaea III was formed. What is now North and South Carolina were firmly attached to the western edge of North Africa, and the great mountain-building compression of the previous 200 million years had finally ended. The newly formed Appalachians may have reached heights approaching 26,000 feet (8,000 m), but then they became vulnerable to the ravages of the erosional forces of ice, gravity, wind, and most importantly running water. Today, the

Years Before Present	Era	Period	Epoch	Event
10 thousand	CENOZOIC	Quaternary	Holocene	Barrier Islands Form
			Pleistocene	Sea levels rise and fall during Ice Ages Carolina Bays are formed Native Americans arrive
1.6 million		Tertiary	Pliocene	Sea level rises Formation of Orangeburg Scarp
5.3 million			Miocene	Sandhills begin to form Gulf Stream develops off coast
23 million			Oligocene	Appalachians continue to rise Sea level continues to fall Increased Erosion of Piedmont and Blue Ridge
37 million			Eocene	Sea level rises to highest point thick layer of marine sediment deposited over entire coastal plain
58 million			Paleocene	Earthquakes are common
66 million	MESOZOIC	Cretaceous		Worldwide extinction of dinosaurs Cape Fear Arch begins to form
144 million		Jurrasic		North America begins drifting westward giving birth to modern Atlantic Ocean as Carolinas pull away from Africa
208 million		Triassic		Rift Basins form as Pangea III begins to rupture along the east coast. Dinosaurs begin to arrive in Carolinas
245 million	PALEOZOIC	Permian		End of Alleghenian Orogeny Blue Ridge shoved 160 miles Northwest causing extensive overthrusting
286 million		Pennsylvanian		Alleghenian Orogeny
320 million		Mississippian		Erosional period between orogenies
360 million		Devonian		Possible Acadian Orogeny in Blue Ridge
408 million		Silurian		Subduction zone island arc welded to North America forms Charlotte Belt
438 million		Ordovician		Taconic Orogeny: Inner Piedmont is accreted to North America
505 million		Cambrian		Carolinas located near the Equator
570 million	PRECAMBRIAN			Pangea I and II form and subsequently rift apart. Proto-Atlantic Ocean (Iapetus) is created and destroyed. Grenville Orogeny begins formation of Appalachians 1.2 billion years ago Earth forms 4.6 billion years ago
4.6 billion				

Table 2.1. The Geologic Time Scale and associated events in the formation of the Carolinas (modified from Murphy, 1995).

highest peaks of the Appalachian chain found along the North Carolina-Tennessee border barely reach 6,600 feet (2012 m).

The most recent stage in this cataclysmic game of bumper cars began about 200 million years ago when Pangaea III split apart. As in the case of the destruction of Pangaea II, a rift formed along the eastern seaboard of the present day United States. When Europe and Africa drifted east, they left behind land that had become attached to North America, including all of present-day Florida and southern Georgia and Alabama. Pieces of the North America plate had also become attached to Europe and Africa, forming what is now Scotland, northern Wales and Africa's Mauritania (Murphy 1995). As the continents drifted apart, the present day Atlantic Ocean was born and is to this day continuing to widen. The tectonic forces that had thrust the Appalachians up were now gone and the erosional forces of running water and wind began to wear down the great mountains. Material washing off of the eroding Appalachian Mountains continues to be carried by the many rivers and streams of the Carolinas, which deposit their loads of sand, silt, and clay along the continental margins forming the broad Coastal Plain. The result of these continental collisions and subsequent eons of erosion are three great physiographic provinces that comprise the Carolinas: the Coastal Plain, Piedmont, and Appalachian Mountains (Figure 2.3).

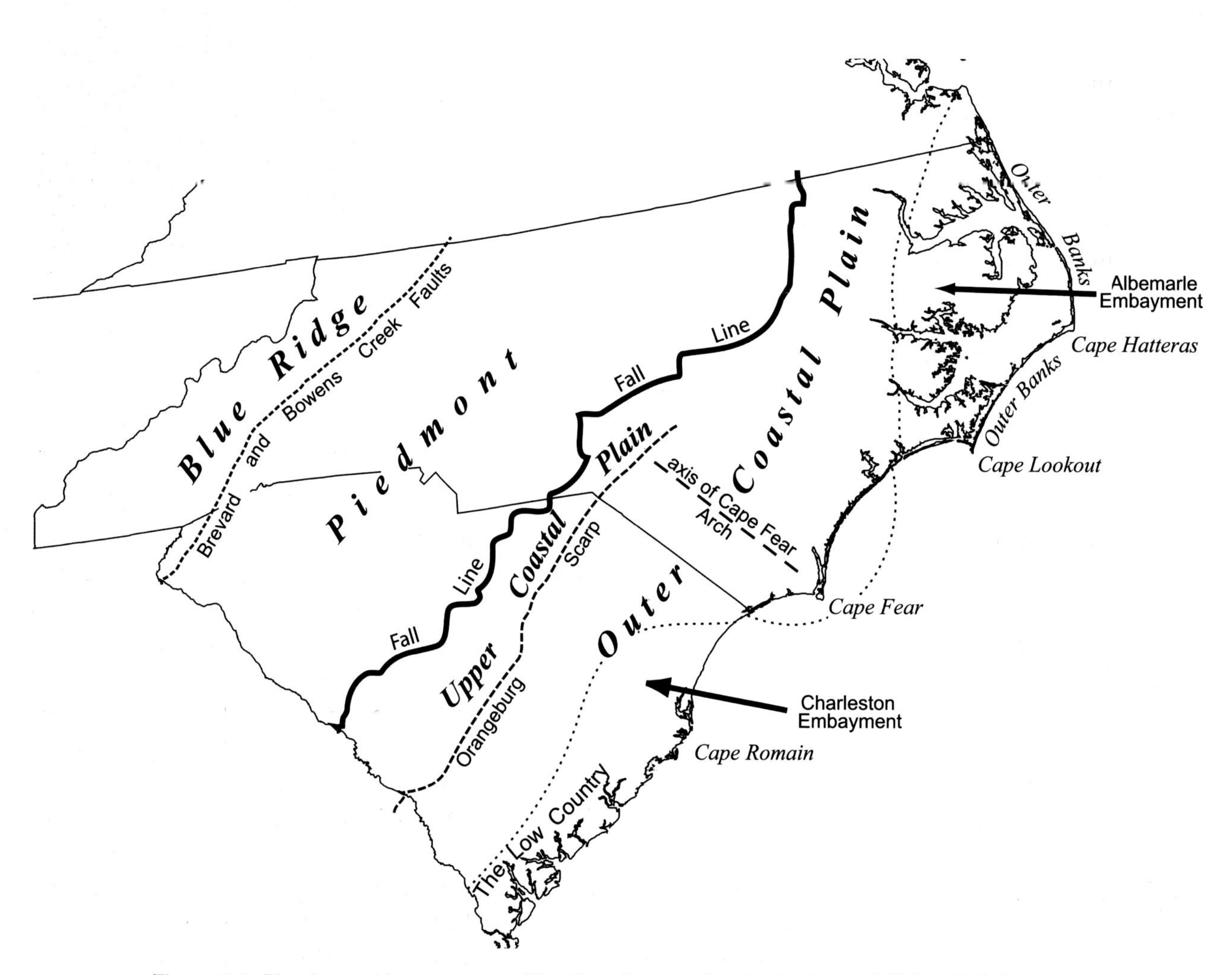

Figure 2.3. Physiographic provinces of the Carolinas and major features defining their borders.

The Physiographic Provinces

The Coastal Plain

Fundamentally, the Coastal Plain is a broad platform of resistant crystalline (pre-Mesozoic igneous and metamorphic) rock which enclose scattered areas of tilted sedimentary rock found in Triassic-Jurassic age basins (Horton and Zullo, 1991). The entire platform is buried beneath layers of Cretaceous and Cenozoic sediments eroded from the highlands to the west. In the west, the basement crystalline rock emerges from its blanket of sediment in a long sinuous line called the Fall Line. The eastern boundary is simply where the sea encounters the soft sediments today. That border is ever changing; at various times the entire Coastal Plain has lain beneath the Atlantic Ocean, while at other times it has extended far seaward of its present position. The broad underlying platform does not simply dip gently from the Fall Line to the sea. Instead, like a piece of plywood left out in the rain, it has several up-warps or arches and down-warps, referred to as embayments.

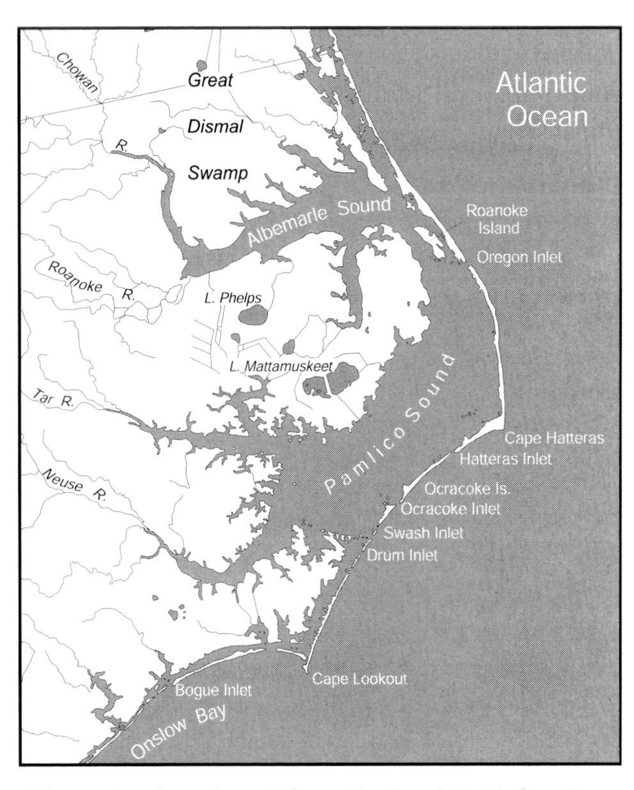

Figure 2.4. Sounds and Outer Banks of North Carolina.

The Coastal Margin

Where the sea meets the Carolina Coastal Plain is one of the most beautiful, fragile and ephemeral landscapes of North America. In North Carolina, the sea first washes ashore on the sands of a series of low-lying barrier islands known locally as the Outer Banks (Figure 2.4). These large elongated sand bars stretch over 200 miles (320 km) from the Virginia border to Bogue Inlet near Swansboro. The Outer Banks form two large cuspate forelands, the northern one having its apex at Cape Hatteras and the southern one at Cape Lookout. The area between the Outer Banks and the mainland varies from just a few miles to over twenty. Known as sounds, these microtidal (tidal range in this region is less than 3 feet, or 1 meter) estuaries are a rich and varied ecological region. Here, freshwater from the Neuse, Pamlico and Albemarle rivers mixes with the saltwater of the Atlantic. A series of seven tidal inlets provides the only passage from the sea into the sounds (Photo 2.1). From north to south, these inlets are Oregon, Hatteras, Ocracoke, Drum, Barden I, Beaufort and Bogue. From historical and geological records, it is clear that these inlets repeatedly fill with sediment and new inlets open elsewhere; over thirty such inlets have been documented since the arrival of Europeans to the region (Fisher 1962, 1967).

Low gradient beaches that change their form from winter to summer, and after each passing storm, characterize the seaward side of the islands. Storms, waves and longshore currents running parallel to the beach are the most important processes affecting the erosion, transportation, and deposition of sediment along North Carolina's barrier coastline (Moslow and Heron, 1994). Near both Cape Hatteras and Cape Look-

Photo 2.1. Drum Inlet. The pattern of sand deposition reflects the role that tides play in filling the sounds of the Carolinas (courtesy U.S. Dept. of Agriculture).

out, a series of sub-aqueous shoals extends out to sea at an angle roughly 90 degrees to the barrier islands. At places, these shoals are less than ten feet (3 m) beneath the surface and have been the bane of navigators for centuries. The nearly 600 shipwrecks testify to the reputation of the Outer Banks as "the graveyard of the Atlantic." At low tide, the water behind the barrier islands drains out, exposing the salt marshes and shallow mud flats that dominate the landward side of the Outer Banks. At high tide, the flow of water reverses and water pours into the sounds, flooding the marshes and mud flats.

Typical of barrier islands formed in a wave-dominant environment, the islands are slowly migrating toward the mainland. Waves born of storms periodically wash over the low-lying islands, pushing the islands ever closer to the mainland (Photo 2.2). At the same time, rivers and streams entering the calm protected waters

Photo 2.2. The pattern of the sand (light areas are sand deposits without a vegetative cover) is a result of storms that literally wash over the low lying barrier islands. Storms including hurricanes and winter Nor'easters push the barrier islands ever closer to the mainland (courtesy U.S. Dept. of Agriculture).

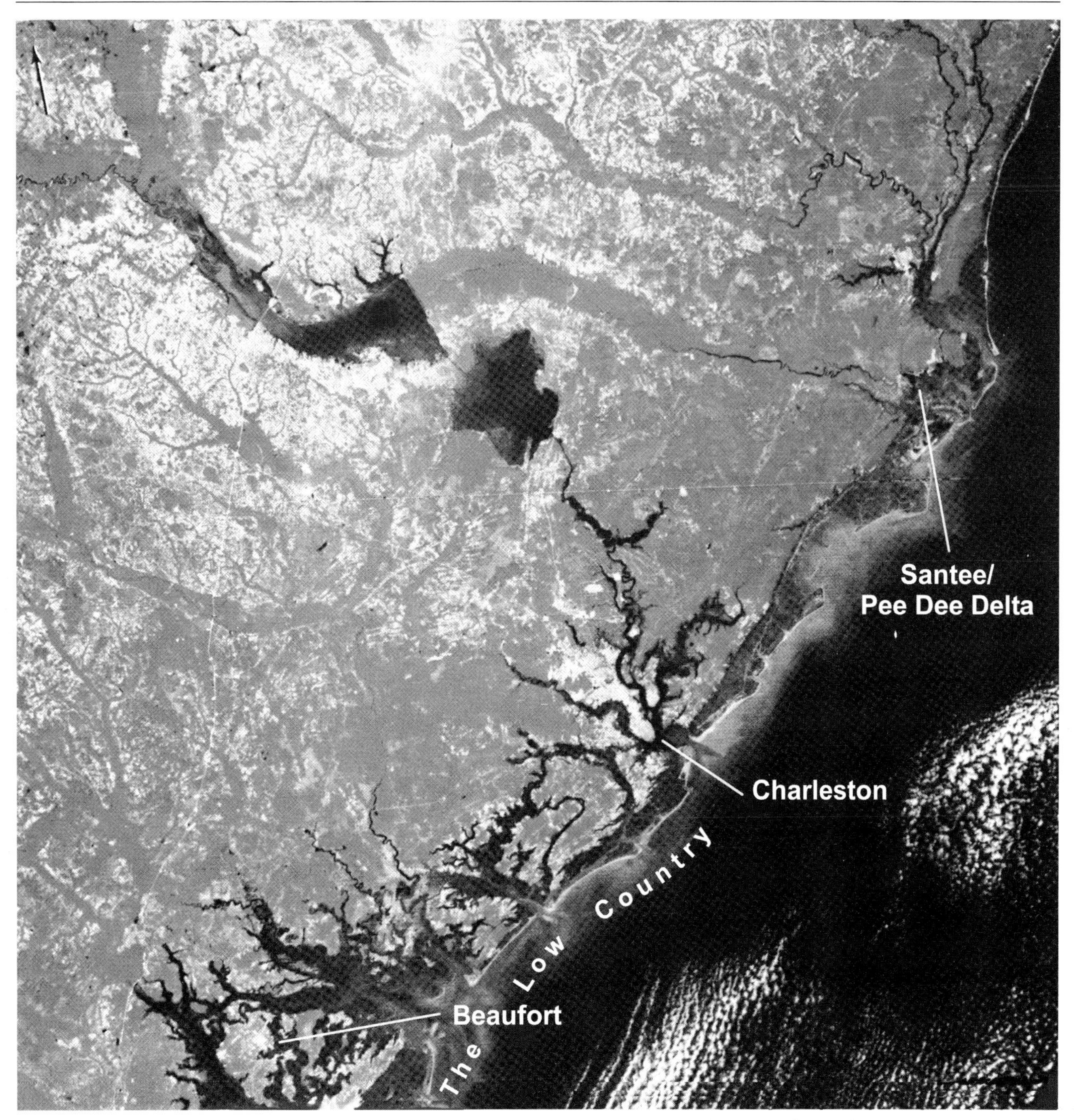

Photo 2.3. This infrared Landsat satellite image shows the many flooded river valleys of the South Carolina coast (courtesy NASA).

of the sounds deposit their loads of silt and sand and thus the mainland grows closer and closer to the Outer Banks.

South of Bogue Inlet to Debidue Island, South Carolina, welded barrier islands dominate the area. Here, the barrier islands are separated from the mainland by only a narrow strip of wetlands. Inlets are far more frequent than are found along the Outer Banks; however, they are very shallow and filled with shifting sand. Part of the coastline in this region has no protecting barrier islands at all. Such is the case for Carolina Beach near Wilmington, N.C. and for much of the Myrtle Beach area known as the Grand Strand. These mainland beaches are comprised of sand deposited during the last Ice Age.

South of Debidue Island, the coast is dominated by the Santee/Pee Dee River delta. This is the largest river delta in the eastern United States. The Santee and Pee Dee rivers have their headwaters in the mountains to the west while the Cape Fear rises in the Piedmont. All three of these rivers have a series of river terraces, where they cross the Cape Fear Arch. Such terraces indicate the slow uplift of the region that continues today. Soller (1988) estimates that the rate of uplift along the Cape Fear Arch in some areas to be as much as 18 feet (6 m) per 100,000 years, while in others as little as 2.5 inches (6 cm) per 100,000 years. Near Cape Romain, the surface begins to down warp and, as a result, the sea is eroding inland at a phenomenal rate of 6 meters per year (Hayes, 1994). As the sea moves inland, stream valleys are flooded, and small ridges and former sand dunes become isolated as islands in a classic submergent coastline (Photo 2.3). This area, known as the "low country," stretches to the Georgia border and beyond. It is a labyrinth of tidal flats, salt marshes and lagoons. Here is found the greatest tidal range along the Carolina coast, averaging over 9 feet (3 m). At low tide, great expanses of mud flats are exposed only to disappear under the rising tides.

The Outer Coastal Plain

The Coastal Plain can be divided into two sections, the upper (inner) and outer plains. The outer Coastal Plain, or Tidewater region, is a relatively topographically monotonous surface, containing large areas of swamps and numerous lakes indicative of poor drainage conditions. Averaging less than twenty feet above sea level the Outer Coastal Plain is comprised of marine sediments deposited as world sea levels rose and fell. These fluctuations during the late Cenozoic period were a result of the cyclic expanding and subsequent melting of glacial ice caps. At one time, one-third of the world was covered by glacial ice, miles in thickness. Today, only ten percent of the earth is covered by a much thinner ice cap. As ice caps grow, sea level drops; thus, at the height of the Pleistocene, or ice ages, significantly more of the continental shelf was exposed. For the last 15,000 years, world sea level has been rising as the glacial ice caps have steadily melted. Today, only Antarctica and Greenland have a significant ice cap. The result has been a progressive encroachment of the Atlantic over the Carolina Coastal Plain. This pattern has been repeated many times over the Coastal Plain; ancient barrier islands, beach ridges, and the remains of former lagoons and salt marshes lay stranded miles inland from the present day coastline. Undoubtedly, the most mysterious features found on the Coastal Plain are the thousands of elliptical depressions known as Carolina Bays. Thomas Ross has long studied these strange features and the following discussion of their formation and origins is derived from his extensive work. Ross (1996) has described Carolina Bays as shallow, elliptical depressions whose long axis is oriented northwest-southeast. Most bays are filled with soil that is darker than the surrounding landscape and are partially surrounded by a sand rim, making them highly visible from the air, reminiscent of a cratered landscape (Photo 2.4). But at the surface, the rims appear as low, broad ridges that enclose the depression, causing them to be barely noticeable at ground level. Carolina Bays are primarily found in North and South Carolina, though they can be found as far north as southern New Jersey and as far south as northern Florida (Johnson 1942; Prouty 1952).

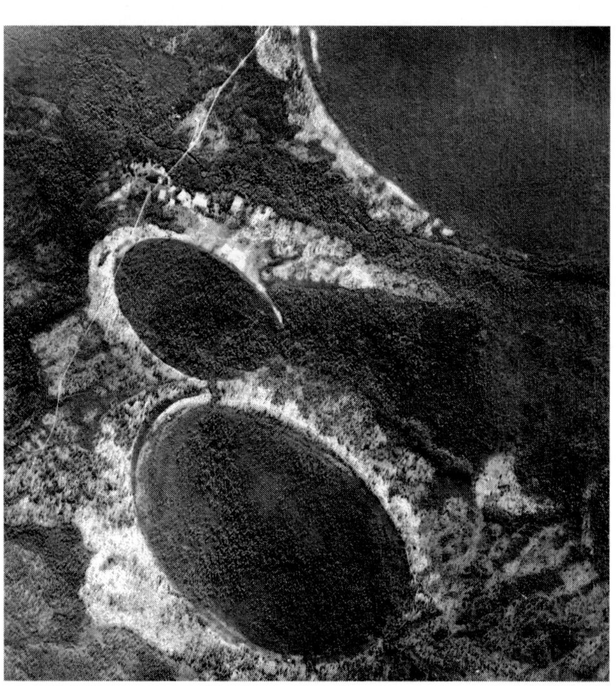

Photo 2.4. Aerial photograph of several Carolina Bays in Bladen County, N.C. (courtesy U.S. Dept. of Agriculture).

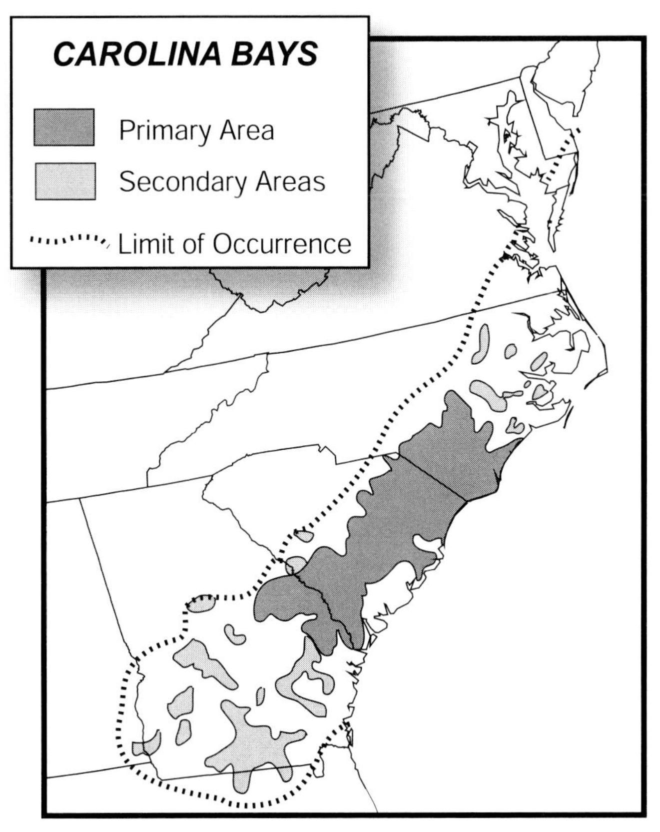

Figure 2.5. Distribution of Carolina Bays (after Ross, 1996).

(See Figure 2.5.) Prouty (1952) estimated that there are as many as 500,000 bays located on the Atlantic Coastal Plain; however, other authors dispute this figure, indicating that there are far fewer (Corbett, 1989; Bennett and Nelson, 1991; and Ross, 1994).

Carolina Bays are ancient lakebeds, most of which have dried up over the past several thousand years. However, after periods of heavy precipitation most bays collect runoff, which is ponded for several hours or days above the zone of saturation. A dozen or so bays have a permanent source of water, the largest being lakes Mattamuskeet, Waccamaw and White. But many others are swampy or wetland areas that contain water for extended periods of time.

The term Carolina Bay is derived from the numerous sweet, red, and loblolly bay trees growing in and around the bays, not because they are maritime bodies. Carolina Bay vegetation is not restricted to bay trees, however. It is not uncommon to find vegetated lakes, grass-sedge prairies, and cyprus-gum swamps in bays. Other bays support pond and loblolly pine and, in drier portions, scrub oaks.

The bays are relict features dating from between 6,000 to 40,000 years before the present. Since Toumey first described their elliptical shape in 1848, many theories of origin have been proposed. The idea that they are the result of a chance encounter with a passing black hole or comet, which showered the east coast with large ice fragments, was initially well received. However, in recent years those ideas have fallen out of favor for a more earthly theory that the bays were formed by a combination of terrestrial factors. These factors include artesian springs creating surface ponds that were subsequently shaped by wind action. Thom (1970) maintains that thousands of years ago shallow ponds and lakes on the sandy Coastal Plain were subjected to prevailing southwest winds, which caused the bays to lengthen along the northwest-southeast axis, and for sand rims to develop on the leeward side of the lake or bay. Kaczorowski (1977) also argued that bay origin was closely tied to ponds and lakes, but differed from Thom in that he claimed that bay development was not restricted to sandy coastal plains, but could occur on silt and clay, as well as sand, a conclusion refuted by Gamble, Daniels, and Wheeler (1977).

The Upper Coastal Plain

The upper Coastal Plain lies between the Fall Line to the west and the Orangeburg Scarp to the east. The Orangeburg Scarp is a wave-cut feature indicating that some five million years earlier, during the early Pliocene, it was the shoreline. As the sea lapped against what would become the base of the Orangeburg Scarp, it deposited sand, clays derived from decaying marsh vegetation, and shell marls over the outer Coastal Plain. Referred to as the Duplin Formation, these recently deposited near-shore marine sediments coat the western portion of the outer Coastal Plain, making it much flatter than the upper Coastal Plain to the west. The most distinctive features of the upper Coastal Plain are the Sandhills. As described by Cabe and Reiman (1996), the Sandhills is a topographically distinc-

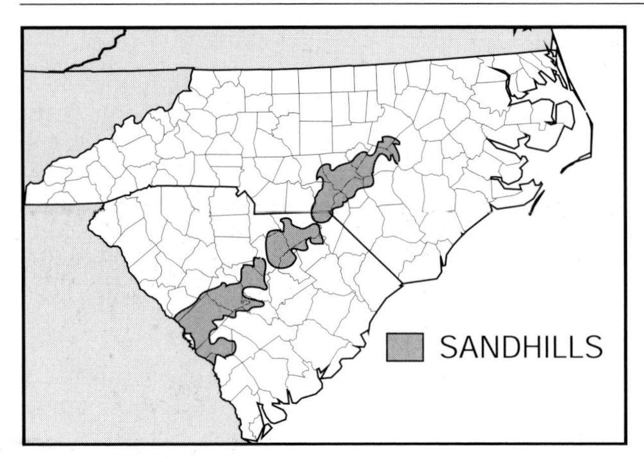

Figure 2.6. Extent of the Sandhills region (after Cabe and Reiman, 1996).

tive area of North and South Carolina, ranging from broad, nearly flat ridges to rolling hills (Figure 2.6). Compared to the Duplin Formation to the east, the sediments comprising the Sandhills are much older. The two major geologic units of the Sandhills are the Cretaceous age Middendorf Formation and the Eocene age Pinehurst Formation. The Middendorf sediments are fluvial in nature and were deposited about 90 million years ago by meandering streams; the Pinehurst sediments are marine sediments deposited in a near-shore environment about 45 million years ago, when world sea level was extremely high. At that time, virtually the entire Coastal Plain would have been beneath the sea.

The younger Pinehurst Formation overlies the Middendorf Formation in much of the Sandhills area and is typically found on the hilltops. Because of the high permeability of these sandy sediments, precipitation is easily absorbed and seldom runs over the surface. As a result, these broad ridges retain their original depositional character and are not highly dissected. However, in some areas, the Pinehurst sediments have significantly higher clay content that greatly lowers the permeability of the soil. In these areas, surface run-off is much more important, creating a more rolling topography.

The Piedmont

While the Coastal Plain is primarily a depositional landscape covered with relatively young sediment, the Piedmont is almost entirely an erosional landscape exposing very old rock units. The Piedmont extends from northern New Jersey to central Alabama. In the Carolinas, the Piedmont is a broad plateau sloping downward from the front of the Blue Ridge to where it slides beneath the younger sediment of the Coastal Plain. Its eastern boundary is marked by the Fall Line, and to the west, some 125 miles (200 km) away, the Brevard Fault and its northern extension, the Bowens Fault, separate it from the Blue Ridge Province. The underlying rock of the Piedmont is primarily metamorphic with scattered intrusions of igneous plutons, and the enclosed Triassic basins, which are made of younger sedimentary layers. Across the Piedmont, the surface has been finely dissected, though not very deeply. As noted by Rodgers (1970) "The entire Piedmont province is covered with a residual mantle of thoroughly weathered material, called saprolite, very uneven in thickness though generally increasing southward; in many places indeed such material extends well below the level of the streams, as shown in mines and in borings for dam foundations." The extent and depth of the saprolite is testament to the tremendous length of time that the region has been subjected to uninterrupted erosion. Perhaps the most recognizable features of the Piedmont are the scattered monadnocks, resistant remnants of ancient mountains that stand well above the general level of the Piedmont plateau (Photo 2.5). Monadnocks include Pilot Mountain, Hanging Rock and Crowders Mountain in North Carolina and Little Mountain, Glassy Mountain and Paris

Photo 2.5. Pilot Mountain, N.C. an erosional remnant or monadnock composed of metamorphic quartzite and gneiss (courtesy Robin Lynch Bennett).

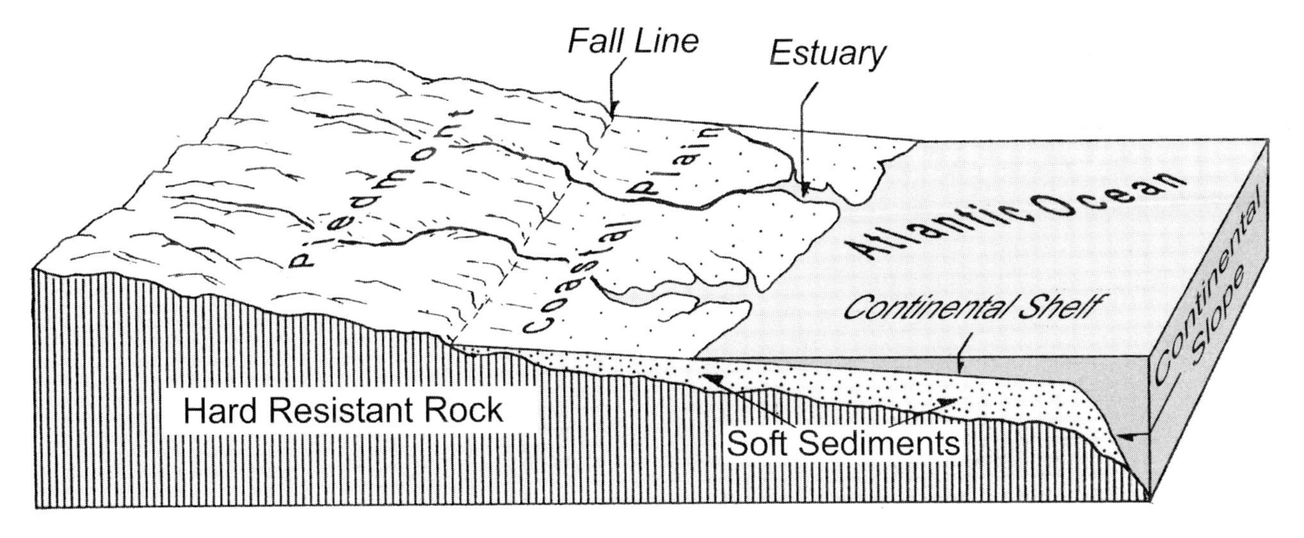

Figure 2.7. Development of the Fall Line (after Bennett and Patton, 2006).

Mountain in South Carolina. Not only are the monadnocks a result of differences in the erodibility of the local bedrock, but also so are most variations in relief of the Piedmont.

When rivers flowing across the resistant igneous and metamorphic rock of the Piedmont move on to the Coastal Plain, they immediately begin to erode downward more rapidly as they encounter the soft sedimentary deposits found there (Figure 2.7). The result is a sequence of rapids. Cities developed near the rapids because they provided an early form of cheap energy—first to power gristmills, and then later to drive turbines producing electricity. In addition, this was often the farthest point upstream that ships could navigate and, thus, they had to unload their cargo for transfer to wagons or rail. As a result, a string of cities developed up and down the Fall Line (Fall Zone). Trenton, Philadelphia, Baltimore, Washington, and Richmond developed along the northern edge where the Fall Line is most distinct. In North Carolina, the Fall Line is just barely visible, but still important enough to influence the settlement of Roanoke Rapids, Raleigh, and Fayetteville. In South Carolina, the Fall Line becomes more pronounced, helping to determine the location of Columbia.

Geologists have long considered the mountains of western North America to be formed when island chains and small micro-continents riding on the eastward-moving Pacific lithospheric plate collided with, and subsequently were compressed upon, the edge of the westward moving North American Plate. These slivers added to our western coastline are referred to as accreted, or exotic, terrane because their origin was far from their current resting place. The tremendous compressional force necessary to suture land masses together resulted in a great deal of metamorphism, deformation, faulting and thrusting-up of the pre-existing rock. Today, we suspect that a similar series of events played out along the eastern shore of North America beginning some 1.1 billion years ago. At that time, the world was divided into two large continental masses, Laurentia and Gondwanaland. Gondwanaland contained the land that would become Africa, South America, Antarctica, India and Australia, and Laurentia comprised the remaining world's landmasses including the ancestral North America. As a part of Laurentia, the ancestral North America began drifting eastward toward Gondwanaland, overriding the intervening oceanic crust of the ancient Iapetus Ocean. As it moved eastward, island archipelagos, bits of the sea floor, and continental fragments riding on the subducting oceanic crust were added to the continental margins.

Inevitably, Laurentia and Gondwanaland closed the Iapetus Ocean and collided head on. The tremendous pressure and heat of that collision metamorphosed much of the rock all along the continental margins—sandstone to quartzite, shale to slate, granite to gneiss, and so on.

Geologists believe that this scenario of Laurentia rifting apart forming a proto-Atlantic Ocean, only to later have it close again as Europe and Africa collided with the ancestral North America, played out several times. The periods of continental collision were times of mountain building as rock layers were folded and thrust upward by the great compressional forces. These episodes, known as orogenies, were also periods when the pre-existing rock was metamorphosed by the pressure and heat of continental collisions. Four orogeny episodes have been identified in the Piedmont and the mountains to the west. They are the Grenville, which occurred in the late Precambrian (1.1 billion years ago), the Taconic during the late Ordovician (470 to 440 million years ago), the Acadian in Devonian time (380 to 340 million years ago), and the most recent, the Alleghenian, which began in the late Carboniferous and reached its peak during the Permian (330 to 270 million years ago) (Zullo and Horton, 1991). Each of these collisions accreted new terrane to the edge of the continent; today, those fragments of accreted crustal material are reflected in the various metamorphic belts of the Carolinas.

Piedmont Belts

The rock underlying the Piedmont has long been divided into a series of geological belts, that is, regions of similar rock having a similar geological history (Figure 2.8). These belts run roughly parallel to one another from the southwest to the northeast. The belts are composed primarily of metamorphosed rock, indicating a landscape that was subject to the intense pressure and heat that only great mountain building events are capable of supplying. The easternmost of the belts, the Eastern Slate Belt, contains metamorphosed volcanic and sedimentary rock formed during the Taconic Orogeny. Intruding through the Eastern Slate Belt are much younger granitic bodies, called plutons, probably created when magma welled upward during the more recent Alleghenian Orogeny. The Raleigh Belt running from the city of Raleigh northward to

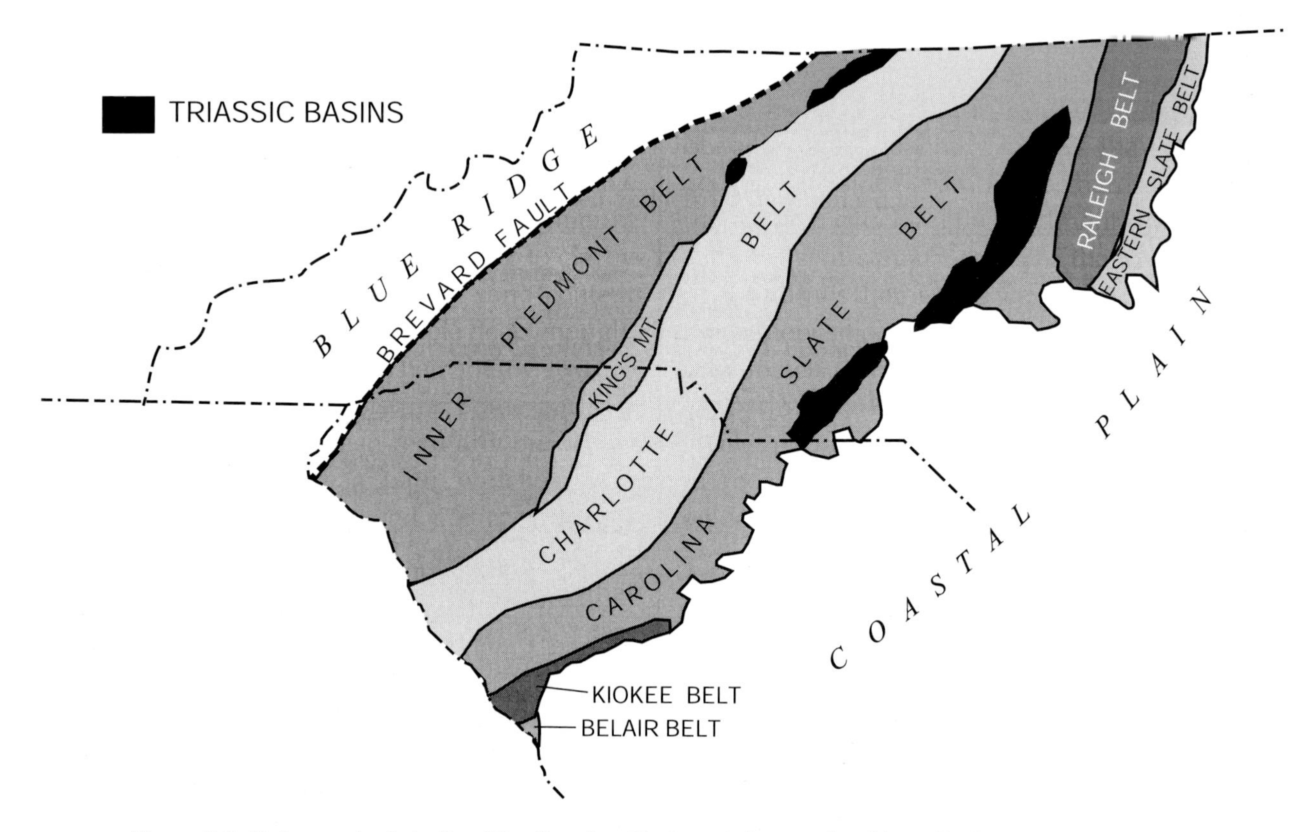

Figure 2.8. Major geologic belts of the Carolina Piedmont. (generalized from Butler, 1991 and others)

the Virginia border is mainly gneiss and schist metamorphosed from granite during the Alleghenian Orogeny. The largest of the belts, the Carolina Slate Belt, is composed almost entirely of chlorite and biotite grade metamorphic rock resulting from the Taconic Orogeny. As in the Eastern Slate Belt, numerous granite plutons became imbedded throughout this belt during the subsequent Alleghenian Orogeny.

The rising magma that cooled and formed these plutons is believed to mark an ancient subduction zone. It is thought that the lighter, less dense and much thicker North American plate overrode the thin oncoming oceanic plate. In this process, the oceanic crust was forced down into the asthenosphere where it melted. Since the oceanic crust was quite dense it sank into the depths of the asthenosphere. A small percentage of the melting crust was quite light and attempted to rise toward the surface like a balloon released at the bottom of a swimming pool. In some cases, the magma flowed through fractures and faults formed when the rock layers could no longer withstand the incredible strain of the great continental collision. Rising rapidly through these passages, the magma would have poured out onto the surface as lava flows and been ejected as volcanic eruptions. Some of the rising magma cooled and solidified before reaching the surface, forming intruded plutons. If the pluton came near enough to the surface, it would force the overlying rock layers to bulge upward in great arches. In several places on the Piedmont, the granitic cores of these ancient arches have become exposed, as overlying rock layers have eroded away. Such is the case for Stone Mountain and Medoc Mountain in North Carolina and Table Rock, Caesar's Head, Forty Acre Rock, and Liberty Hill in South Carolina. The low-lying Uwharrie Mountains, just to the east of Albemarle, N.C., represent an area where the magma reached the surface in a series of explosive volcanic eruptions. In the Uwharries, one finds lava flows, lava domes, and plugs (Conley and Bain 1965; Seiders 1981).

As the magma inched upward through the pre-existing rock, the contact between the incredibly hot molten material and the enveloping rock resulted in the mineralization of gold and other metals through extensive hydrothermal alteration of the surrounding rock. The first gold mine in North America, the Reed gold mine, began operation in 1802 in Cabarrus County, North Carolina. Three years earlier, Conrad Reed, the young son of a German immigrant farmer, John Reed, found a yellowish rock reportedly weighing seventeen pounds while fishing in Little Meadow Creek on the family property. Not knowing what he had found, John Reed reportedly accepted $3.50 for the unusual rock, offered by a Fayetteville jeweler. Reed later demanded and received a fair price for the nugget, after he learned that the jeweler had sold it for thousands of dollars. The story of the find and the jeweler's attempted larceny spread quickly, sparking the first gold rush in the United States (Stine and Stine, 1996). The most productive of the Carolina gold mines was the Haile mine in Lancaster County, South Carolina, which produced over 300,000 oz. of gold (Bell, 1986). Until the discovery of gold at Sutter's Mill in California, more gold was being taken from the Carolina Slate Belt and the neighboring Charlotte Belt than from any other region of the country.

Like its neighbor to the east, the Charlotte Belt was also metamorphosed during the Taconic orogeny and includes a great number of plutons and the remnants of volcanic eruptions. It appears that the Charlotte Belt is the core of an ancient volcanic island archipelago that collided with, and was subsequently added to, the margins of Laurentia. Ash blown out of volcanoes erupting during the collision blanketed much of the Charlotte Belt, as well as the nearby Carolina Slate Belt and Kings Mountain Belt. In areas near volcanic eruptions, the ash was still incandescent when it fell to the surface, and the hot particles stuck together, creating a hard resistant rock called a welded tuff. Outcrops of welded tuff are found throughout the Charlotte Belt. The boundary between the Kings Mountain Belt and the Charlotte Belt is poorly differentiated. The rock in the Kings Mountain Belt underwent a lower-grade metamorphism than did the rocks of the Inner Piedmont Belt to the west. This means that minerals in the Kings Mountain Belt are more stable at the surface and take longer to chemically weather. The result is that the narrow Kings Mountain Belt is erosionally more resistant and, thus, topographically higher than the Inner Piedmont Belt (Murphy, 1995). The western edge of the belt is marked by the Kings Mountain shear zone and its southern extension, the Lowdensville shear zone.

In the western Piedmont, the rock of the Inner Piedmont Belt, again primarily metamorphic, appears to be a series of highly eroded thrust sheets. A thrust sheet is a compressional feature created as rock first fractures; then, one fragment is pushed up and over the other. (See Figure 2.2.) Some geologists believe that the Inner Piedmont Belt represents a piece of North America that was torn from the continent during the Precambrian and then reattached during the Taconic Orogeny. Others believe that it is a piece of exotic terrane that was pushed up onto the North American coast during the Taconic Orogeny (Murphy, 1995). The Brevard and Bowens Faults mark the western edge of the Inner Piedmont Belt and the beginning of the Blue Ridge Province.

The Brevard Fault is a one-to-two mile wide zone of highly shattered rock. It extends from Virginia to Alabama and clearly separates the Blue Ridge from the Piedmont. The degree to which the rock has been crushed, and the extent of its depth and width, testifies to the incredible power and movement that created this thrust fault (Murphy, 1995).

Triassic Basins

Eventually, Gondwanaland and Laurentia sutured together, forming the great supercontinent Pangea III. Pangea III lasted some 100 million years and then began to split apart, starting some 250 million years ago during the Triassic Age. As it did, the crust near the split or rift was stretched and thinned. This crustal extension would have been similar to what is occurring in the Great Rift Valley of eastern Africa today. From outer space, the African Rift Valley appears as a 4,000-mile long scar on the continent's surface. It marks the line where Africa is literally ripping apart. Earthquakes are common, and portions of the crust slump down, forming a long series of valleys beneath ragged scarps as eastern Africa attempts to tear itself away from the rest of the continent. In the wet tropical rainforest of central Africa, great lakes, including Victoria, Albert, Tanganyika, and Malawi, have filled the rift valleys. In the drier northern reaches of the rift, seasonal streams flow in before evaporating beneath the hot tropical sun. The valleys become sediment traps, as streams flowing into these basins empty their load of pebbles, sand, silt and clay. The lakes also collect sediment. Some sediment originates from decaying vegetation, some is wind-blown material, and still other is the sand and silt carried by small in- flowing streams. The mix of all of these materials makes the fine lacustrine (lake) muds that ooze up between a swimmer's toes.

A series of rift valleys of Triassic age (225 million years ago), trending from the southwest to the northeast, are found along the eastern seaboard of the United States. In North Carolina, several of the Triassic Basins are exposed on the surface, including the Dan River-Danville, Durham, Sanford, Wadesboro, and Davie County Basins. In South Carolina, the Florence and South Georgia Basins lie beneath more recent sediment deposits of the Coastal Plain. In addition, several more rift basins have been identified beneath the shallow water of the continental shelf off the coast of the Carolinas (Olsen, et al 1991). Like their contemporary African counterparts, the Triassic Basins of the Carolinas were ideal sediment traps. At the bottom of the basins is found pebbly sandstone, indicative of fluvial (river) deposition. Above that are fine-grained, organically rich sediments, with many fossils indicative of lacustrine (lake) deposition. At the top of the basins is found a mix of lacustrine and fluvial sediments, indicating the final period of deposition when the basins had virtually been filled with sediment, the lakes were disappearing and small streams flowed across the surface.

The Blue Ridge Province

The Appalachian Mountains are a chain of low, but steep, mountains that run for nearly 2,000 miles (3,200 km) from Newfoundland in

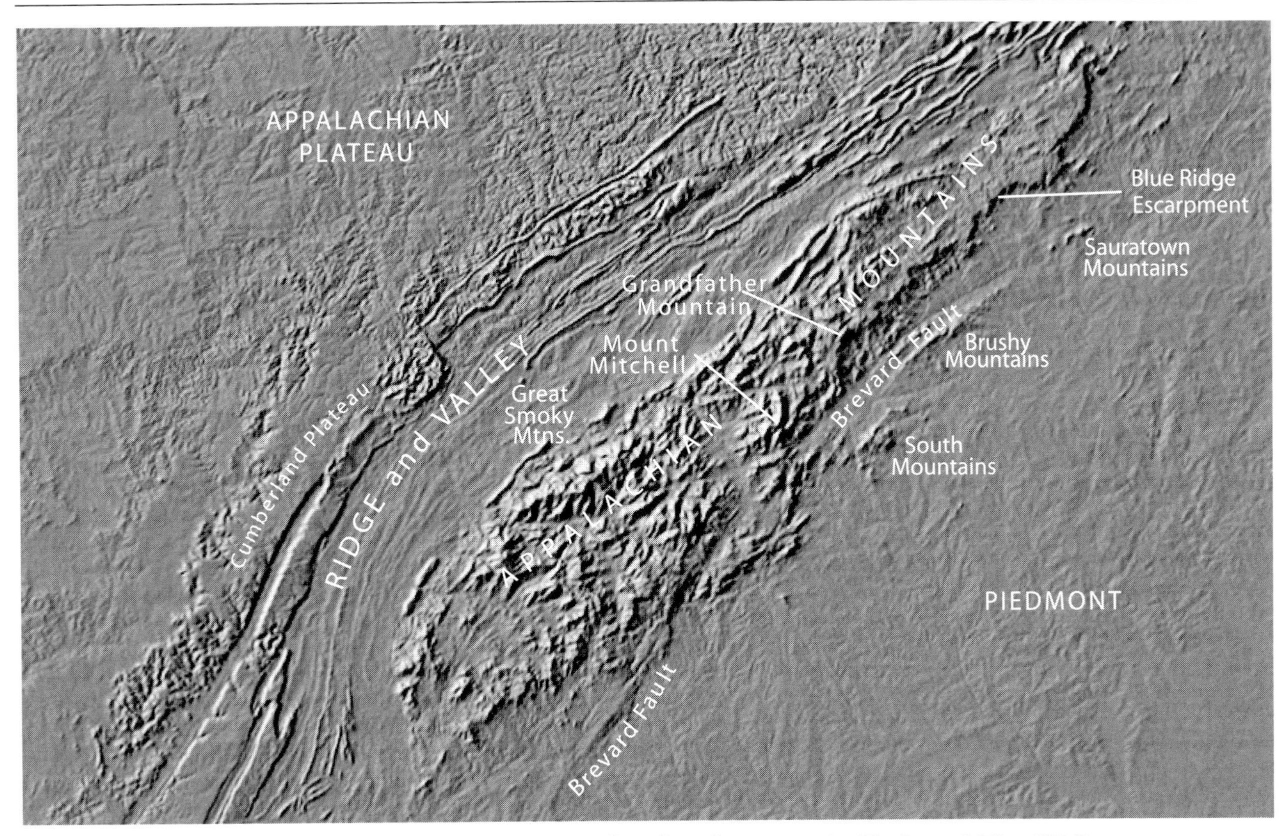

Figure 2.9. The mountains of western Carolina. (base map by Thelin and Pike, 1991).

the Gulf of St. Lawrence to central Alabama. Traditionally, the Appalachians have been divided into four physiographic provinces—the Piedmont, Blue Ridge, Folded Appalachians also known as the Ridge and Valley, and Appalachian Plateau. Only the Piedmont and Blue Ridge Provinces are found within the Carolinas. The Blue Ridge Province is composed of resistant metamorphic and igneous crystalline rock and extends for over 600 miles (965 km) from southern Pennsylvania to northern Georgia. It reaches its widest extent of about 65 miles in North Carolina and just barely clips the northwestern corner of South Carolina. The name Blue Ridge, in addition to being the term used for the entire province, is also the name given to a specific range within the province. Two other ranges, the Unaka Mountains (including the Great Smoky Mountains) and the Black Mountains, compose the Blue Ridge Province in North Carolina (Figure 2.9). Some of the highest and most famous peaks in the Appalachians are found in the Blue Ridge Province, including Mt. Mitchell, the loftiest peak in the Eastern United States at 6,684 feet (2,037 m), Clingman's Dome on the Tennessee border in Great Smoky Mountains National Park which rises some 6,642 feet (2024 m), Grandfather Mountain in the Blue Ridge range at 5,984 feet (1,824 m) and Sassafras Mountain, the highest point in South Carolina, standing 3,554 feet (1,083 m) above sea level. The Blue Ridge Province is noted for its very straight and high eastward-facing scarp, that marks its border with the Piedmont Province (Shimer, 1972). This dramatic scarp reaches a maximum elevation of 4,000 feet (1,220 m) near Boone, N.C. To the west in Tennessee, lies the Folded Appalachians composed of a series of folded sedimentary layers.

As was found in the Inner Piedmont Belt to the east, the Blue Ridge is composed of a series of thrust sheets that were pushed up and over other layers of rock, like a pack of playing cards, when Africa and Europe collided with North America for the final time during the Alleghenian Orogeny. When the great thrust sheets making up the Blue Ridge Province were pushed toward the continental interior, they in turn forced the sedimentary rock to the west to fold, as if someone had pushed a loose carpet up against a wall. The resulting sequence of low parallel ridges and valleys marks the western edge of the Blue Ridge Province. Much of the rock material of the Blue Ridge thrust sheets, originally on the continental margins of ancestral North America when the collision with Africa and Europe occurred, was heavily metamorphosed during the Alleghenian Orogeny. On the other hand, the rock comprising the Folded Appalachians, being much further into the interior of the continent and away from the greatest pressure of the collision zone, was little deformed and remains almost entirely sedimentary. Those sedimentary layers are significantly softer than the crystalline rock of the Blue Ridge, and so they have been more thoroughly eroded. The result is that the Blue Ridge Province is topographically a much higher region than the Folded Appalachians.

In several places in the Blue Ridge, the overthrust sheets that were pushed up and over other layers of rock have eroded through in small local areas so that the underlying layers are now exposed. This creates an unusual stratigraphic sequence of older rock lying above younger rock. Called windows (or fensters), they have been invaluable in understanding the tectonic events that created the Appalachians and the Blue Ridge. Important windows are found in the "coves" of Great Smoky Mountains National Park and at Grandfather and Stone Mountains. The Sauratown Mountains on the Inner Piedmont Belt are also an example of a thrust sheet window; however, in this case, the underlying rock brought to view through the window is basement rock from the Grenville Orogeny, over 1.1 billion years old. This is some of the oldest exposed rock in the southeastern United States.

Unlike their northern counterparts, the Southern Appalachians were never subjected to continental or alpine glaciation. Instead, the erosional work was accomplished by the disintegration and decomposition of the rock layers through a series of chemical and mechanical weathering agents. Mechanical processes include the freezing and thawing of water in fine cracks, the growth of plant roots in joints and fractures, the expansion and contraction of rock due to diurnal and seasonal temperature changes, and the process of unloading, or pressure release fracturing. In this latter process, rock, which is at depth and under tremendous pressure, expands as the overlying rock gradually erodes away. Because rock is brittle, it cracks during the expansion. Exposed granite plutons found in the Blue Ridge Province are particularly vulnerable to this type of weathering. Called "exfoliation domes," they fractured by unloading, with the resulting thin concentric layers eroding away like layers peeling off a large onion (Photo 2.6). Notable exfoliation domes include Stone Mountain, Looking Glass Rock, and the Mt. Airy pluton (today one of the world's largest open-faced granite quarries).

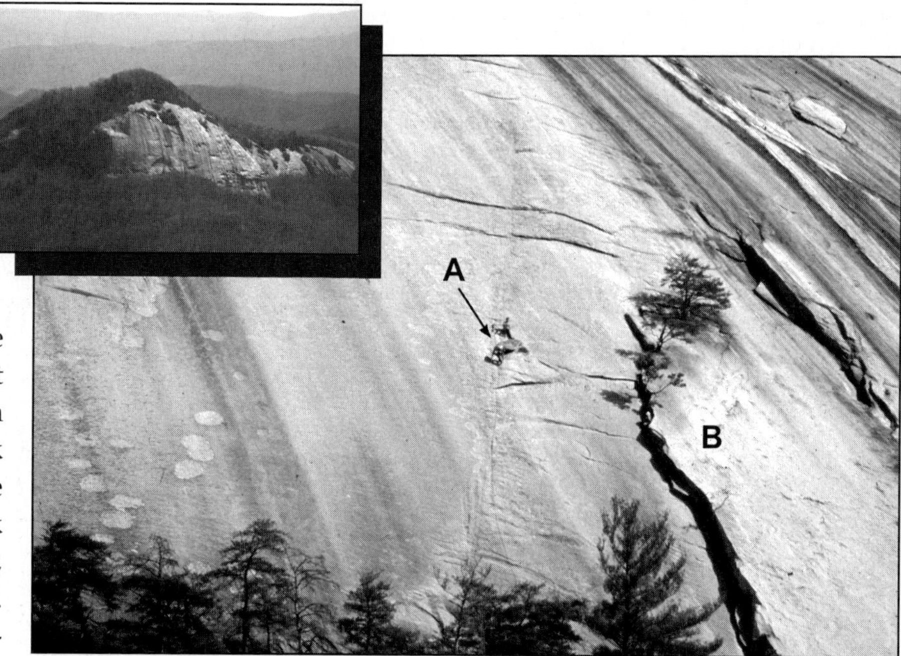

Photo 2.6. Stone Mountain, N.C. Letter "A" indicates two climbers making the ascent of this spectacular exfoliation dome. Letter "B" refers to the large sheet of granite that is "peeling off" the mountain side. This form of erosion is typical of these exposed granitic plutons.

The most important chemical process at work on the surface of the Carolinas has been the solution of minerals by water, particularly by rainwater, which absorbs small amounts of carbon dioxide as it falls through the atmosphere. The combination of water and carbon dioxide produces a weak form of carbonic acid that is particularly effective at dissolving calcium-based material. Much has been written about how the strength of this acid rain has been increasing, a result of the burning of fossil fuels which releases carbon dioxide into the air.

In the Blue Ridge Province, once the surface rock had been broken, flowing water became the dominant erosional agent. The characteristic steep, smooth, and rounded look of the mountains of the Carolinas is typical of a landscape shaped by running water.

The River Systems

Through the eastern section of the Blue Ridge runs the divide that separates streams flowing southeastward across the Piedmont and Coastal Plain to the Atlantic Ocean from those that flow northward or westward as part of the Mississippi River drainage system, ultimately emptying into the Gulf of Mexico. The streams, which flow across the Piedmont and Coastal Plain, are the largest of the Carolina rivers. They include the Roanoke, Cape Fear, Neuse, Yadkin/Pee Dee, Santee, and Savannah (Figure 2.10). The overall direction of their courses reflects the southeastward slope of the Piedmont and Coastal Plain. The drainage pattern of the eastern streams is mostly dendritic, meaning that the pattern of their tributaries is similar to that of branches

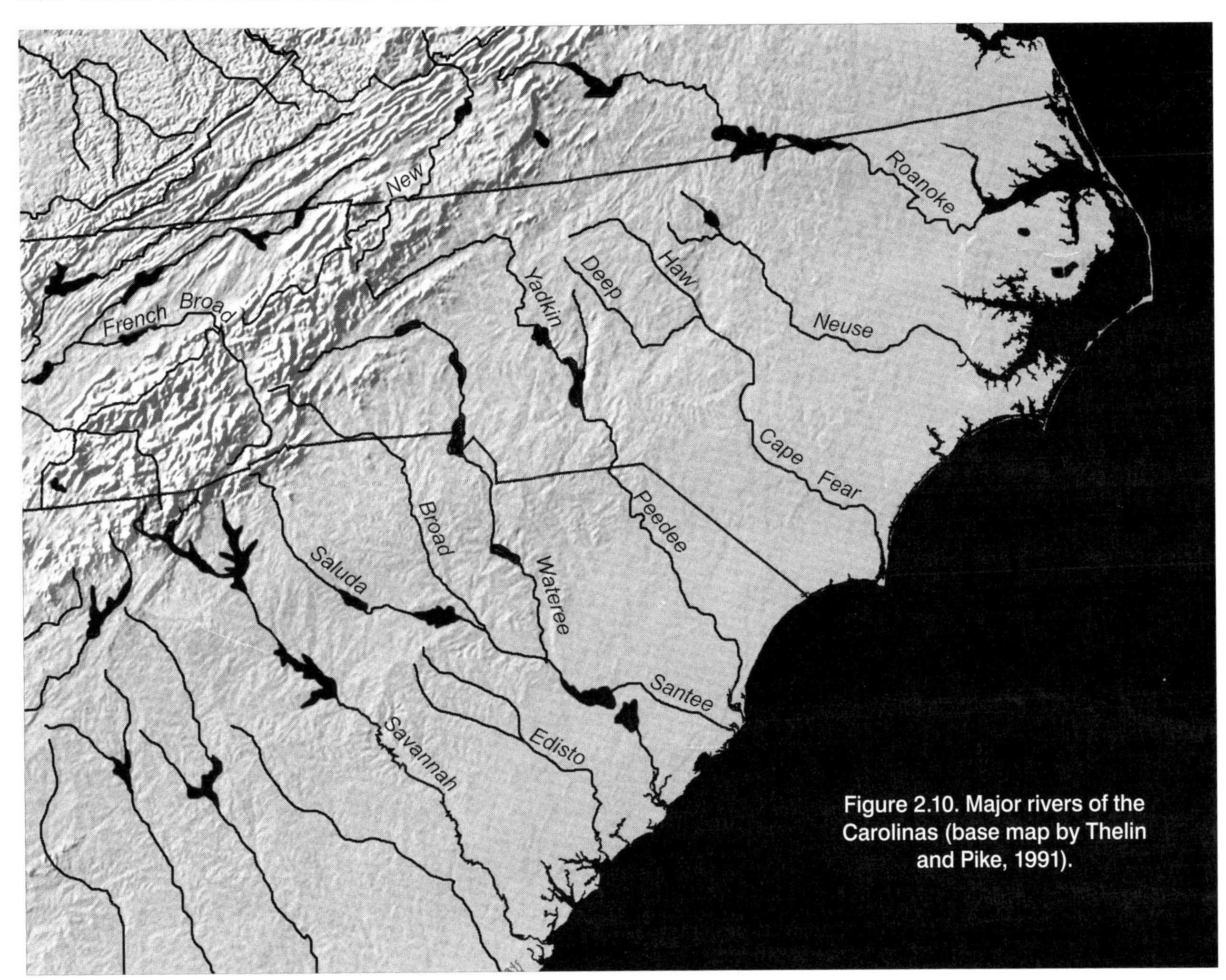

Figure 2.10. Major rivers of the Carolinas (base map by Thelin and Pike, 1991).

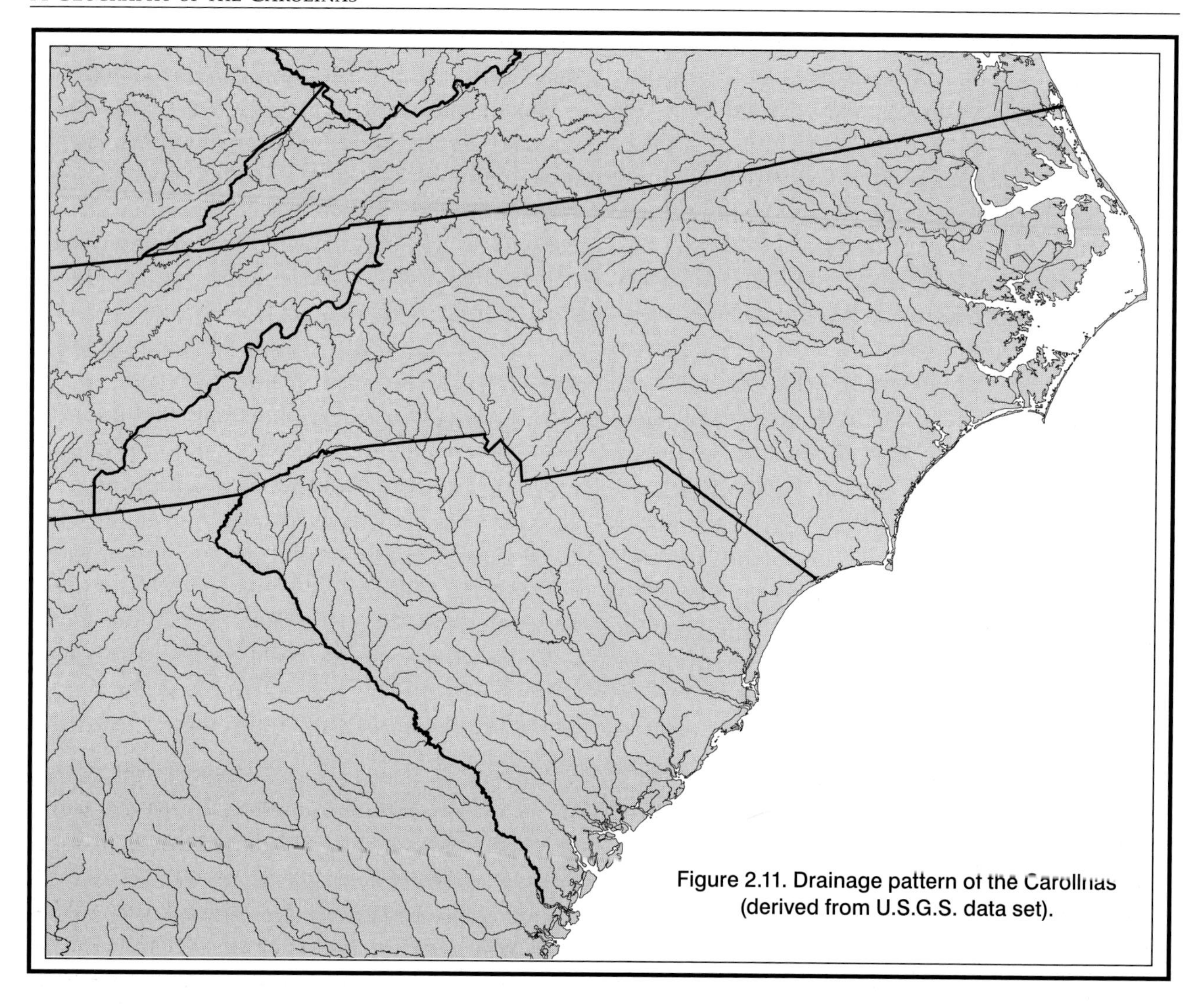

Figure 2.11. Drainage pattern of the Carolinas (derived from U.S.G.S. data set).

joining the main trunk of a tree (Figure 2.11). One would not expect this type of pattern on a surface that slopes as steeply as the Piedmont plateau. What the dendritic pattern suggests is that the streams' courses were established when the Piedmont plateau was worn even lower than today and did not tilt to the same degree. On this relatively flat topography, dendritic drainage would naturally have developed. The branching pattern may have become permanently etched into the landscape as the Piedmont slowly began to rise. Pavich (1985, 1986) suggested that this ongoing uplift of the Piedmont is a result of the weight material lost by erosion. Called isostatic adjustment, the concept is much like that of unloading a ship. As material in the ship's hull is removed, the ship becomes lighter and so rides higher in the water. In the case of the Piedmont and Blue Ridge, as material erodes off and is deposited in the sea, the portion of the lithospheric plate that they are riding upon rises. Pavich also suggests that streams of the Piedmont are now flowing across a landscape in equilibrium where the amount of material being removed is roughly balanced by the amount of uplift.

Two major streams drain the western portion of North Carolina—the French Broad, which flows into the Tennessee before moving onto the Ohio and Mississippi Rivers, and the New River, whose water flows north to the Gauley and Kanawha Rivers before emptying into the Ohio. A 26.6 mile (42.8 km) section of the New River in Ashe and Allegheny Counties, was designated a national Wild and Scenic River in 1976. While this river undoubtedly deserves this designation, it is interesting to note that supporters of the designation frequently cited the superlative age of the New River as a reason for its preservation as a free-flowing stream (Schoenbaum, 1979). The promotion of the antiquity of the New River was so effective that the label of "oldest river" in North America and second oldest in the world, following the Nile, appears to have become an accepted fact by 1977 (Mayfield and Morgan, 1996). The primary evidence to support the idea that the New is a very old river involves the fact that the New River is an unusual stream in that it flows northward across three physiographic provinces of the Appalachians—the Blue Ridge Province, Folded Appalachians, and Appalachian Plateau. In this scenario, the New is believed to be an antecedent stream, meaning that it flowed in its present course before the existence of the present topography. The original surface through which the New flowed was hypothesized to have been a region reduced by erosion to a relatively flat featureless area called a peneplain. The next step was a very gentle uplift of the Appalachians allowing the New to cut down and thus become incised into the surface. As the land continued to rise, the New continued to react by cutting ever downward through the underlying folded structure (Johnson, 1931). As noted by Marie Morisawa (1989), Johnson's theory is appealing in its simplicity, but it suffers from a lack of evidence. Most scholars of Appalachian geology and hydrology agree that the New undoubtedly is a very old river; however, they find no evidence that the New is any older than other streams that traverse the Appalachians (Smosma, 1977; Lessing, 1986). As Mayfield and Morgan write, "it is difficult to understand why the New River was transformed from an old river to the 'oldest river' in the 1970s." (Mayfield and Morgan, 1996)

Conclusion

The geologic history of the Carolinas is extremely complex, and our understanding of it is continuously in a state of flux. As new information is unearthed, some concepts will be discarded and others will be formed to account for the new information. However, the general framework appears to have been established. The evidence is clear that the Appalachians are an old and heavily eroded mountain chain formed by a series of continental collisions. These great collisions compressed the surface, thrusting great layers of rock upward and metamorphosing the pre-existing rock of the Piedmont and the Blue Ridge Province. Those collisions also caused molten magma to intrude into the underlying basement rock, where in some cases it cooled forming plutons, or in other cases found cracks and faults that provided passage to the surface where the lava spilled out, forming volcanic peaks. For the last 200 million years, the east coast has been a passive margin. During this time, little tectonic activity has been taking place. The primary geologic events occurring have been the wearing down of the mountains and Piedmont by erosion and the subsequent transportation and deposition of the sediment onto the coastal plain by rivers. Located on the Piedmont are numerous remnants of the greater mountains that once existed there. These monadnocks exist because they are composed of rock that is harder or less fractured than the surrounding rock.

The Coastal Plain—that broad relatively flat surface—has been the recipient of the sediment eroding off of the highlands to the west and of deposits laid down when the sea has periodically covered it. More or less of the Coastal Plain has been exposed, in recent geologic history, as world sea levels rose or fell—dependent on the degree of the thermal expansion of ocean water and the amount of water held in the earth's glacial ice caps. The coastal margin is the most ephemeral of the Carolina landscapes. A single storm can create a new inlet in the barrier islands or close an existing one. The strings of islands along the Carolina shores are ever moving; some grow larger, while others are quickly disappearing.

There are states that have more awe-inspiring canyons, higher mountains, and larger rivers, but few states can claim the variety and beauty of landscapes that are the Carolinas. It is both a primeval landscape and a recent landscape produced by the unimaginable force of continents colliding and the everyday drifting of grains of sand by waves lapping the shore.

References

Bell, H. III. 1986. Some Geological Aspects of the Haile Gold Mine, Lancaster County, South Carolina. In *Volcanogenic Sulfide and Precious Metal Mineralization in the Southern Appalachians*. Knoxville: University of Tennessee, Department of Geological Sciences, Studies in Geology 16, 124-145.

Bennett, D. G. and J. C. Patton. 2006. *The United States and Canada: A Systematic Approach.* Dubuque, Iowa: Sheffield Publishing Co.

Bennett, S. H. and J. B. Nelson. 1991. *Distribution and Status of Carolina Bays in South Carolina.* Nongame and Heritage Trust Publications. Columbia, S.C.: Wildlife and Marine Resources Department.

Cabe, S. and R. Reiman. 1996. Sandhills of the Carolinas. In *Snapshots of the Carolinas: Landscapes and Cultures,* ed. D. G. Bennett, Washington D.C.: Association of American Geographers, 83-86.

Conley, J. F. and G. L. Bain. 1965. Geology of the Carolina Slate Belt West of the Deep River Wadesboro Triassic Basin, North Carolina. *Southeastern Geology* 6: 117-118.

Corbett, J. 1989. Our Mysterious Bays, Their Origin and History. *Bladen Weekly,* 13 December, 1.

Fisher, J. J. 1962. Geomorphic Expression of Former Inlets along the Outer Banks of North Carolina. MS Thesis, University of North Carolina at Chapel Hill.

_____. 1967. Development Pattern of Relict Beach Ridges, Outer Banks Barrier Chain, N.C. Unpublished Dissertation, University of North Carolina at Chapel Hill.

Gamble, E. E., R. B. Daniels, and W. H. Wheeler. 1977. Primary and Secondary Rims of Carolina Bays. *Southeastern Geology* 18: 199-212.

Hayes, M. O. 1994. The Georgia Bight Barrier System. In *Geology of Holocene Barrier Island Systems.* ed. R. A. Davis, Berlin: Springer-Verlag.

Horton, J. W. and V. A. Zullo. 1991. An Introduction to the Geology of the Carolinas. In *The Geology of the Carolinas: Carolina Geological Society Fiftieth Anniversary Volume,* eds. J. W. Horton and V. A. Zullo, Knoxville: The University of Tennessee Press, 1-10.

Johnson, D. W. 1931. *Stream Sculpture on the Atlantic Slope: A Study in the Evolution of Appalachian Rivers.* New York: Columbia University Press.

_____. 1942. *The Origin of the Carolina Bays.* New York: Columbia University Press.

Kaczorowski, R. T. 1977. *The Carolina Bays: A Comparison With Modern Oriented Lakes.* Technical Report No. 13-CRD. Columbia: Coastal Research Division, Department of Geology, University of South Carolina.

Lessing, P. 1986. Geology of the New River Gorge. *Mountain State Geology* 1986: 48-55.

Mayfield, M. W. and J. T. Morgan. 1996. The New River. In *Snapshots of the Carolinas: Landscapes and Cultures,* ed. D. G. Bennett, Washington D.C.: Association of American Geographers, 87-90.

Morisawa, M. 1989. Rivers and Valleys of Pennsylvania, Revisited. *Geomorphology* 2: 1-22.

Moslow, T. F. and S. D. Heron. 1994. The Outer Banks of North Carolina. In *Geology of Holocene Barrier Island Systems,* ed. R. A. Davis, Berlin: Springer-Verlag.

Murphy, C. H. 1995. *Carolina Rocks!* Orangeburg, S.C.: Sandlapper Publishing Co.

Olsen, P. E., A. J. Froelich, D. L. Daniels, J. P. Smoot, and P. J. Gore. 1991. Rift Basins of Early Mesozoic Age. In *The Geology of the Carolinas: Carolina Geological Society Fiftieth Anniversary Volume,* eds. J. W. Horton and V. A. Zullo, Knoxville: University of Tennessee Press, 142-170.

Pavich, M. J. 1985. Appalachian Piedmont Morphogenesis: Weathering, Erosion, and Cenozoic Uplift. In *Tectonic Geomorphology: Proceedings of the 15th Annual Binghamton Geomorphology Symposium*. Eds. M. Morisawa and J. T. Hack, Boston: Allen and Unwin, 299-319.

_____. 1986. Processes and Rates of Saprolite Production and Erosion on a Foliated Granitic Rock of the Virginia Piedmont. In *Rates of Chemical Weathering of Rocks and Minerals*, eds. S. M. Colman and D. P. Deither, New York: Academic Press, 551-590.

Prouty, W. F. 1952. Carolina Bays and Their Origin. *Bulletin of the Geological Society of America*, 63: 167-224.

Rodgers, J. 1970. *The Tectonics of the Appalachians*. New York: John Wiley & Sons.

Ross, T. E. 1996. Carolina Bays: Coastal Plain Enigma. In *Snapshots of the Carolinas: Landscapes and Cultures*, ed. D. G. Bennett, Washington, D.C.: Association of American Geographers, 77-81.

Schoenbaum, T. J. 1979. *The New River Controversy*. Winston-Salem, N.C.: John F. Blair Publisher.

Seiders, V. M. 1981. *Geologic Map of Asheboro, North Carolina and Adjacent Areas*. U. S. Geological Survey Miscellaneous Investigations Series Map I-1314, scale 1:62,500.

Shimer, J. A. 1972. *Field Guide to Landforms in the United States*. New York: Macmillan Publishing Co.

Smosma, R. A. 1977. Geological Facts about the New River. *Mountain State Geology*, December: 51.

Soller, D. R. 1988. Geology and Tectonic History of the Lower Cape Fear River Valley, Southeastern North Carolina. U. S. Geological Survey, Professional Paper 1466-A.

_____ and H. H. Mills. 1991. Surficial Geology and Geomorphology. In *The Geology of the Carolinas: Carolina Geological Society Fiftieth Anniversary Volume*, eds. J.W. Horton and V. A. Zullo, Knoxville: University of Tennessee Press, 290-308.

Stine, L., F. and R. S. Stine. 1996. *An Archaeological Search for a Blacksmith Shop, Site 31Ca29, Reed Gold Mine State Historic Site Cabarrus County, North Carolina*. Raleigh, N. C.: Gold History Corporation and the Department of Cultural Resources, Division of Archives and History, Historic Sites Section.

Thelin, G. and R. Pike. 1991. *Landforms of the Conterminous United States*. United States Geological Survey Map 1-2206.

Thom, B. G. 1970. Carolina Bays in Horry and Marion Counties, South Carolina. *Geological Society of America Bulletin* 81: 783-814.

Chapter 3

CLIMATE, VEGETATION AND SOILS

Gregory Carbone[*] And John Hidore[**]

[*]University of South Carolina at Columbia

[**]University of North Carolina at Greensboro

North and South Carolina are nearly midway between the equator and the North Pole. This location between the extreme heat of the tropics and the cold of the polar region results in a climate that is one of moderate temperatures and, yet, has seasonal weather that can be hot in summer and cold in winter. Prevailing winds from the south and southwest bring rainfall much of the year. The climate throughout the region is similar in seasonal patterns of temperature, moisture and weather systems. The climate varies primarily because of differences in topography and proximity to the Atlantic Ocean.

Landforms provide the most visible expression of the physical geography of the Carolinas. The Appalachian Mountains (or Blue Ridge Province), Piedmont, Sandhills, Coastal Plain , and Tidewater regions are the stage on which physical processes occur. Climate, vegetation, and soils are the actors on this stage and their stories are interconnected. The elements of climate—temperature, moisture, and sunlight—influence vegetation. They also influence the rate of decay of plant litter and the transport of materials important for soil development. Vegetation supplies organic material and nutrients to the soil and influences climate by altering the moisture and energy budgets. Soils provide nutrients for vegetation and their texture governs moisture supply available for plants. Since climate is often regarded as the driving factor of physical geography, we will first describe the climate of the Carolinas, followed by a discussion of broad vegetation and soil patterns, and an examination of the interaction among the three using case studies that illustrate the five topographic regions within the Carolinas.

Climate

Hot, humid summers, mild winters, and ample precipitation throughout the year characterize the climate of the Carolinas. The climate is shaped by several factors, including latitude, prevailing pressure and wind systems, the southern Appalachian Mountains, and the proximity of a warm ocean current. To understand their effects, let us examine these factors and the resulting patterns seasonally.

Summer

Summers are hot in the Carolinas; maximum summer temperatures frequently exceed 90°F—the result of intense solar heating. *Mean* July temperatures range from 80° F on the south coast to 65° F in the mountains (Figure 3.1). Daily high temperatures are moderated along the coast where sea breezes draw cooler air inland from the Atlantic Ocean. Because temperature drops with altitude, the mountainous parts of the Carolinas are also cooler.

Summers in the Carolinas are humid. The moisture-laden air is due to two factors. One is the proximity of the region to two large moisture sources—the Atlantic Ocean and the Gulf of Mexico. The second is the wind system, which carries moisture from these large expanses of

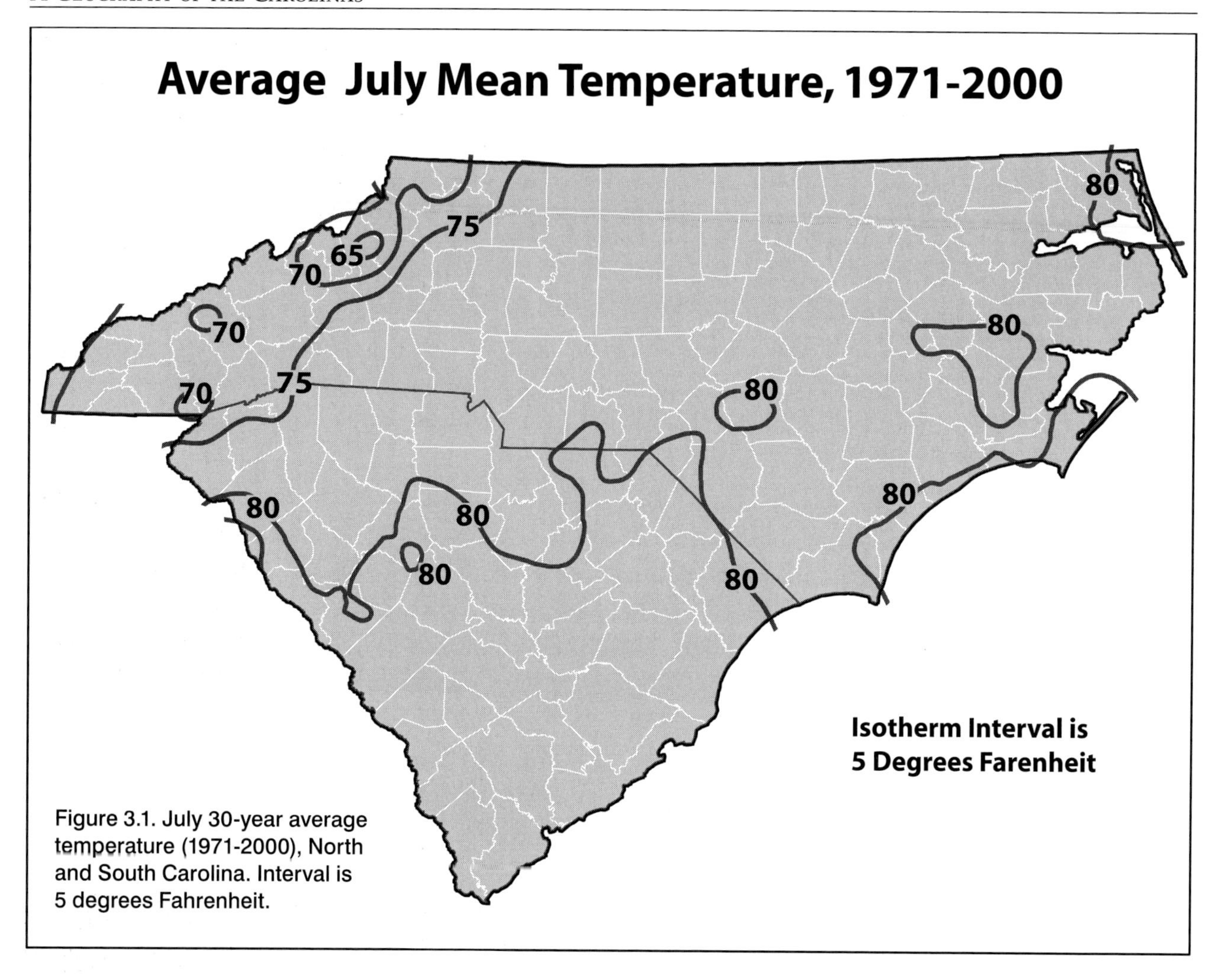

Figure 3.1. July 30-year average temperature (1971-2000), North and South Carolina. Interval is 5 degrees Fahrenheit.

water over the Carolinas. These winds, typically from the south or southwest, are often part of a semi-permanent subtropical high pressure cell in the Atlantic called the Bermuda High (Figure 3.2). The large quantity of moisture in the atmosphere provides an abundant source for precipitation. Annual precipitation ranges from approximately 40 inches in the northern Piedmont to over 80 inches in the southern Appalachians (Figure 3.3). This precipitation is spread fairly evenly through the year, but results from mechanisms that differ seasonally.

Most summer precipitation comes from thunderstorms that develop as intense surface heating forces warm, moist air aloft. Precipitation from these thunderstorms can be very intense, but usually of short duration. Thunderstorms are essential in sustaining agricultural crops during the summer when temperature and evaporation rates are high. During some years, the Bermuda High moves westward, establishing itself over the Southeast. When this happens, drier air develops, vertical lifting weakens, and clouds and precipitation are suppressed. In some summers, this latter process results in drought which damages crops.

Tornadoes

Severe thunderstorms produce heavy rainfall, frequent lightning, hail, and, on occasion, tornadoes. While the tornadoes that develop in the Carolinas tend to be less severe than those that form in the Great Plains, they produce casualties and cause considerable damage. Tornadoes have struck in all months of the year, but the majority occurs between March and June when the land surface begins to warm and warm, moist air clashes with cold polar air invading from the northern interior part of North America. Tornadoes most frequently develop between noon

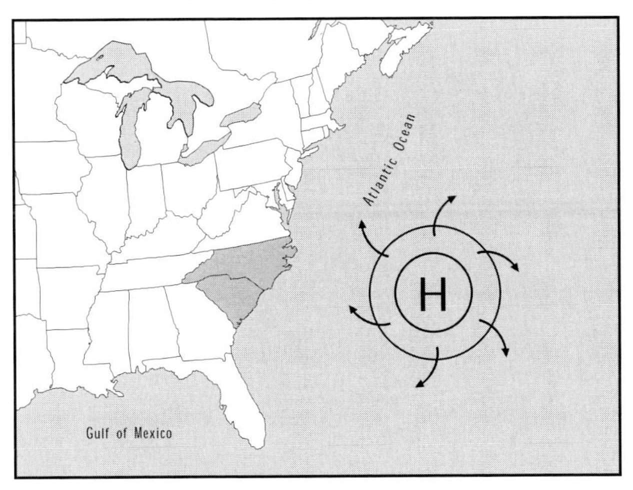

Figure 3.2. Location of Bermuda High

and 10 PM, peaking between 4 and 6. This is the time of maximum atmospheric heating and instability. On the ground, the average width of a tornado in the Carolinas is about 150 yards and the average length of travel is 8.1 miles. A tornado scale is the Fujita scale. In this scale, the less intense storms are considered class F1 tornadoes. The intensity increases to F5 for the most severe storms. Ranked by the Fujita scale, the most frequent class of tornadoes are in the F1 category. While tornadoes have been recorded in all classes, few category 3, 4, or 5 tornadoes have been recorded. Fewer tornadoes occur in the Appalachian Mountains than in other parts of the Carolinas, and there is a general increase in frequency as one moves from the mountains to the coast, probably due to increased daytime heating and increased atmospheric moisture. Since 1975, there has been an increasing number of tornados annually. It is unclear if this is an actual trend as a result of global warming, a matter of chance, or a result of more complete reporting recently.

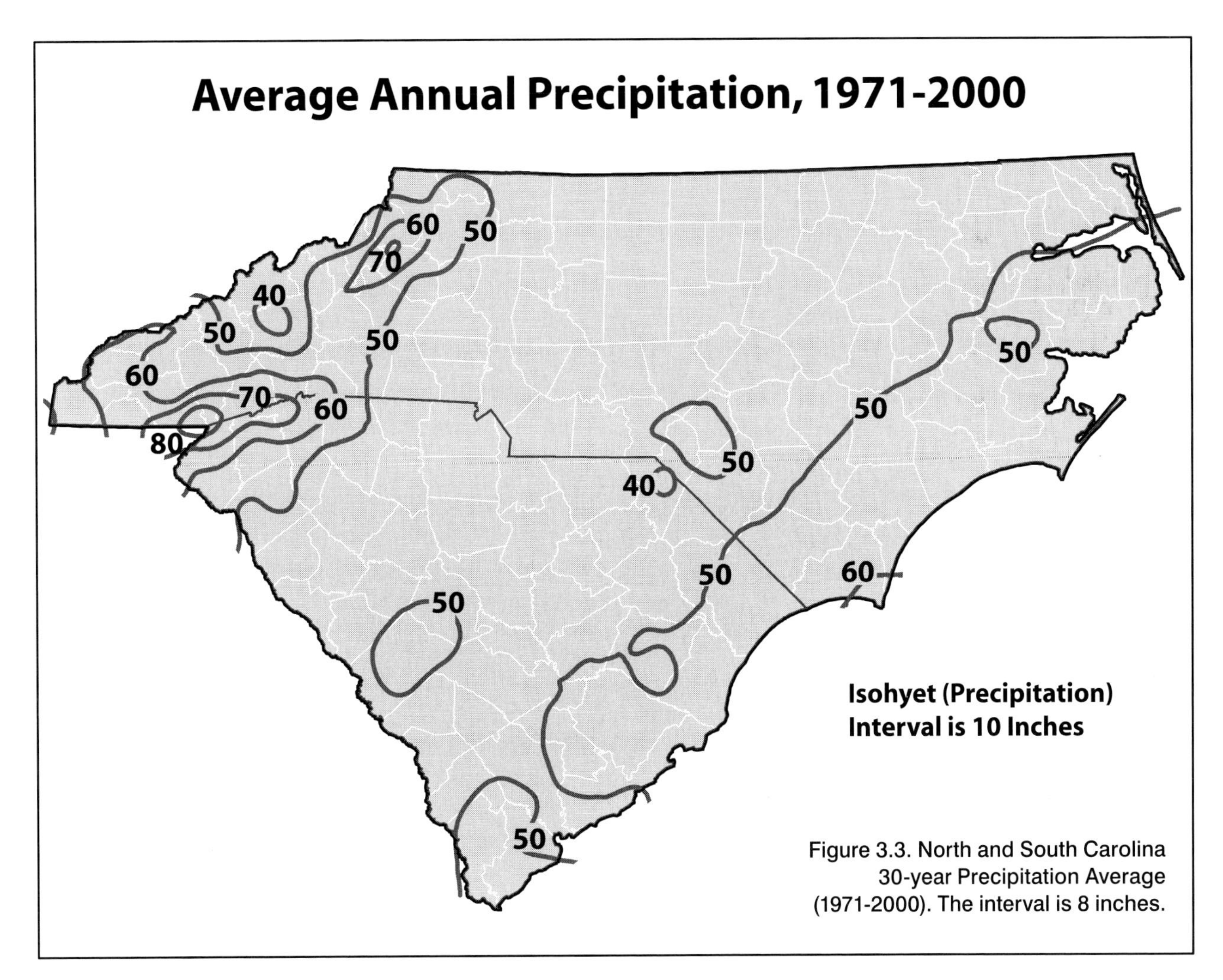

Figure 3.3. North and South Carolina 30-year Precipitation Average (1971-2000). The interval is 8 inches.

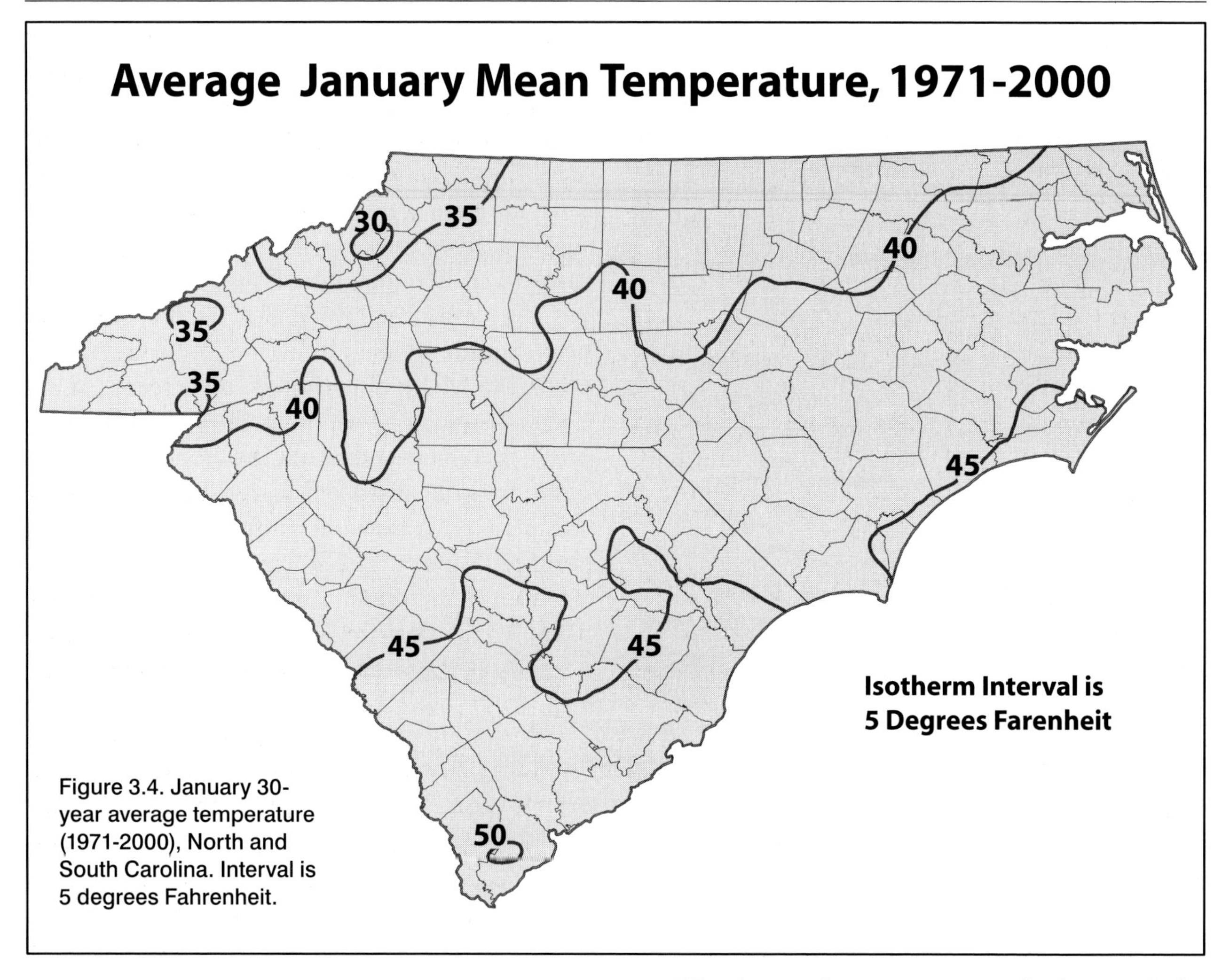

Figure 3.4. January 30-year average temperature (1971-2000), North and South Carolina. Interval is 5 degrees Fahrenheit.

Hurricanes

Tropical cyclones, or hurricanes, provide another source of rain in the warmer months. When they, or their remnants, strike the Carolinas, they can drop several inches of rain over a short period of time. These storms develop as low pressure systems, originating either in the Trade Winds west of Africa or over the Caribbean Sea. The path of those developing west of Africa often takes them in a westerly and then northerly direction, directed by circulation around the Atlantic subtropical high. These storms may strike North America anywhere from Florida to New England. Since the Carolinas lie midway along this coastline, they are often in the strike zone. Hurricane season extends from early June to late November, but peak strike time is from mid-August to mid-September when ocean temperatures in the source regions are warmest.

Hurricane damage can result from winds, flooding, heavy rainfall, and storm surge. The threat each hazard poses depends on a variety of factors including the intensity of the storm, the resulting storm surge height, and the time it makes landfall with respect to natural tides. Damage occurs to both natural and human environments. Hurricanes can alter the shoreline, killing wildlife and/or damaging their habitat. While improved forecasting has reduced human casualties in recent decades, extensive building in coastal areas has made the human landscape more susceptible to damage. Buildings, transportation networks, and utilities are all vulnerable. Damage and casualties are not limited to the coastal zone. For example, Hurricane Hugo, in 1989, caused significant damage from the South Carolina coast into Canada. Casualties directly related to the storm occurred as far north as Virginia. Flood damage from Hurricane Floyd extended across all of eastern North Carolina in 1999.

Winter

As with any location outside the tropics, winters in the Carolinas are cooler than summers because of shorter daylight hours and less intense solar radiation. With the solar noon sun approximately 30° above the horizon on December 21-22, solar intensity is only half that on the summer solstice. In addition, the Atlantic subtropical high weakens and becomes less persistent, limiting the delivery of mild subtropical air. Despite these circumstances, the region is warmer than it could be. Mean January temperatures are 50°F along the south coast, and 34°F in mountainous areas (Figure 3.4). Several factors contribute to the relatively mild winter temperatures. First, the warm Gulf Stream along the Atlantic coast supplies energy to the overlying atmosphere. Second, the Southern Appalachians limit the flow of air from interior portions of the continent and alter its character when it does pass over the mountains. The mountains are high enough to retard and, in some cases, prevent the intrusion of cold winter air masses. When air does move over the range and sinks on the leeward side, it is compressed and warmed.

Midlatitude Lows & Nor'easters

Midlatitude lows are very large storm systems resulting from the convergence of warm moist air from the Gulf of Mexico or the Atlantic Ocean with cooler and drier air from the interior United States and Canada. Usually originating east of the Rocky Mountains and moving eastward with the prevailing winds, these systems move across the Carolinas every five to seven days in the winter. They can bring prolonged periods of drizzle or steady rain. Cold fronts also occur with these midlatitude lows and lifting along these fronts often produces thunderstorms. In the winter, when the storms pass far south and temperatures are cold enough, these storms can produce snowfall.

The Nor'easter, or Hatteras low, is an intense type of midlatitude low. While typically less intense, the Nor'easter is to winter what the hurricane is to summer. These storms sometimes develop in the western Gulf of Mexico off the Texas coast or just east of the southern Rocky Mountains. The storms move eastward along the Gulf Coast and then turn northward along the Atlantic Coast. They are characterized by extremely low pressure, high wind velocities, high waves and storm surges, severe thunderstorms, and heavy precipitation.

The Blizzard of 1993 was such a Nor'easter. The storm began as a weak low-pressure system over the western Gulf of Mexico, intensified rapidly in the Gulf, crossed the panhandle of Florida, and moved northeastward over the mid-Atlantic states. Heavy precipitation fell in the western parts of the Carolinas as snow, and over the remainder of the area as rain. Flooding was widespread as the coast experienced a three-foot storm surge with high waves on top the surge. Since the storm lasted for many hours, part of the time the storm surge and waves built on top of high tides resulting in major beach erosion and structural damage. Because of their size, intensity, and frequency, Nor'easters have had a major impact on the Carolinas.

The Physiographic Regions of the Carolinas

Having established the basic similarity of climate over the Carolinas, it is useful to examine the environmental differences which occur within the region. The varying landscapes in the Carolinas provide a visible expression of the physical geography. As noted in Chapter 2, there are five distinctly different physiographic provinces and sub-regions. Each is a product of different geological origins and environmental processes. The three major physiographic provinces are the Coastal Plain, Piedmont and Blue Ridge. In addition, there are two major physiographic sub-regions. These provinces and subregions differ more in terms of vegetation and soils than in climate. The following is a brief discussion of the basic vegetation and soil characteristics and processes. In turn, there are case studies of each of these five provinces and sub-regions.

Vegetation and Soils

The Carolinas' warm and humid subtropical climate supports luxuriant vegetation, but pine forests dominate much of the landscape. Before European settlement, nearly the entire Coastal Plain was covered in pine forests. The dominant species of the Coastal Plain was long-leaf pine, but other pines including scrub, short-leaf, and loblolly were also found. Pines abound because of their competitive advantage in a region with relatively infertile soils and frequent fire. While pines predominate on the Coastal Plain, hardwood forests thrive in river valleys and swamps on the Piedmont, and extensively at the higher elevations of the Blue Ridge. On the Coastal Plain, savannah grasslands maintained by fire exist, and expanses of marshes can be found in the Tidewater region.

Current vegetation patterns reflect not only adaptation of plants to specific environmental conditions and disturbance, but to the influence of human activities. In particular, agriculture and urban and residential development have affected vegetation in the Carolinas. In the following material, however, the focus is on the natural vegetation patterns resulting from climate variations (due to latitude, elevation, or proximity to the coast). In the case studies, there is an examination of the impact of local influences on vegetation, such as topography, soil type, and depth of the water table. These regional and local factors, together, determine which plants will succeed.

Soil is a mixture of mineral and organic matter that covers the surface of the land. It is the milieu in which vegetation grows, and is, thus, the base for the vast agricultural enterprises in these states. Soils vary a great deal over the region. Basic to soil formation is the nature of the underlying rock or locally deposited sediment (parent material). Some parts of the region are covered by sand and other sediments, and other parts are underlain by very hard rock. As a result of the differences in parent material, the soils vary in mineral content.

The type of soil is also affected by climate and other factors. The warm, moist climate accelerates the decomposition of mineral and organic matter. The soils on older land surfaces, such as the Piedmont, have undergone extensive decomposition (ultisols). In areas of younger surfaces, such as river sediments (entisols) or steep mountain slopes (inceptisols), soil development has not proceeded nearly as far. The vegetation growing in the soils also affects soil development. Vegetation supplies the organic matter to the soil through decomposition of plant litter. The type of vegetation determines the kind and amount of litter placed on the surface or in the soil (i.e. pine needles, leaves, grasses). The combination of mineral and organic matter determines soil fertility.

Case Studies

To illustrate how climate, vegetation, and soils vary across the Carolinas, and how these three components interact, five geographically separate counties will be used as case studies (Figure 3.5). These counties represent the five physiographic regions—the Tidewater (Beaufort County, SC), the Coastal Plain (Bertie County, NC), the Sandhills (Kershaw County, SC), the Piedmont (Guilford County, NC), and the Blue Ridge Province (Haywood County, NC). While there are general similarities among climate, vegetation, and soils in the five counties, a variety of broad-scale forces create differences among them. Local causes, such as relief, drainage and geologic history result in variation within each county.

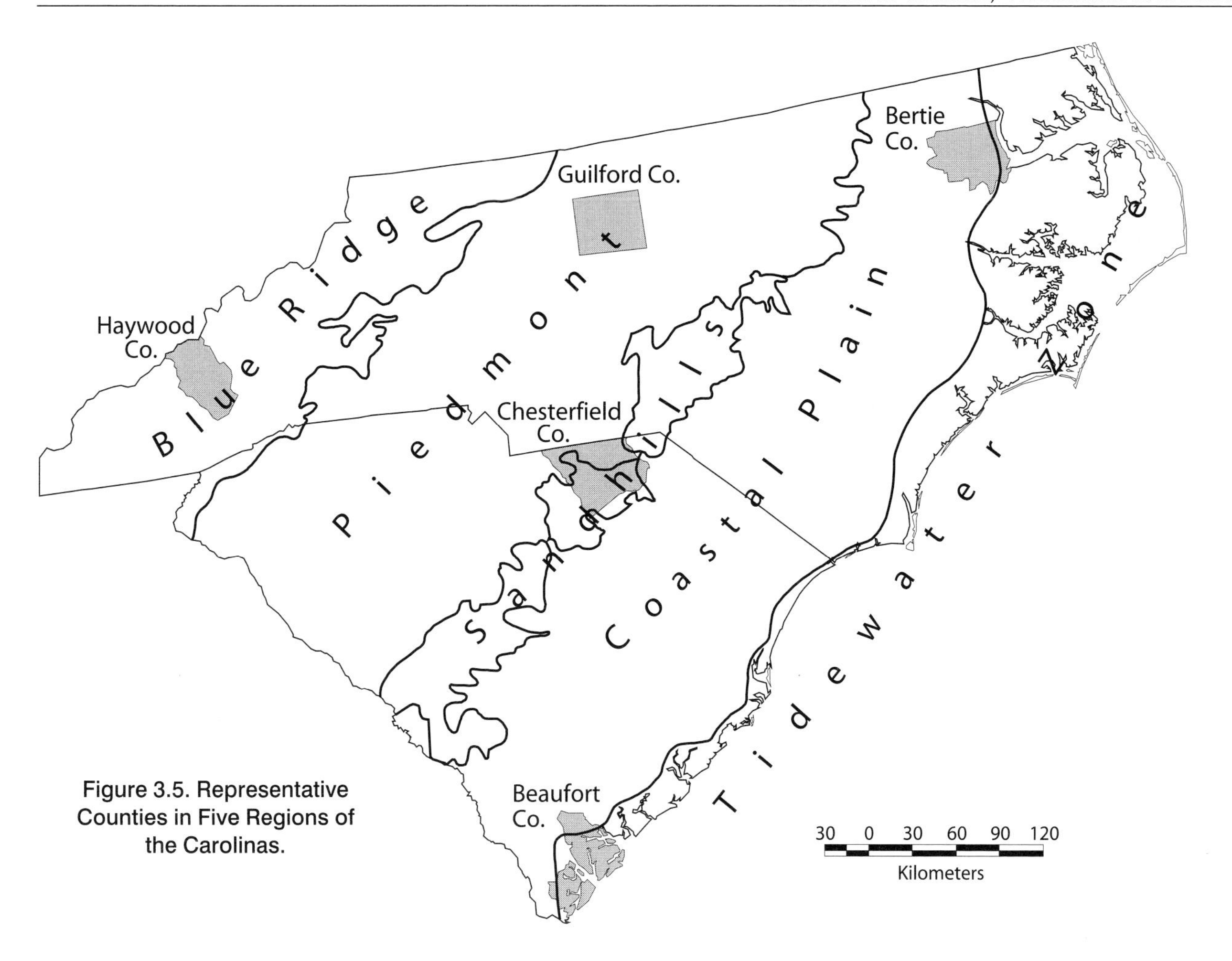

Figure 3.5. Representative Counties in Five Regions of the Carolinas.

The Tidewater: Beaufort County, South Carolina

The Carolinas have 488 miles of Atlantic coastline. The deltas, tidal inlets, barrier islands, and beaches comprise a distinctive physical landscape. Distance to the sea is responsible for most variation in this landscape. We use Beaufort County, South Carolina to illustrate this coastal setting.

Climate

The proximity of the Atlantic, and especially the Gulf Stream, moderates the daily and seasonal temperature range in Beaufort County. Warm coastal waters buffer the region from low minimum temperatures during the winter months. The average minimum January temperature is warmer than inland locations at the same latitude. This moderating effect extends the average frost-free season to 266 days, from March 5 to November 27. While the influence of the Atlantic is less dramatic during summer, sea breezes draw relatively cool air from the ocean that reduces daily maximum temperature along the coast.

In Beaufort County, average annual precipitation (51 inches) differs little from other non-mountainous parts of the Carolinas. However, its seasonal precipitation pattern shows noticeably higher rainfall during summer months due to the greater incidence of thunderstorms and the influence of periodic hurricanes along the coast. Moisture-laden hurricanes can produce over ten inches of rain, so their occurrence can inflate average precipitation during the tropical cyclone season.

Vegetation and Soils

Vegetation in coastal regions responds to the maritime climate, as well as to the influences of water table depth, soil development, and salt concentration. In Beaufort County, three general categories characterize the coastal vegetation—forests, marshes and dunes. The *pine forests*, characteristic of the Carolinas, develop in most inland parts of Beaufort County. The mix of species in the county largely reflects differences in the water table. Pond pine and slash pine, with a dense understory of saw palmetto, sassafras, and sweet bay magnolia are found in the wetter areas. On drier ridges, longleaf pine is found. *Semitropical magnolia forests* also develop inland. Such forests include magnolia, live oak, laurel oak, and beech trees along with numerous vines and epiphytes, such as Spanish moss, that attach to trees. These species block enough light to limit understory growth. Closer to the coast, *maritime forests* develop with live oak, palmetto, and slash pine—species that have a high tolerance for salt spray from the ocean and strong winds.

The pine, semitropical magnolia, and maritime forests of Beaufort County grow in sandy, rather permeable soils that developed in marine sediments. Often terraces mark abandoned shorelines. The soil layers, or *horizons*, are weakly defined because of sandy parent material. Species that tolerate relatively dry conditions (e.g., longleaf pine) occupy the higher ridges where soil development extends two to three feet deep and the water table is several feet below the surface during most of the year. Species that require more water (e.g., those of magnolia forests) grow in lower areas and where the water table is closer to the surface. In the immediate coastal zone, maritime forests develop on wind-driven sands that were deposited on ridges and troughs paralleling the current coastline. Again, the relative relief of these areas determines the locations of particular species.

Fresh water, brackish water, and salt marshes form a continuum along tidal rivers, from inland locations to the ocean. Marsh vegetation must tolerate periodic inundation and saturated soils (Photo 3.1). One common response to this environment involves development of a strong, dense root structure. Grasses, rushes and cattails grow furthest inland. In salt marshes, grasses that endure high concentrations of salt water dominate, accompanied by fleshy glasswort. Other species can occur depending on the duration and depth of inundation. Marshes in the coastal zone are important breeding grounds for aquatic and marine fauna.

Soils in the flood plains and tidal marshes of the coastal zone developed in silty and clayey marine sediments. They are occasionally flooded

Photo 3.1. Spartina grasses in expansive salt marshes. These species tolerate salt water inundation with each high tide.

by fresh or brackish water, or daily inundated by salt water with the tides. Commonly, these soils consist of a silty, clay-loam surface layer that is underlain by clay and clay loam. Greenish-gray clays form beneath the soil surface layer as a result of waterlogging and limited oxygen.

Dunes present perhaps the harshest environment for vegetation. Here plants must tolerate limited soil moisture, saline soils, wind desiccation, and salt spray. While sea oats dominate most dunes, a variety of grasses and sedges are also found (Photo 3.2). Behind the protection of the dune, a variety of salt-resistant plants grow, such as groundsel, Virginia creeper and greenbrier. The creation and maintenance of dunes is important to human settlement in the coastal zone, since these features buffer against coastal erosion and storm surge. The soils in dunes exhibit little, if any, development because materials are very coarse and are regularly deposited and eroded by wind and water, and vegetation is relatively sparse.

The Coastal Plain: Bertie County, North Carolina

The Atlantic Coastal Plain is one of the more clearly defined regions of the Carolinas. It is part of the continental shelf that has been above sea level for a long period of time. It slopes gradually towards the Atlantic Ocean and is a nearly level physiographic province. The long straight stretches of highways and railroads which cross it are visible evidence of the uniformity and flatness of the region. The outer edge of the region has been under water since the last continental ice sheet began to melt, and even the Coastal Plain itself was under water until recent geologic time. The sediments that cover the region are mainly unconsolidated or partly consolidated sand, silt and clay. In some areas, a thin layer of limestone lies beneath the surface. Movement of surface water is slow because the land is so flat. Groundwater is very close to the surface and provides the main water supply for the region.

Photo 3.2. Sea oats dominate sand dune vegetation

Climate

Like most of the Carolinas, Coastal Plain summers are hot and humid. Summer temperatures average in the high 70s. Winter temperatures average in the 40s. Precipitation is spread fairly evenly throughout the year and averages 45 to 50 inches per year, slightly less than the immediate coast. In most years, precipitation is abundant enough to meet agricultural needs. Prevailing winds are from the south and southwest as in the rest of the state. Snowfall is infrequent and remains on the ground only a few days each year. In Bertie County, the growing season averages 191 days, or over six months.

Vegetation and Soils

Before European settlement, most of the Coastal Plain consisted of pine forests. Today, much of the region remains in woodland, typically fast-growing loblolly pines that sustain a thriving lumber industry. These pines tolerate the relatively sandy, nutrient-limited soils that form in the marine sediments of the Coastal Plain. Despite its pervasiveness, Coastal Plain vegetation is not limited to pine forests. Since European settlement, the region has supported extensive agriculture. Agricultural fields, of course, occupy the best soils—those that are well-drained and have a loamy surface layer with a moderate amount of organic matter. The subsoil is loamy to clayey and light yellowish-brown. When sand content is high enough and the water table deep enough, these soils are productive. Their loamy texture helps retain some soil moisture. Some parts of the Coastal Plain have been regularly burned to produce savannas—areas with widely-spaced trees, limited understory, and a carpet of herbaceous plants and grasses. Soils in savannah landscapes often have a dark organic layer that results from decaying grass roots. Where grasslands have occupied the land for some length of time, the underlying soils are among the richest on the Coastal Plain.

Hardwood stands exist in some of the drier areas. These stands include several species such as American elm, yellow poplar, sweetgum, hickory, red maple, willow oak, white oak and southern red oak. Understory trees in these environments are mainly dogwood, sourwood, American holly, and sassafras. Hardwoods also occupy the floodplains of major river valleys with species that include sweetgum, overcup oak, sycamore, cottonwood, and loblolly pine. Holly, hackberry and swamp dogwood are found in the understory of the floodplains. Portions of the Coastal Plain are poorly-drained swamps. Flooding regularly occurs and bald cypress and water-tupelo are the dominant species. The soils along drainageways are poorly drained, acidic, and have high organic content due to the slow decomposition of a saturated environment. Similarly, the soils found in cypress-tupelo swamps are deep and poorly drained. They usually have a dark gray loamy surface layer which is rich in organic material. The subsoil beneath this has increasing amounts of firm clay.

The Sandhills: Chesterfield County, South Carolina

The Sandhills form a discontinuous, narrow tract oriented from southwest to northeast across the Carolinas' mid-section from Augusta, Georgia to near Raleigh, North Carolina. This region of rolling hills generally follows the Fall Line, a geologic division separating the hard igneous and metamorphic rocks of the Piedmont from the less-resistant sedimentary rocks of the Coastal Plain. The Sandhills themselves were formed during two major depositional periods—the Cretaceous, approximately 90 million years ago, and the Eocene, approximately 45-50 million years ago. During the former period, rivers deposited sands and gravels in and around their channels and finer material on flood plains. During the latter, rivers deposited sediments from the Piedmont onto a coastal delta (sea levels were considerably higher than today) where ocean waves and wind reworked them. The coarsest material was left near the wave action, while finer materials were deposited offshore. The unique character of the Sandhills' soils and vegetation, as well as variation within the region, results largely from the patterns of these ancient sand and clay deposits and the subsequent processes that have reshaped them.

Climate

Chesterfield County, in northeastern South Carolina, serves as an example of the Sandhills. The long-term weather records in Cheraw, South Carolina show typical subtropical humid climate conditions. Mean January temperature is 39° F with average morning lows of 27° F and average afternoon highs of 51° F. Mean July temperature is 78° F with morning lows averaging 67° F and afternoons highs averaging 89° F. The mean first fall frost date occurs on November 1, and the mean last spring frost is on April 2, providing an average 212-day frost-free season. Average precipitation is 49 inches; the wettest month is July (5.74 inches) and the driest month is November (2.90 inches).

Vegetation and Soils

Despite having temperature and precipitation patterns similar to the rest of the Carolinas, vegetation patterns in the Sandhills differ from those elsewhere. Because the coarse-textured soils found here do not hold water very well, vegetation must tolerate relatively dry conditions. The driest environments are found on broad ridges where soils have developed from deep, sandy marine deposits. Water permeates through these coarse-textured soils rapidly and leaches nutrients from their upper layers. The resulting droughty, nutrient-poor soils severely restrict vegetation. The xerophytes that do survive in this environment have developed specific adaptive strategies. Longleaf pine adapts by growing a deep taproot in its first few years to draw moisture from the deep water table. Because its regeneration depends on periodic fires to release seeds from its cones, it was the dominant tree species in the Sandhills before the control of fire. The turkey oak, a common understory tree, adapts to the harsh environment by growing deep-lobed, waxy leaves to limit transpiration (Photo 3.3). It also orients its leaves vertically to reduce the absorption of sunlight. Shrubs and herbaceous plants survive the dry environment using a variety of adaptive strategies. The narrow leaves of rosemary and the thick, waxy leaves of dwarf huckleberry reduce transpiration loss. Wire-grass, a common clump grass, survives by folding its slender leaves, producing woody cells and minimizing living tissue. It also has an extensive root system. Some annuals, such as the wire plant, grow and flower during the spring and early summer and enter the dormant seed stage during the warmest parts of summer when evapotranspiration rates are highest. Prickly pear, yucca and other succulents found across the Sandhills store water in plant tissues. As the coarse, droughty soils of the uplands control vegetation type, the resulting vegetation influences soils development. The relatively sparse vegetation cover reduces plant litter returned to the soil. This lack of organic material retards soil development allowing only drought-tolerant species to grow.

Where coarse-grained deposits are thinner and where original marine sediments included finer material, soils with greater moisture-holding capacity have developed. Long-leaf pines

Photo 3.3. A typical Sandhills landscape. Long-leaf pine, turkey oak, and bunch grasses growing on ancient beach terraces.

grow faster and taller in these environments and share the landscape with loblolly pine and hickory, along with post, blackjack and bluejack oaks. Even black and water oaks can survive in wetter spots. Understory trees include dogwood, sassafras, blackgum, and persimmon. Even more diverse plant communities can develop along hillsides where shallow clay layers perch water just below the surface and make moisture available to plants. Denser vegetation stands produce increased plant litter. As organic material breaks down, nutrients and finer sediments are added to the soil which further increases moisture-holding capacity. Moister areas also support a wider variety of shrubs covering a greater portion of surface area (e.g., wax myrtle, sparkleberry, and hawthorn) and herbs (e.g., thistle, dogfennel, and blazing star). Clumps of wire grass are thicker and more closely spaced.

Even in the otherwise xeric Sandhills, wetlands develop where the water table intersects the surface. These flooded soils lack the oxygen necessary for proper decomposition, so partially-rotted plant material accumulates in a thick layer. In floodplains, the arrival of new alluvial material often makes these young soils quite thick. The deep, poorly-drained soils provide a suitable environment for water-tolerant trees such as pond pine, sweetgum, blackgum, yellow-poplar, water oak, sweet bay, water tupelo, water ash, and bald cypress. Dense stands of evergreen shrubs and vines dominate the wetland understory. Species include fetterbush, holly, cane, blueberry, swamp azalea, and greenbrier. A variety of ferns (e.g., cinnamon fern, and sensitive fern) rushes, and sedges typically dominate the ground cover.

The Piedmont: Guilford County, North Carolina

The Piedmont is an area of rolling hills stretching from Virginia to Alabama. These hills are carved from a plateau that rises from about 600 to 1500 feet above sea level. This region lies between the Appalachian Mountains and the Coastal Plain and consists of a very mature stream-dissected upland. The underlying bedrock is mainly metamorphic and igneous rock of ancient age. Most of the bedrock in the Piedmont date from Paleozoic and Precambrian time. Guilford County, North Carolina provides an example of the Piedmont landscape.

Climate

Summer weather in the Piedmont is much the same as that for the Carolinas as a whole. Winter temperatures average about 40° F in Greensboro, the county seat of Guilford County and summer temperatures average about 76° F. The mean frost-free season is from April 11 to October 26—198 days. Mean annual precipitation is 44 inches, and it is nearly evenly distributed through the year. November is typically the driest month (2.96 inches) and July the wettest (4.51 inches). Snowfall is a minor element in the weather averaging about 11 inches annually, with most of this occurring from only one or two events. Typically, there are only four days each year with snow cover. Relative humidity averages 55 percent. Possible sunshine is greater in summer (64%) than in winter (54%) as a result of fewer mid-latitude cyclones in summer. Prevailing wind direction is from the south and southwest, bringing moisture from the Gulf of Mexico and the Atlantic Ocean.

Vegetation and Soils

The Piedmont of North and South Carolina was mainly forested before European settlement. In Guilford County, forest or woodland covers the majority of the land surface today, but the amount of land devoted to forest is decreasing rapidly with urban development and the growth of major transportation networks. The original forests consisted of a mix of hardwoods—mostly oak and hickory—and evergreens—loblolly, short-leaf and Virginia pine. Primary understory species are dogwood and sourwood. While floodplain forests do exist in the Piedmont, they are narrower and less extensive than those found on the relatively flat Coastal Plain. Red gum, tulip-poplar, red maple, and river birch are found in these floodplains (Photo 3.4).

A typical Piedmont soil is well drained,

forming in residual material weathered from old metamorphic and igneous rock. As a result, texture is finer than that found in the Carolina Coastal Plain or Sandhills and permeability is less. The typical fine sandy-loam soils are found on broad smooth divides between streams. Their surface layer is dark grayish-brown about three inches thick with relatively low organic matter. The subsoil is approximately two feet thick and consists of sandy clay-loam and clay. The seasonal water table is high, sometimes within one to two feet of the surface.

In Guilford County soils developed largely under a hardwood forest. These trees take up elements from the subsoil and add organic matter to the surface by depositing leaves, roots and, eventually, the entire woody structure. Organic material decomposes fairly rapidly in the environment because of mild temperatures throughout the year and abundant precipitation. The latter hastens dissolving and removal of elements, such as calcium, magnesium, potassium, and phosphorus. Organic compounds are also absorbed rapidly by existing vegetation leaving the soils acidic and low in mineral salts.

Areas with the most productive Piedmont soils traditionally have been used for crops and pasture. But extensive use has resulted in severe soil erosion. Prior to settlement, soil erosion was near or below the rate of soil formation. Early European settlers in the Piedmont (around 1700) found the streams to be clear of sediment, even in times of high water. These settlers cleared forests and planted crops on the same fields year after year. Corn, cotton, tobacco, and other row crops grown during the nineteenth century hastened soil erosion. During the four to five month growing season, rows were tilled regularly to reduce weeds. Intense summer rains displaced soil. Erosion continued during the winter when the land was bare and remained unfrozen. By the time of the Civil War, much of the Piedmont was suffering from severe land degradation, and by the 1920s, the region was one of the most severely eroded areas of the United States. This continued until the conservation movement of the 1930s.

Erosion reduced soil fertility in the Piedmont by stripping away the rich organic layer and incorporating subsoil into the surface layer. Rain that fell on the exposed yellow to red clayey subsurface sealed it with an even harder layer that was more impervious to water. The resulting decreased infiltration and increased runoff led to common flash flooding. The original topsoil that disappeared into rivers delivered alluvial material to floodplains forming mineral-rich soils upon deposition.

Photo 3.4. Mixed forest in the Piedmont consists of a mixture of Virginia pine, tulip poplar, oak, and other native hardwoods.

The Blue Ridge: Haywood County, North Carolina

The Blue Ridge Province is the region of the Carolinas with the most variable climate, vegetation and soils. The terrain consists of rugged mountains, including the Great Smokies, deep valleys, and areas of intervening hills between the mountains and the larger valleys. Climate, vegetation, and soils vary throughout the Blue Ridge, largely in response to elevation which ranges from about 1,400 feet to nearly 6,600 feet. Twenty mountain peaks exceed 5,000 feet. In many areas, the high mountains shade the intermediate and lower mountains, as well as the valleys between them. Many sites, including even south facing slopes, are relatively cool. Haywood County, located in the western North Carolina mountains west of Asheville, is representative of the region. It includes part of Great Smoky Mountains National Park, Pisgah National Forest and the Blue Ridge Parkway.

Climate

The climate varies with elevation and with exposure to solar radiation and prevailing winds. At Canton, North Carolina (elevation, 2662 feet), the January temperature averages about 34° F and the July temperature averages about 70° F. The mean date of the first frost in the fall is October 9, and the last date for freezing temperatures is May 2. The average number of days above freezing is 159—over three months shorter than at Beaufort! Temperatures and length of growing season decrease significantly with elevation.

Portions of the southern Blue Ridge Province receive the greatest amount of precipitation of any place east of the Mississippi River (Figure 3.3). The Blue Ridge, which rises out of the Piedmont in Georgia and extends northeastward, intercept the winds from the Gulf of Mexico and from the Atlantic Ocean. The lifting of the warm and moist air over the southern end of these mountains results in greatly enhanced orographic precipitation. In fact, the name Blue Ridge comes from the bluish cast given to the mountains by the nearly perpetual haze that exists over them. The haze consists of the moisture being carried to and over the mountains, as well as that given off by evapotranspiration from the vegetation. The amount of precipitation in the mountains is largely a function of elevation. Highest values (over 85 inches) are found near the highest peaks of the southern end of the range. Precipitation is relatively abundant all year with mean monthly precipitation being only slightly higher than in winter. October and November are the driest months, averaging slightly less than three inches per month. Summer precipitation tends to come in large part from convective thunderstorms. The highest region, now mainly consisting of Great Smoky National Park receives its name from the vertical development of streams of cloud as the air becomes unstable as it rises along the flanks of the mountains. The high mountain areas (above 4,800 feet) receive an unknown amount of moisture from condensation of cloud droplets directly on the vegetation and ground surface. The same is true in the winter when condensation is in the form of rime ice. Snow accumulation can be extensive depending on elevation and temperatures in any given year. The high mountain regions are quite cold and windy in the winter and receive much of their winter precipitation in the form of snow. At Canton, snowfall averages only 8.5 inches, but higher elevations receive an average of over 55 inches.

Vegetation and Soils

Most of the Blue Ridge Province, like the rest of the Carolinas was originally covered with woodland or forest. Like climate, vegetation varies with elevation and is generally associated with high peaks, intermediate slopes and lower flood plains. Generally, spruces and firs dominate the highest elevations, while deciduous hardwoods are increasingly found in the intermediate and lowest elevations.

There are six main forest types found in this region: spruce-fir, yellow pine, eastern white pine, cove hardwoods, upland hardwoods, and northern hardwoods (Photo 3.5). The spruce-fir forests occur mainly above 4,800 feet. While extensive in pre-settlement time, the forests are

Photo 3.5. Mixed pine and hardwood forest of the lower Blue Ridge Mountains. The hardwood forest is gradually replacing the evergreen forests which originally existed.

now of limited extent due to a combination of factors including deforestation by logging and paper interests, fires, acid rain, and insect infestation. Red spruce is the dominant species but is declining as is Fraser fir. Both plants are being challenged by a combination of infestation of balsam wooly adelgids and acid precipitation. In summer, these elevations receive as much as 10 inches of moisture in the form of interception of cloud droplets. These cloud particles are extremely acidic, much more so than the precipitation that falls from the clouds. There are intrusions of hardwood trees, but they tend not to be very successful.

The soils on the side slopes and ridges above 4,800 feet are subject to extremely cold temperatures and high winds. Since they are on steep slopes, they are generally very well drained despite consisting of mainly fine soil particles. Metamorphic or igneous bedrock is found within 30 inches of the surface and stones and boulders are scattered over the surface. Soil profiles are generally thin. The surface layer is up to four inches deep and the subsoil up to 34 inches deep. The steep slopes also cause rapid runoff, especially in bare areas. The potential for frost action is moderate in these soils when water is retained within the profile during winter.

The most extensive and most varied forests occur in the lower elevations (below 4,800 feet). Two of these forests are evergreen—the yellow pine and the eastern white pine forests. The yellow pine forest occurs generally on poor soils on ridge tops and on the drier and warmer south-facing slopes. Pitch pine, short-leaf pine, and Virginia pine are the dominant species of yellow pine. The eastern white pine forest grows in a wide variety of sites. It is a rapidly growing species, found on most sites except those with the best soils. It can be grown on almost any site, except at high elevations. Following forest removal, it quickly invades and eventually becomes the dominant species until hardwoods recover.

The cove hardwood forest is found mainly on the best, most productive soils. These soils are deep, warm, and moist. Common species are black cherry, eastern hemlock, eastern white pine, American basswood, yellow buckeye, white ash, sweet birch, northern red oak, and white oak. The upland hardwood forests are found on sites intermediate in productivity between the cove hardwoods and the yellow pine. These sites contain a predominance of different hardwoods from those of the cove hardwoods sites.

Many of the narrow flood plains and terraces in the Blue Ridge are cultivated. The main crops are tomatoes, burley tobacco, silage corn, pasture, and ornamental crops. While not extensive, soils found on alluvium are deep and well-drained. Where soils are deepest (up to about 50 inches) they tend to be subject to considerable erosion which limits use primarily to woodland.

Summary

The climate of the Carolinas is one of generally moderate temperatures and precipitation. Summers are hot and humid. Most summer precipitation comes from convective showers and is generally sufficient to support field crops without supplemental irrigation. Winters are cool, but do not have extremely cold temperatures. Typical daily low temperatures are in the 20s and 30sF. Only a few days each year average near or below 0°F. Winters are also humid with more midlatitude cyclones and more prolonged periods of precipitation than in summer.

Within the region, the major variations in climate are due to proximity to the Atlantic Ocean and to differences in elevation. Sites near the ocean have warmer winter temperatures and cooler summer temperatures. There is also more precipitation along the coast than in most inland locations. Elevation strongly influences the climate of the Appalachian Mountains. In the mountains, both temperature and precipitation vary with elevation and exposure to prevailing winds.

Across the Carolinas vegetation and soils vary with the drainage characteristics of the site. Virtually the entire area was originally covered by forest. Forest communities differed from place to place largely because of soil wetness and elevation. Much of the region remains in second growth forest today. The forests consist of more pine species on the drier sites and more hardwoods on the wetter sites. Some uplands are poorly drained and are not well suited to agriculture. The same is true for some of the alluvial soils along the river bottoms. Better drained soils are found in some sections of uplands and throughout the area on valley slopes. At the turn of the twenty-first century, forest, pasture and cropland are found intermixed, largely depending on soil characteristics.

References

Barry, J.M. 1980. *Natural Vegetation of South Carolina*. Columbia: University of South Carolina Press.

Kovacik, C.F. and J.J. Winberry. 1987. *South Carolina: The Making of a Landscape*. Columbia: University of South Carolina Press.

United States Department of Agriculture, Soil Conservation Service. 1980. *Soil Survey of Beaufort and Jasper Counties, South Carolina.*

United States Department of Agriculture, Soil Conservation Service. 1990. *Soil Survey of Bertie County, North Carolina.*

United States Department of Agriculture, Soil Conservation Service. 1995. *Soil Survey of Chesterfield County, South Carolina.*

United States Department of Agriculture, Soil Conservation Service. 1979. *Soil Survey of Edgecomb County, North Carolina.*

United States Department of Agriculture, Soil Conservation Service. 1997. *Soil Survey of Haywood County, North Carolina.*

United States Department of Agriculture, Soil Conservation Service. 1994. *Soil Survey of Davidson County, North Carolina.*

United States Department of Agriculture, Soil Conservation Service. 1977. *Soil Survey of Guilford County, North Carolina.*

Chapter 4

SETTLEMENT GEOGRAPHY OF THE CAROLINAS BEFORE 1900

John J. Winberry* and Roy S. Stine**

**University of South Carolina at Columbia*

***University of North Carolina at Greensboro*

The human geography of the Carolinas can be traced back more than a dozen millennia. During that time, many different societies have crossed the face of the region, from the first occupants, the Paleo-Indians, to the most recent Hispanic and Asian immigrants, who contribute to an increasingly complex population that lives in the Carolinas today. This chapter, however, will discuss the story of settlement only to 1900; but that story includes the Native Americans, who created the landscape prior to the European encounter; the Europeans and Africans, who forged the plantation landscapes of the Coastal Plain and Piedmont provinces; and the Germans and Scotch-Irish, whose small farmsteads molded the Back Country of the Carolinas. It also involves institutions, such as the cotton plantation of the antebellum period, tenancy and small cabins of the post-Civil War era, the arrival of the textile industry, and new transportation patterns during the last decades of the nineteenth century.

We can identify five major stages, which could be referred to as watersheds, in that long history. These critical events divided one period of settlement from another and significantly altered the landscape of the Carolinas. The first was the arrival of the Native Americans, perhaps 12,000 years ago; the second, associated with village settlement, was the introduction of agriculture, about 3,000 years ago; the third was the encounter of Native Americans and Europeans on the coast of the Carolinas, nearly five hundred years ago; the fourth was the Revolution and American political independence at the end of the eighteenth century; and the fifth was the Civil War and its aftermath.

ARRIVAL OF THE NATIVE AMERICANS

No evidence exists of pre-human populations in the New World, and the overwhelming preponderance of data, from archaeology to molecular biology to DNA testing, indicates the evolution of modern humans (Homo sapiens sapiens) in Africa, perhaps 150,000 years ago. From that hearth, they moved northward into the Middle East and Europe, eastward to South, East, and Southeast Asia, and northeastward through Central Asia to Siberia, being well-established in all these regions by about 30,000 years ago. Those in Siberia looked across the Bering Strait to North America, contemplating unknowingly their movement into what would be called the New World hundreds of thousands of years later—but how were they to make the crossing? And when did they do it?

The story of the human migration into the Americas began some 2 million years ago, when the world entered the Pleistocene Geologic Epoch or Ice Age. Lasting until about 11,000 years ago, this period was characterized by great sheets of ice or glaciers that covered the northern areas and many of the mountains of Europe, Asia, and North America. Also, sea level fell 400-450 feet below what it is today, exposing at different times a land bridge that connected extreme western Alaska and Siberia. In time, it was covered by vegetation, and humans literally walked across it into the Americas, probably 20,000-22,000 years ago (Meltzer, 1997, 754; Gibbons, 1998, 1306), although this date is debated.

These early humans spread quickly through the ice-free parts of the Western Hemisphere, arriving in the Carolinas perhaps 12,000 years ago. By then, they had developed the New World's Paleo-Indian tradition, which included especially the Clovis point. This fluted three-inch long blade, worked on both sides to form a sharp edge and hafted to a spear shaft, was used in hunting and probably for other purposes. The new migrants to Carolina faced a changing climate that presaged the end of the Pleistocene Epoch. The pine-spruce vegetation complex was largely replaced by beech and hickory-dominated forests, and the Pleistocene mammals—such as mammoth and great bison—disappeared about 10,800 to 11,000 years ago, victims of the changing ecology, but also of the successful hunting by the Paleo-Indian peoples (Goodyear et al., 1989). Paleo-Indian settlement typically "occupied both riverine and inter-riverine environments... [and] terrace settings along major drainages" (Anderson, 1992a, 14). Few intact Paleo-Indian sites exist, so little really is known about this stage of the Carolinas' prehistory. It is assumed that the Paleo-Indian peoples did hunt the great herd animals, relying on stealth and cooperation to approach them and thrusting spears from close quarters to disable and ultimately kill them. They also hunted smaller animals and gathered plant materials to provide for their subsistence.

By about 10,000 years ago, the Carolinas entered the Holocene, or Recent Geologic Epoch; warmer and wetter conditions prevailed, the floral complex changed to an oak-hickory forest, and the Pleistocene mammals had been replaced by a modern fauna, including the white-tailed deer. This correlated with the appearance of the Archaic phase of prehistory, a very successful hunting and gathering economy that used new technologies and strategies to support a larger population. The Archaic peoples occupied the Piedmont and Coastal Plain and exploited a wide selection of foods—hunting, fishing, and gathering nuts, berries, fruits, roots, and other plant materials. Their occupied sites clustered on swamp margins and the slopes overlooking riverine flood plains as they took advantage of the ecotones (the boundaries between adjacent ecologic complexes) and avoided low-lying, poorly drained areas. Their "central-based wandering" subsistence strategy involved "logistically provisioned seasonal base camps during the winter and a series of short-term foraging camps throughout the remainder of the year" (Anderson, 1992b, 36). Groups tended to move toward the coast in early spring, into the Inner Coastal Plain and Lower Piedmont by early fall, and back to the fall line base in winter.

The Introduction of Agriculture

The Late Archaic, beginning about 4,500 years ago, foreshadowed the second great watershed of settlement, the arrival of agriculture. This would support larger populations and provide the subsistence base for further cultural advance. Individual settlements were larger during this period and probably moved from a band-level to a tribal political organization (Anderson et al., 1981, 92-94). Settlement became more permanent, and pottery first appears in the Carolinas in riverine and coastal sites that exploited shellfish. Of greatest significance, though, is the appearance of the "Eastern Agricultural Complex," the limited cultivation of certain native plants, such as the sunflower, marsh elder, chenopodium, sumpweed, maygrass, and knotweed (Hudson, 1990, 58).

The Woodland stage of prehistory was an extension of the Late Archaic because of its ce-

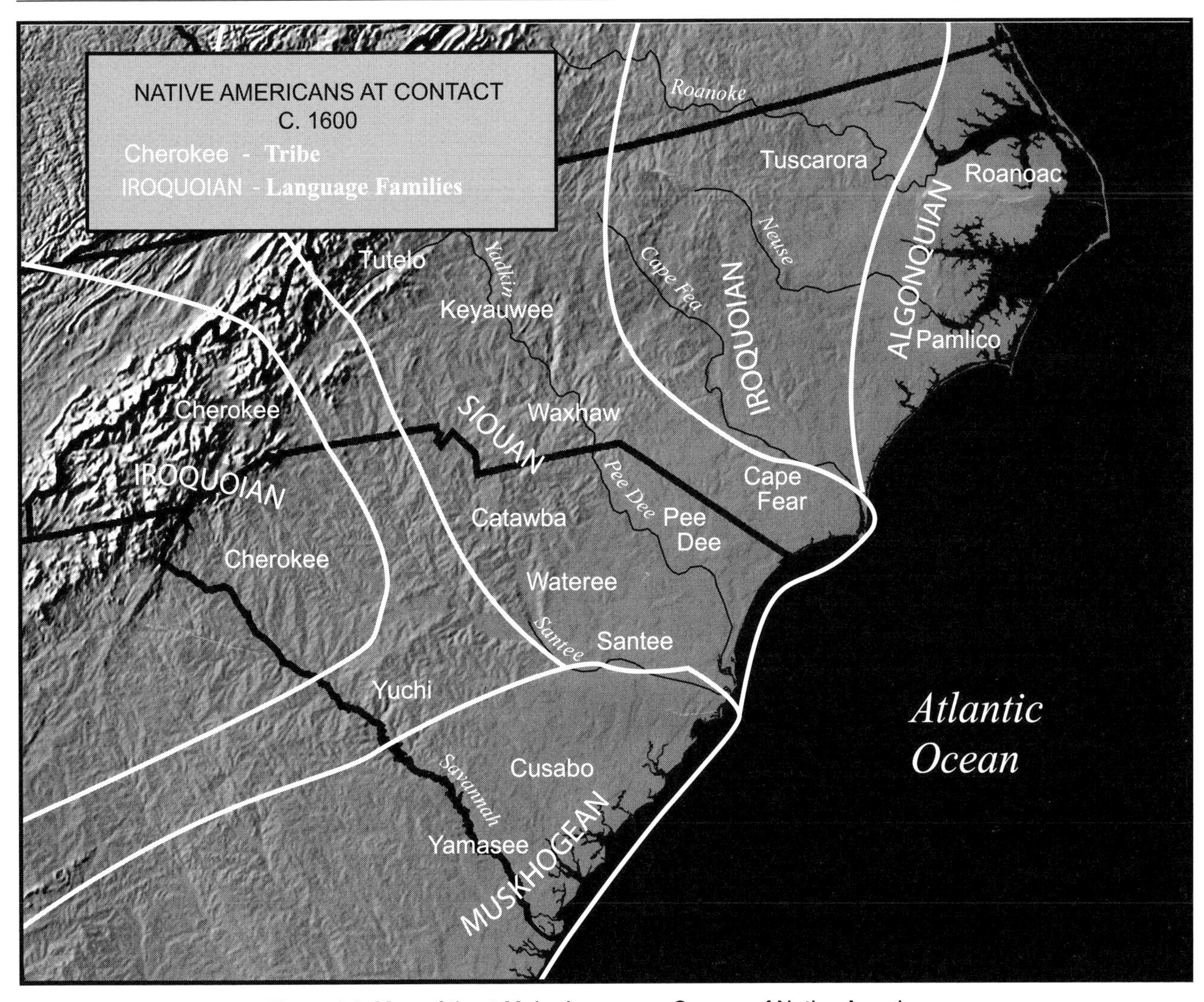

Figure 4.1. Map of the 4 Major Language Groups of Native Americans

ramics and continued reliance on the "Eastern Agricultural Complex," but it also marked the second watershed in the settlement geography of the Carolinas. It involved some significant differences, including the appearance of the bow-and-arrow and new, smaller point types that allowed individual stalking and longer-range killing of animals. Most important for its impact on the settlement landscape, however, was the arrival of tropical crops from Mexico that included initially the bottle gourd and squash and much later corn and beans. The Woodland peoples still relied on hunting and gathering for subsistence, but their settlements were oriented now to the lowlands along rivers where farmers successfully planted the rich and productive soils of the bottomlands. Based on this economy, the houses and villages of the Woodland period took on a permanence and stability not seen earlier.

About 900 years ago, around A.D. 1100, a group of people referred to by some as "the invaders" crossed the Savannah River near present-day Augusta and moved along the Fall Line into the Carolinas (South, 1976, 17). Today, we know these people as the Mississippian Culture, the phase of prehistory that immediately preceded and approximated the arrival of the Europeans. The full-blown Mississippian itself apparently had its origins in the middle Mississippi Valley at a 5 1/2-square-mile site known as Cahokia (near present-day East St. Louis, Illinois). From Cahokia, the Mississippian spread in all directions, but it had a particularly strong presence in the Southeast, establishing large settlement and ceremonial centers at Moundville, Alabama, and Ocmulgee and Etowah, Georgia. When the Mississippian reached the Savannah River, however, it had lost much of its force and was characterized more as a frontier settlement. Its small villages, with only one or two mounds, were characteristically surrounded by a palisade apparently for protection against the displaced Woodland peoples who preceded them. Because the Mississippian sites had no earlier levels of human occupation and ethno-historical records suggest that they spoke a language different from that of surrounding tribes, it is assumed that they were intrusive populations. The Mississippian Culture moved along the Fall Line of South Carolina and North Carolina; its villages were situated on bluffs, overlooking the fields on the rich soils of the riverine floodplains. The easternmost site of the Mississippian tradition was Town Creek near present-day Mt. Gilead, North Carolina, established about A.D. 1300 (Ferguson, 1971; Anderson, 1989).

In the spring of 1540, the ruthless Spanish explorer Hernan de Soto arrived at Cofitachequi, one of the major Mississippian sites in the Carolinas, probably located on the Wateree River just south of present-day Camden (DePratter, 1989, 134-135). De Soto's chroniclers gave their impressions of the settlement and its people, its apparent wealth, and the political and trade influence that it had over the region (Hudson, 1990, 68-70). They also recorded other items that suggested Cofitachequi's contact with the Vasquez de Ayllon colony of 1526 on the coast of South Carolina, to be discussed below, the site of which supposedly was but two days away. The Mississippian settlements soon disappeared (the last mention of Cofitachequi in the literature, for instance, was in 1673), but the Native American population in the Woodland stage of development remained significant until European diseases destroyed village after village.

Four major language families were represented in the Carolinas at the coming of the Europeans (Figure 4.1), and it must be noted that they also were migrating into, out of, and back and forth across the region (Hudson, 1976; Milling, 1969; Swanton, 1979). One Iroquoian-speaking tribe, the Cherokee, was related to Indian nations in New York and was spreading into the northwestern corner of South Carolina from highland Tennessee and North Carolina. The other Iroquoian people, the Tuscarora, were well established in the Inner Coastal Plain of north central North Carolina, some distance from the Cherokees. Separating them were the Siouan speakers, the most important of which were the Catawbas, who had come from the Midwest. They settled in eastern South Carolina and the southern coastal plain of North Carolina, reaching into Piedmont South Carolina at least by 1680. They were blood enemies of the Cherokees, and the two tribes were separated by an unpopulated "No-man's Land" between the Wateree/Catawba and Broad Rivers in South Carolina.

On the northern coast of North Carolina, around Albemarle and Pamlico Sounds, was the southern extension of Algonquian speakers, who also had ties to Indian tribes in New York. They were the first to meet the English in the Roanoke colony and were friendly to the settlers during much of the colonial period. Another group of Algonquian speakers, not shown on the map, referred to initially as the Savannahs and later as the Saludas, settled in the area of present-day McCormick and Edgefield counties in the early 1600s to trade with the Spanish settlement of St. Augustine. They departed the area in the early eighteenth century but left their names for two major rivers in South Carolina. Along the lower coast of South Carolina were the Muskhogean speakers who were related to tribes across the Southeast, including the Creeks and Chickasaws.

EUROPEAN DESCRIPTION OF THE NATIVE AMERICANS AT CONTACT

Within about 30 years after the arrival of Christopher Columbus in the Bahamas, Europeans were exploring the coast of the Carolinas. In 1522, two ships under the commands of Francisco Gordillo and Pedro de Quexos arrived on the coast of South Carolina after sailing north from the Spanish colony of Hispaniola and through the Bahamas in search of Indian slaves for the sugar plantations of the Caribbean. They captured a large number of Indians, including a bright young man who later was baptized and given the name Francisco de Chicora. He impressed the Spaniards with his descriptions of what some have referred to as the land of Chicora and its surrounding territories. Many of these stories were pure fantasy, but he and the European explorers provide early information on the Native American populations of the Carolinas (Quattlebaum, 1956, 103-120).

In 1524, Giovanni Verrazzano arrived in the area of present-day Cape Fear, sailed southwestward, perhaps along the Grand Strand of South Carolina, and then turned around to sail up the North Carolina coast on a journey that eventually took him to Nova Scotia (Sauer, 1971, 52-57). He was the first to describe the Algonquian and Siouan Indians of the Carolina coast. In 1566-67 and again in 1567-68, De Soto's exploration of the interior of South Carolina was complemented by Juan Pardo's journeys into the interior and visit to Cofitachequi (Hudson, 1990). While their descriptions of the Native American cultural patterns were limited, the chronicles of Pardo's journey do provide important insight into the political geography of the pre-Columbian populations of the Carolinas.

In 1562, the French attempted a colony on Parris Island off Port Royal, South Carolina, and later on the St. John's River in Florida. Although both failed, they did leave us narrative descriptions and Jaques LeMoyne's drawings of the settlements and ways of life of the coastal peoples. These were engraved and published in 1591 by Theodore de Brys. While they are generally accurate representations, scholars have pointed out the addition of a number of distinctly European elements to the indigenous ways of life (Sauer, 1971, 206-211). Another collection of illustrations was by John White of the failed Roanoke Colony in 1586. His drawings, describing the Algonquian populations and their ways of life along the coast of North Carolina, also were printed and remain a visual record of Native American culture at contact.

The Indian populations of the Carolinas lived in villages of probably 75-100 people, although some of the larger settlements were easily twice that size. Many of them were palisaded for defense, but this artifact began to disappear by the eighteenth century. Among the Siouan people of the interior, the houses were circular wigwams, as described by John Lawson, "built of bark and round like an oven" (Swanton, 1979, 410). John White's drawings of the Algonquian village of Secotan in North Carolina records a long house shaped like a quonset hut with an oblong floor plan. As described by Thomas Hariot, these houses were "made of small poles, made fast at the tops in round form after the manner as is used in many arbors in our gardens of England" (Sauer, 1971, 261). In the late eighteenth century, William Bartram described the houses of the Cherokees as made of logs with shake roofs, a type probably adopted from European pioneers (Milling, 1969, 12-13).

The Native-American houses were commonly surrounded by vegetable gardens, while the larger fields in which the three major crops—corn, beans, and squash—were grown outside the village. Plants associated with the Eastern Agricultural Complex were cultivated as well, and the Indians also gathered wild plants including chestnuts, hickory nuts, and acorns; the roots of greenbrier, arrowhead, and Jerusalem artichoke;

fruits such as persimmons, pawpaws, crab-apples, blackberries, and blueberries; and plants like cabbage palmetto and Indian hemp, which were used for both food and fiber (Kovacik and Winberry, 1989, 61). The Indians also adopted fruits from the Europeans; and orchards of figs, apples, and peaches were found in Indian villages.

Native Americans continued to rely on hunting as an important source of food and used the bow and arrow and blowpipe before adopting the gun from the Europeans. Deer was the most favored meat, but wild turkeys and bear, the latter for its meat and grease, and a number of smaller animals were also hunted. Apparently, fire was an important hunting strategy to create open areas, and travelers referred to savannas (large grassy openings in a forested region) in the interior that probably were formed and preserved by fire, although these also were called Indian old fields. Along rivers and on the coast, the Indians complemented their subsistence with fishing, using weirs, fish traps, nets, and spears.

The European impact on the Indian populations was devastating and included diseases that virtually wiped out whole tribes and a hunger for land that drove the Native Americans out of the Carolinas. In 1682, the Catawba had an estimated population of about 4,600, but, by 1775, it had been reduced to 400. The collapse actually was worse, though, because many smaller Siouan-speaking tribes, whose numbers had declined precipitously, had been added to the Catawbas over that same period (Swanton, 1979, 104-105). In 1721, the English finalized a treaty with the Cherokees that recognized Cherokee control of much of western South Carolina, an area that came to be known as the Northwest Frontier. A series of wars and subsequent treaties, however, reduced the size of that territory until the Treaty of Holston, in 1791, pushed the Cherokee completely out of South Carolina and out of much of North Carolina, as well. These two tribes still have reservations in the Carolinas, though. The Cherokee control the 44,000-acre Qualla Boundary and adjacent land areas in highland North Carolina (Jones, 1996, 33-36), while the Catawba have a small 630-acre reservation in Lancaster County, South Carolina. The latter tribe, however, just concluded a large legal settlement with the state and federal governments over a previous 225-square-mile grant illegally taken from them in 1840.

Coming of the Europeans

The arrival of the Europeans marked the third of the watersheds of settlement. While their initial impact seemed minimal, their introduction of disease, a new culture, a new economy, and two new peoples (European and African) would have significant impacts on the settlement geography of the region. Three European nations struggled for dominance of the Carolinas during the early colonial period—the Spanish, the French, and the English. As already mentioned, the earliest European explorers of the Carolinas were the Spanish navigators Francisco Gordillo and Pedro de Quexos, who sailed along the South Atlantic coast, probably making landfall in the area of present-day Winyah Bay in June 1521. The Spanish chronicles suggest that this area was known as Chicora, but doubtfully was the actual name for the location or the Indians living there, and it does not appear on maps today (Hoffman, 1990, 11). One other place name, however, did have a long-term impact. Gordillo apparently assigned it to the estuary of a large river near present-day Port Royal, South Carolina, but the documents are not clear on exactly when the naming took place. "Santa Elena" does show up on the Juan Vespucci map of 1526 (Cumming, 1998, 69, 107) and persists today in its Anglicized form, Saint Helena. It is considered one of the earliest European place names in North America (Sauer, 1971, 192).

In 1526-27, Lucas Vasquez de Ayllon attempted what was perhaps the first European settlement in the Carolinas. Called San Miguel de Gualdape, its location is uncertain because no archaeologic remains of it have yet been found to support the documentary arguments. Some

The Fort at the English Settlement at Manteo

Historical marker (above) showing a reconstruction of the fort and excavated earthworks (left).

Photo 4.1. The fort at the site of the first English Settlement at Manteo

scholars have placed it as far south as the coastal islands of Georgia (Hoffman, 1990, 70-71), while others have suggested it was located at present-day Georgetown and Winyah Bay (Quattlebaum, 1956, 21-23). It could be argued that if Cofitachequi were on the Wateree and two days away (according to the DeSoto chroniclers), the most logical site would be near the Santee River or Winyah Bay. The colony failed within a few months and was abandoned.

During this period, France became the second European power to show interest in the Carolinas as, noted above, Giovanni Verrazzano explored the South Atlantic coast in 1524. He sailed along the outer banks of North Carolina but never ventured into Albemarle or Pamlico Sounds, although they were visible across the dunes. In 1562, the French under Jean Ribaut attempted a settlement, this time in the area of present-day Beaufort, South Carolina. It was established on Parris Island, in a deep embayment that the French called Port Royal because of its magnificent harbor (which probably was the estuary that the Spaniards had earlier named Santa

Elena). The French were interested in controlling the lucrative trade with the Indians, but they also wanted to establish a naval base/fortification near the Spanish convoy route to prey upon the silver-laden treasure galleons. The settlers built Charlesfort for protection and established a small village, but the colony was unsuccessful and failed within a year.

The Spaniards by this time were alarmed by the French presence in the territory which they claimed as their own and directed a fleet under the command of Pedro Menendez de Aviles to sail to Santa Elena and establish a base to protect the convoy route. Because a new French settlement, Fort Caroline, had been built on the St. John's River in Florida, Menendez de Aviles established his initial base just to the south of it, at St. Augustine, before capturing the French fort. He then continued on to Santa Elena, arriving there in 1566. The Spaniards found the site of the first French settlement on Parris Island in 1564 and established their town and fortress, apparently building it on the foundation of the French fort itself. Santa Elena had a mixed history as the capital of Spanish Florida, sometimes larger and more important than St. Augustine and sometimes not even occupied. By 1586, however, Spain began to withdraw from the margins of its empire, and Santa Elena became expendable. The Spaniards destroyed the fort, burned the town, and retreated to St. Augustine (DePratter and South, 1995, 5-11).

The last of the European powers to occupy the Carolinas—and the one that would become dominant—was England. Interested in extending its empire into the subtropical parts of the Americas and participating also in the Indian trade, England attempted its first settlement on Roanoke Island, North Carolina, in 1585. (See Photo 4.1) Like its Spanish and French predecessors farther to the south, the settlement did not last and was abandoned by the summer. A passing English ship under the command of Sir Francis Drake returned the survivors to England. A second colony, dispatched in 1587 under the leadership of Governor John White, also failed; but it became one of the most famous colonies in the settlement history of the Americas. Known today as the "Lost Colony," it had an inauspicious beginning when its governor had to return to England to secure additional supplies. By the time Governor White could return to the colony in 1590, he found only the abandoned fort and the inscription "Croatoan" on one of the trees, perhaps referring to a nearby friendly Indian tribe with whom the colonists may have settled after abandoning Roanoke Island (Sauer, 1971, 265).

The earliest permanent settlement of what today is North and South Carolina was by individual farmers moving south out of Virginia into the area just north of Albemarle Sound and east of the Chowan River in present-day North Carolina. The Virginia colony had been founded in 1607, and, as it grew, its inhabitants spread into the interior and eventually southward. The actual date of the first settlers in North Carolina is unknown, but records indicate that they began arriving no later than the early 1660s (Robinson, 1979, 79) and perhaps as early as 1653 (Powell, 1977, 21). By 1694, the northern province of the Carolina colony had an estimated population of 2,000, almost all of it located north of Albemarle Sound (Brown, 1948, 63).

Despite this early focus of settlement on Albemarle Sound, it was another area, Cape Fear and the Charles River (now the Cape Fear River), that was considered the site to complement the earlier English settlements of Virginia (1607) and New England (1620/1630). It was the location of two attempts at settlement in the mid-1660s, neither of which was successful. In 1662, the Massachusetts Bay Colony sent William Hilton to explore the area of Cape Fear for a potential settlement. Based on his favorable report, a group of New Englanders sailed to the Cape Fear area in the late winter of 1663, but soon abandoned the colony. Some have suggested the failure was due to poor lands, contradicting Hilton's positive assessment. In fact, the departing settlers left a posted message to that effect, giving the Charles River a very bad reputation among the English. But another reason may have been the inclusion of the area within the Carolina Land Grant of 1663 given by Charles II to the Eight Lords Proprietors. Acquiring the vast territory that included what was to be all of the Carolinas and Georgia and part of Florida (territories that also were claimed by the Spaniards who threatened from St. Augustine), the Lords Proprietors intended the promulgation of a rigorous model of settlement based on the Fundamental Constitutions, to be discussed below, that left no room for Puritan self rule (Lee, 1965, 33-34). Realizing that they would not be able to govern as they wished, the colony abandoned the Charles River.

A second attempt at settling Cape Fear occurred after the establishment of the Proprietary colony of Carolina and had its origin on the island of Barbados. William Hilton played a role here as well, hastening to the island to clear up misperceptions regarding the quality of the land along the Charles River and, more important, to restore his own reputation. He explored the coast of the Carolinas for the Barbadians, from Port Royal to Cape Fear, reporting the latter area to include "as good tracts of land, dry, well wooded, pleasant and delightful as we have seen any where in the world" (Lee, 1965, 38). In May 1664, a group from Barbados established Charles Town on the Charles River about 20 miles above the mouth. Within two years, its population had grown to 800 and the colony was supported by a successful agriculture and Indian trade. But problems began to arise which resulted in a steady out-migration and an end to immigration of new settlers. One matter involved the stringent land distribution policies of the Proprietors, which they sought to base on agglomerated townships, similar to those in New England, and the required payment of quitrents on both good- and poor-quality lands. Settlers felt that these regulations were both inefficient and unfair. There also were problems related to shortages of supplies that the Proprietors had promised to provide, and the hostile relations with previously friendly Indians was the last blow to the disintegrating colony (Lee, 1965, 44-50). By 1667, Charles Town had failed, joining a long list of failed settlements in the Carolinas going back 140 years! But plans were underway to establish a new colony to the south, near Port Royal, one that carried the same name—Charles Town—but which would survive to prosper.

After the failure on Cape Fear, the colonists on Barbados still faced the problem of finding new settlement areas because of the population pressure on the island. They sent William Hilton on another exploration of the Carolina coast, but this time to the south of Cape Fear. Visiting the area of Port Royal, he remarked on the magnitude of the harbor and the richness of the land.

"The ayr is clear and sweet, the Countrey very pleasant and delightful: And we could wish that all they that want a happy settlement, of our English Nation, were well transported thither" (Kovacik and Rowland, 1973, 334).

In April 1670, one of the ships carrying settlers from England and Barbados put into Port Royal. For many reasons, not the least of which was the strong encouragement by the Chief of the Kiawah, an Indian tribe located farther up the coast, the settlers sailed north in search of a new landfall. They found it on the Ashley River and settled on its western bank, naming the place Charles Town. A decade later, the settlement was moved to a new location on the peninsula formed by the Ashley and Cooper Rivers, giving the city a greater control of the port. By 1671, the number of settlers in Charles Town barely exceeded 400, but it had become the focus of settlement for what was to be South Carolina. By the end of the 1600s, the population of the newly settled colony was about 16,000, approximately half European and half African (Petty, 1943, 20-21).

At this time, the settlement geography of the Carolinas under the English focused on three locations: Albemarle Sound, where a few hundred families had occupied lands along the northeast bank of the Chowan River; Cape Fear, where two colonies had failed within five years and the land now lay open; and Charles Town, where a small settlement had been established north of Port Royal. The two occupied areas were separated by some 400 miles and represented two different ways of life. Recognizing this, the Proprietors in 1691 established a deputy governor for the northern province of the colony; and, in 1712, separate governors were appointed for North and South Carolina.

The Fundamental Constitutions and the Settlement Process

The eight Lords Proprietors received a second, revised charter for the new colony of Carolina in 1665, which covered a vast territory south of Virginia, extending this time as far as present-day central Florida. Largely under the leadership of Anthony Ashley Cooper, later the Earl of Shaftesbury, who took great interest in the colony, the Proprietors established an outline for settlement called the "Fundamental Constitutions." Drafted largely by Ashley Cooper's secretary, John Locke, it advocated an agglomerated settlement focused on the township, arguing that it was more efficient and safer than scattered farmsteads. The document also sought to recreate a kind of medieval Europe in the new colony with local nobility, leetmen, and the quitrent.

The territory of Carolina was to comprise 480,000-acre counties, each of which was to be subdivided into 40 properties of 12,000 acres. Eight of these properties, called seignories, were to be granted to the Proprietors, and another eight were to be distributed to the local Carolina nobility—the landgraves and caciques. The remaining 24 properties were to be distributed to freeholders or yeoman farmers. The lowest in this social hierarchy were the leetmen, the equivalent of medieval serfs, who held no property but would instead provide labor for the local nobility. Each county also was to contain a town that would complete the agglomerated model. Despite the attempts of the Proprietors, the model never was adopted, although counties were established, land grants made, and some nobility appointed (Clowse, 1971, 15-22).

One of the major problems that the new colony had to overcome was the dearth of settlers, and this required the creation of a positive image for Carolina. In the late seventeenth and early eighteenth centuries, a number of pamphlets intended to burnish that image were published (Greene, 1989). As early as 1672, Carolina was referred to as "one of the best [colonies] . . . that ever the English were masters of," a land that promised "Health, Pleasure, and Profit." Despite such promises, word got back to England that Carolina was not necessarily a paradise and that the colony faced problems like miserably hot and unhealthful summers and bouts with disease that plagued the settlers. Later promotional tracts continued to paint a highly flattering image of the colony, but they also offered practical advice to deal with its problems, recommending a temperate lifestyle and the taking of cool water and rest in the shade during the heat of a summer day, reminding readers that the winters were mild, and strongly advocating that new settlers arrive in late fall, after the sickly part of the year, so that the body could season over the winter and spring and thus be able to meet the rigors of the summer (Merrens, 1969).

The key to the success of the Carolina colony, however, was not the Fundamental Constitutions but the free distribution of land. Land policies varied over time, but all individuals benefited from a liberal program to assure the settlement of the colony (Ackerman, 1977). An individual newly arrived in Carolina or an indentured servant who had completed his service or even someone who had received no land previously could petition the council for a grant of land based on a 50-acre headright for himself and an additional 50 acres for each dependent in the family, including servants and slaves (the headright grants were higher during the early years of the colony). Once the petition was received and verified, the Council issued a warrant for an official survey by the colony's Surveyor General. Once the plat of that survey was complete, it was submitted to the Council, which presented the claimant with the official grant. One obligation that the new grantee incurred was the quitrent, a payment from medieval times that freed the landholder from various obligations to the lord, in this instance the Lords Proprietors. Most of the early land grants were located on streams or rivers, underscoring the role of water in agricultural technology and as an important means of transportation and communication.

By around 1730, the two colonies each held about 30,000-35,000 inhabitants, but geographic distance between their centers of settlement exacerbated the differences already evident in their economic, political, and social geographies. In fact, many scholars have argued that it is incorrect to refer to them as "the Carolinas" in a historical sense, especially during the colonial period, because it implies that the two were alike (Brown, 1948, 61-62), when in actuality they were very different places. These differences laid the foundation for the division into two separate colonies in 1729, when Carolina became a royal colony.

Developments in North Carolina

Expansion of Settlement

The geography of coastal North Carolina is dominated by the Outer Banks, the offshore islands that parallel the coast and block the rivers and sounds from the ocean. The shallow inlets that separated these islands did not allow the easy passage of large ocean-going ships, which seriously hampered foreign trade and the economic development of North Carolina. As noted above, the population of North Carolina was limited to the area of Albemarle Sound during much of the seventeenth century, but, in the early 1700s, settlers spread southward along and beyond the Roanoke, Tar, Neuse, and Trent Rivers (Merrens, 1964, 20). As early as the 1690s, French Huguenot colonists from Virginia settled on the Pamlico River, and, in 1710, a few hundred German-speaking Swiss from Berne and the Rhine Palatinate established New Bern on land between the Neuse and Trent Rivers (Merrens, 1964, 21-22).

One colony that had a significant impact on the demographic diversity of North Carolina was that of the Moravians, a religious group that had attempted unsuccessfully to settle in Georgia but then found a home in Pennsylvania. They received a 10,000-acre gift of land from Lord Granville (the one proprietor who held on to his land after Carolina became a royal colony) in the mid-1700s and set about establishing a colony in north-central North Carolina in the 1750s. In 1765, the town of Salem, the center of Moravian settlement, was laid out (Enscore, 1996, 29-32).

Despite the immigration of different groups, North Carolina was overwhelmingly English in the early colonial period, and Africans probably accounted for less than 20 percent of the population in 1730. This was in sharp contrast to South Carolina where Africans accounted for some 65 percent of the total population. Furthermore, most slaves in North Carolina came from adjacent colonies or perhaps from the West Indies, while South Carolina established a slave trade directly with Africa at least by 1720.

North Carolina also remained a predominantly rural area with barely two percent of its population that could be considered urban in 1775 (Merrens, 1964, 142). The colony had no focus of settlement like Charles Town to the south (Meinig, 1986: 381). The only area that could serve such a role was Cape Fear, which had an outlet to the ocean and access to a hinterland through the tributaries of the river; but as late as 1724-25, it was devoid of European settlement. In the early colonial period, a scattering of small port towns developed in North Carolina, each serving a certain part of the colony (Merrens, 1964, 87). Bath was incorporated in 1705 as the port for Pamlico Sound, while the town of Edenton was founded in 1713 as the port for Albemarle Sound (Figure 4.2). In 1710, as noted above, settlers from the Rhine Palatinate established a settlement between the Neuse and Trent rivers and named it New Bern. But the geographic constraints of the Outer Banks prevented any of these smaller settlements from developing into a significant trading site or becoming a focus for settlement. One report noted,

> There are great Tracts of good land in this Province and it is a very healthy Country, but the situation renders it for ever uncapable of being a place of considerable Trade by reason of a great sound near

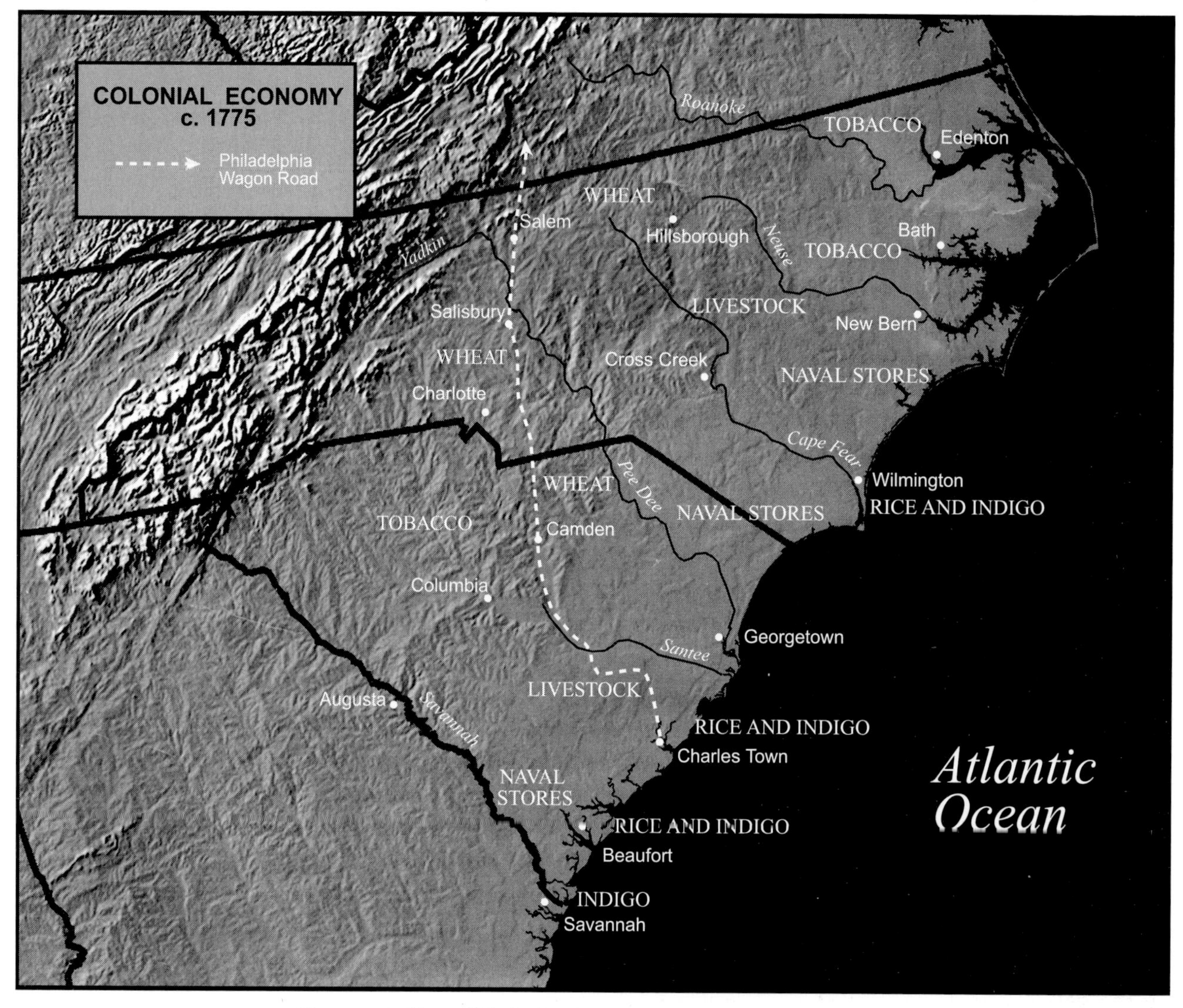

Figure 4.2. Map of Early Settlement and Agricultural Regions

> sixty miles over, that lies between this Coast and the Sea, barr'd by a vast chain of sand banks so very shallow and shifting that sloops drawing only five foot water run great riske of crossing them (Lee, 1965, 66).

For the most part, North Carolina's trade was restricted to Virginia, New England, and the West Indies, and its economy reflected these orientations,

> The little Commerce therefore driven to this Colony is carryd on by very small sloops chiefly from New England, who can bring them Clothing and Iron Ware in Exchange for their Pork and Corn . . . (Lee, 1965, 66-67).

North Carolina was a colony of predominantly small farmers, in contrast to the rice and indigo plantations that dominated coastal South Carolina. The major cash crop in the seventeenth century was tobacco, shipped through Edenton or more commonly sent overland to Virginia for export. Corn and wheat, cattle and hogs, furs and skins from the Indian trade, and eventually naval stores including tar and pitch also played a major role in North Carolina's economy.

By 1726, the settlement of the Lower Cape Fear had begun with the layout of the town of Brunswick on the west bank of the river about 12 miles above its mouth. At that point the river was deep enough for sea-going ships, far enough upriver to provide protection, and still in a position to control the hinterland defined by the river's tributaries. In 1728, the settlement was described as a "dispersed multitude of people residing up and down Cape Fear" (Lee, 1965, 102). Large landholdings, unique for North Carolina, were the basis for a plantation economy that would dominate the area, especially Brunswick and New Hanover counties, for the rest of the colonial period. In 1733, the town that would become Wilmington and replace Brunswick was founded. In time it would be the largest city in North Carolina, but in the mid- to late-colonial period, it shared prominence with a scattering of other urban places.

Urban Places

By the end of the colonial period, there were almost twice as many towns as there had been in 1750, and these included seaports, midland towns, and western towns (Merrens, 1965, 146-155). The three largest seaport towns included Edenton, New Bern, and Wilmington. Situated in the northeastern part of the colony, but limited to the navigation of small ships, Edenton handled exports of staves and shingles, tobacco, and corn, almost all of which were destined for other British colonies on the Atlantic seaboard or the West Indies. The port was hampered not only by the geography of the Outer Banks—the shallow inlets, the size of Albemarle Sound, and the distance to the ocean—but its proximity to the Virginia boundary and location on the sound reduced the size of its hinterland. New Bern grew rapidly during the last half of the eighteenth century, having become the de facto capital of North Carolina in the 1740s, and a nascent road system gave it an access to its hinterland not enjoyed by Edenton. New Bern was at the head of navigation of the Neuse River, but smaller boats could carry goods farther upriver, thus expanding the port's influence into the interior. In the area of Cape Fear, the last quarter of the eighteenth century saw Wilmington become the dominant city and eclipse the older Brunswick. Even as early as 1754, Wilmington's population was three times that of Brunswick (Merrens, 1964, 153). It benefited from easy access to the ocean and a productive hinterland that made it a successful port. Cape Fear was North Carolina's dominant producer of such commercial products as naval stores, lumber, rice, and indigo.

Wilmington's viability as a port also was related to its direct trade ties with the town of Cross Creek (present-day Fayetteville), one of the Midland towns. Acting as an intermediary, Cross Creek directed a great deal of trade by land and water from the Back Country to Wilmington (Robinson, 1979, 179). Located 16 miles upriver from Brunswick, Wilmington offered an intervening opportunity for Back Country merchants seeking an outlet for their products. Similarly situated on the Inner Coastal Plain but north of Cross Creek was another colonial midland town: Halifax. Sited on the Roanoke River and established in 1754, Halifax was at the head of navigation for small boats. It also was close to Virginia and had ties with Edenton, shipping corn, wheat, tobacco, pork, and lumber to that port town. Like Cross Creek to the south, it was an entrepot for the developing Back Country of North Carolina. The western towns included Hillsboro and Salisbury, both of which served as collection points for goods from the Back Country. Located on the Wagon Road from Pennsylvania to South Carolina, they also maintained significant commercial ties with Charles Town and Virginia.

Economic Geography

In the early 1700s, both Carolinas benefited from an economic activity that provided the first major infusion of wealth into South Carolina's colonial economy and would very much define North Carolina's commercial economy through the remainder of the colonial era: the production of naval stores—tar, pitch, and turpentine. In North Carolina this industry began near Albemarle Sound as producers took advantage of the pine forests, but it was the Cape Fear region and the development there of large landholdings that would become the leading producer of naval stores in North America.

As the world's most powerful maritime power, England needed these products for the caulking of ships, preservation of rope and sails, and other uses. In 1702, it was cut off from its traditional sources of naval stores in Scandinavia by the War of Spanish Succession, requiring the Royal Navy to seek new sources for these valuable products. Bounties or subsidies to encourage production in England's American colonies were approved by Parliament to take effect in 1705. The Carolinas, relying on their extensive pine forests, became major producers of these commodities, and naval stores was their leading export during the first decade of the eighteenth century (Clowse, 1971, 170-178). The accumulated wealth was applied to the purchase of slaves to maintain the heavy labor demands of the industry, beginning the introduction of large numbers of Africans especially into the Low Country of South Carolina, but also into the area of Cape Fear.

In 1772, North Carolina exported more than 117,000 barrels of naval stores, representing 63 percent of all the exports of these products from England's North American colonies. The port of Brunswick/Wilmington was the major exporter of tar, while the other North Carolina ports shipped quantities of tar, pitch, and turpentine to New England and the Middle Atlantic states to be re-exported to Europe. Naval stores was the major economic activity for the Cape Fear region especially, and the American colonies dominated English imports of these products. Despite this, the English Navy was a constant critic of the quality of Carolina's tar because of the preparation techniques. Instead of making green tar from trees freshly cut down, Carolina producers relied on already fallen trees for the production of what was called hot tar, which did not require as much labor. They also did not exercise quality control in other areas of the process, which frequently violated the requirements of the bounty.

Cape Fear planters also experimented with rice and indigo, but their production and export were insignificant in comparison with that of South Carolina as the great bulk of labor was directed to the naval stores industry. North Carolina exported only about 500 barrels of rice as late as 1770, when exports from Charles Town were averaging about 130,000 barrels a year (Gray, 1958, 289; Merrens, 1964, 126). Similarly, North Carolina in the mid-1770s exported a couple of thousand pounds of indigo, while South Carolina's exports exceeded 1,000,000 pounds.

North Carolina did have success with the export of grains, both corn and wheat, and flour. Indian corn was almost ubiquitous in the colony, was produced in surplus, and found a good market in New England (Merrens, 1964, 108-111). Wheat had been grown by the first settlers, but its production increased with the arrival of the Moravians and the waves of settlers flowing into the Back country from the northern colonies. Wheat also was exported in the form of flour that could be milled in the Back Country or Midland towns.

Another important product of North Carolina was lumber and forest products, most of which went to the West Indies. Cape Fear was the leading exporter of sawn lumber, which made up some 75 percent of its shipments, and Edenton was the major port for staves, predominantly for barrels and hogsheads, accounting for about half the exports from the colony in the 1770s (Merrens, 1964, 94-95). Cape Fear relied on its extensive forests of pine trees for sawn lumber, and the oak-covered bottomlands around Edenton supported the production of staves.

Since its cultivation, harvest, and preparation required considerable labor, tobacco was one crop associated with slavery, although its production area did not have as great a concentration of slaves as did Cape Fear. The distribution of tobacco warehouses in 1754 suggests that the early production of tobacco was centered in the Coastal Plain northeast of the Tar River, and, apparently, a considerable quantity of North Carolina's crop was marketed just across the boundary in Virginia at Suffolk or Norfolk (Gray, 1958, 231). By the 1760s, tobacco production had spread southwestward toward the upper reaches of the Cape Fear River and its tributaries and probably was introduced into the Piedmont by settlers coming from Virginia. By the early 1770s, some 1.5 million pounds of tobacco were exported annually through North Carolina's ports (Merrens, 1964, 120, 122), and the crop would dominate the Coastal Plain until the arrival of cotton in the nineteenth century.

Developments in South Carolina

Expansion of Settlement

Charles Town's growth and sophistication made it the focus of the Carolinas. Established on the peninsula formed by the Ashley and Cooper Rivers as a walled city in 1680, which it outgrew by 1704, Charles Town was considered "a larger and more elegant center than any of its counterparts in the English West Indies" (Meinig, 1986, 182, 185). It had an estimated population of some 8,000 by 1761. While North Carolina's towns were hampered by geography, Charles Town benefited from its site on a magnificent harbor formed by the estuaries of the two rivers and a situation that made it almost a natural port for the rice, indigo, and other products of the Carolinas during the age of sail. It was located on the western edge of a "circle" sailing route, as ships from England took advantage of the tropical easterly Trade Winds across the Atlantic, sailed up the South Atlantic coast to Charles Town, where they exchanged their cargos, and then caught the Westerlies to sail home. Charles Town also took advantage of the inland waterways behind the fringing barrier islands that extended from Cape Fear southward to the Altamaha in present-day Georgia and the paths and trails that led into the interior of the Carolinas to control much of its trade. As rice became a major export for Carolina, Charles Town accumulated the wealth that led to its golden age, a century of success and wealth that began about 1720 (Rogers, 1980, 3, 8-9).

As early as 1715, an estimated 60 percent of South Carolina's 16,000 inhabitants was African, illustrating the growing importance of the plantation. While South Carolina's colonial population later included French Huguenots, Germans, and Scotch-Irish, it was the Africans who provided the colony's ethnic diversity. Concentrated in the Low Country, they had a significant impact on the cultural and social geography of the coast, bringing religion, dance, song, food, and language to create a rich cultural mosaic. In the early years of the colony, African slaves were imported from the West Indies, but at least by 1720 they came directly from Africa, predominantly from the Senegambia region (present-day Senegal and Gambia) but also from the Grain Coast (Sierra Leone and Liberia) and Gold Coast (Ghana) of West Africa and Angola in Central Africa (Winberry, 1996, 11-15). During the decade of the 1730s, an estimated 20,000 Africans were brought into the colony, and in 1740 the population of slaves was estimated to be 40,000, accounting for at least 65 percent of the total population of South Carolina. The Stono slave rebellion of 1739, however, resulted in a number of restrictions on the slave population and reduced significantly the slave trade.

In 1715, about a half century after its founding, Charles Town continued to dominate the fledgling colony, and almost the entire population of South Carolina was located along streams and inlets between the Santee River to the north (where a number of French Huguenot families had settled) and the Edisto River to the south and about 25 miles into the interior. As the rice industry developed, settlement extended northward into the area of Winyah Bay and looked beyond to Cape Fear (Meinig, 1986, 178). In the early eighteenth century, some new towns were founded that enlarged South Carolina's settlement geography. In 1711, the port of Beaufort was established on the southern frontier of the Carolina colony to provide military protection against the Spaniards in St. Augustine and to serve as an outlet for the Indian trade, livestock activity, and developing naval stores industry that characterized the area (Rowland et al., 1996, 88, 90). The town of Georgetown was established somewhat later in 1731, even though the first petition for a port on Winyah Bay for the rice trade had been presented to the royal officials as early as 1723 (Rogers, 1970, 30, 32).

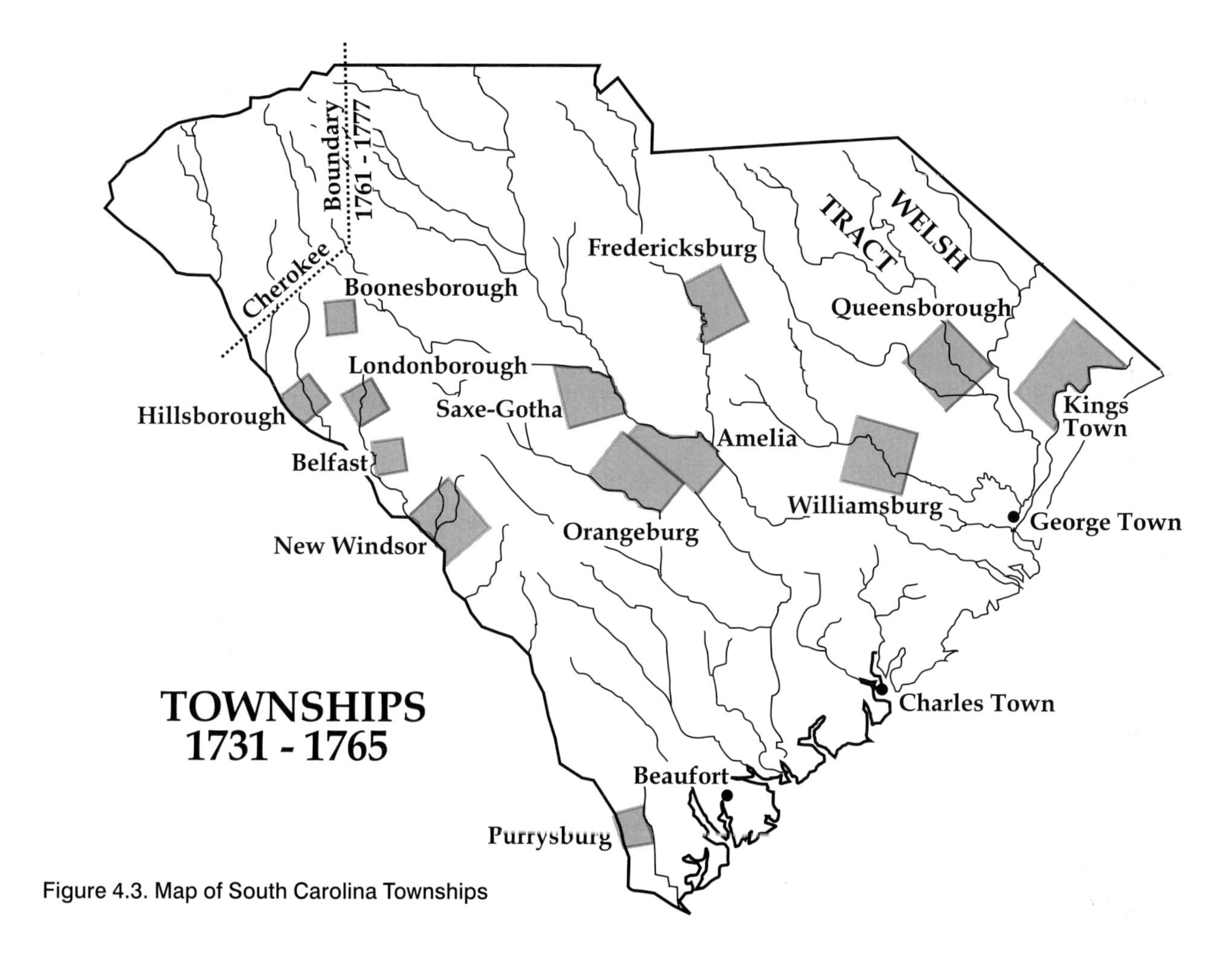

Figure 4.3. Map of South Carolina Townships

What greatly concerned the South Carolina authorities, however, and which underlay a concerted effort to increase population and expand territorial control was the concentration of settlement in the area of Charles Town, the lingering threat of renewed hostilities with the Indian tribes of the interior, and the constant fear of a slave revolt. In 1731, the colonial authorities introduced the Township Scheme, which was intended to encourage European settlement in the colony, provide the essence of a militia for protection, and assure an orderly expansion of settlement in the Back Country. It identified ten sites on major rivers in an arc with a radius of about 100 miles from Charles Town. Two additional townships were supposed to be on the Altamaha, but neither was developed because of the formation of the colony of Georgia in 1730. Another township, Purrysburgh, situated on the lower Savannah River, which was not one of the original ten, became a part of the plan because it was settled about the same time (Figure 4.3).

The remaining eight included sites on the Savannah, Edisto, Saluda, Congaree, Wateree, Black, Pee Dee, and Waccamaw rivers (Meriwether, 1940; Robinson, 1979, 162-174). Each township comprised 20,000 acres of land, and settlers were to claim one town lot and grants of 50 acres of surrounding land for each person in the family, including slaves and servants (Ackerman, 1977). New settlers were also to receive tools and a payment to carry them through the first year, but the funds for these were quickly exhausted. Within a few years, at least a few hundred people had settled in each of the townships with the exception of Queensborough, which was almost a complete failure. One area of great success, just to the north of the poorly situated Queensborough, was the Welsh Tract along the Pee Dee River, which had an estimated population of some 3,000 by

1757 (Robinson, 1979, 167). The township scheme brought diverse settlers to much of eastern and northern South Carolina. Germans settled in and near the Saxe-Gotha Township in present-day Lexington County and toward the north of it in an area today known as the Dutch (corruption of Deutsch) Fork. Welsh Baptists from Pennsylvania settled in the Pee Dee, and French Swiss occupied Purrysburgh. In the 1760s, additional townships—Hillsborough, Boonesborough, and Londonborough—were set up in the former Cherokee territory on the Northwestern Frontier. A colony of French Huguenots settled in the area, providing place names, such as Abbeville and New Bordeaux.

Just as North Carolina had interior settlements tied to its ports, so did South Carolina. The most important Back Country town—Camden—was located on the Catawba Path and the Philadelphia Wagon Road where the Wateree River crosses the Fall Line. It began to develop after the arrival of a group of Quakers in 1750/1751 (Ernst and Merrens, 1973). By the early 1760s, Camden had become a major collection center for the products of the Back Country, including butter, cheese, hemp, flax, and especially wheat, which was milled into flour. Its hinterland in the mid-1770s included all of South Carolina and extended to the upper reaches of the Catawba and Yadkin rivers in the northern North Carolina Piedmont and northeastward into that state's Coastal Plain (Schulz, 1976, 94). The town also had close ties with Charles Town merchants with whom it exchanged the products of the interior. The Charles Town newspaper in 1771 reported, no less than 113 Wagons on the Road to Town, most of them loaded with two Hogsheads of Tobacco, besides Indigo, Hemp, Butter, Tallow, Bees Wax and many other Articles; who all carry out on their Return, Rum, Sugar, Salt, and European Goods (Klein, 1990, 19).

Other Back Country products included in this trade were flour and wheat, and Camden became a major center for milling in the Carolinas. One other important trading center was the settlement of Ninety-Six located on the Cherokee path toward the Northwestern Frontier. It also was within the hinterland of Camden.

Colonial Economy

Charles Town spent the first 40 years of its existence trying to establish a commercial foundation and, like North Carolina, relied on a number of economic activities. The most remunerative of them during the seventeenth century was the Indian trade, and Charles Town merchants followed trade routes across the South to the Mississippi River. Even as late as 1750, some 100,000 deerskins and a small quantity of furs were exported from Charles Town (Meinig, 1986, 184). South Carolina also had success with other endeavors. Because of the mild winter, cattle and hogs were turned out to roam the forests on the Carolina frontier, primarily between the Edisto and Savannah Rivers, in a semi-wild state. They were rounded up periodically and driven to market in Charles Town. Like the livestock trade in the Albemarle region of North Carolina, this was a frontier activity that characterized the Coastal Plain but did not extend very far westward or northwestward into the Piedmont. The great forests that covered the Coastal Plain were also the basis of a forest industry that included the production of planks, shingles, and staves and later of naval stores. All of these activities—Indian trade, livestock herding, and the forestry industry—provided an economic foundation for South Carolina during the last decades of the seventeenth century, but none of them attained the returns that Virginia earned with tobacco.

In 1720, however, South Carolina's Low Country exported 20,000 barrels of rice, which assured the colony a viable commercial base that promised great wealth and the development of its own plantation economy. Rice was initially considered a highland crop, but later it was grown in the swampy regions along rivers. By the 1760s, rice producers began to rely on the tidal system with the extensive construction of dikes that enclosed the fields and the use of trunks (box-like wooden culverts) that allowed the flow of water between fields and streams or canals. This technology flooded and drained the rice fields at different stages of the crop's growth cycle and controlled weeds and other pests. Because of its reliance on the tidal system, rice production was restricted to specific locations along rivers that experienced the tidal effect. Fields could not be too far upriver because of the weakening of that effect, but they also had a downriver limit to assure that the fields would not be inundated with salt or brackish water during high tide. The rivers flowing into Winyah Bay and the Santee River delta supported large concentrations of plantations because the powerful streams had flows that created considerable tidal interaction. On the other hand, the rivers south of Charleston were largely blackwater rivers with a limited flow, and they supported fewer plantations (Hilliard, 1978, 101-104).

Rice plantations had heavy labor demands, which were satisfied by the importation of tens of thousands of African slaves to cultivate and harvest the rice and to maintain dikes and trunks. But the Africans provided knowledge as well as labor. Many of the slaves, being from West Africa where rice was grown widely, were already familiar with the crop, and their knowledge undoubtedly benefited the planters. West Africans also used technologies very similar to the trunk system, and it is possible that they deserve at least part of the credit for its success in the colony (Carney and Porcher, 1993; Wood, 1974).

A second crop that attained considerable importance in the Carolinas and especially South Carolina was indigo. The plant produced a blue dye in great demand by the developing textile industry in England. Brought from the West Indies in the early 1740s, the plant was successfully raised and harvested in South Carolina in 1744. This achievement traditionally has been associated with Eliza Lucas, but she had assistance from some French Huguenots, already familiar with the textile trade, who had settled in Charles Town and its environs. North Carolina planters also achieved a successful crop and produced and exported a small amount of indigo in the 1740s.

In the rice-producing areas above Charles Town, indigo production was a complement to rice. Not only could it be grown on the higher ground above the rice fields that had been carved out of the swamps, but its annual cycle required labor at times when it was not needed for rice production. Indigo was grown all along the coast and also at a number of inland locations because its low-bulk, high-value nature could bear high transportation costs. In the Sea Islands off Beaufort, it was cultivated on small plantations and was the only commercial crop (Rowland et al., 1996, 167). It has been argued that the Parliamentary bounty of six pence per pound, effective in 1749 and removed at the time of the Revolution, was the cause of the crop's success and ultimate failure, while others cited eighteenth-century wars, which made the high-value indigo a better export commodity than the bulky rice in the face of high shipping charges.

The bounty undoubtedly encouraged Carolina planters to try indigo, and the wars made it a desirable crop because of its high value; but it was a steady market demand in England that explained indigo's success in South Carolina. Throughout the colonial period, however, South Carolina indigo had a marginal reputation because planters took a number of shortcuts in the processing of the dye (Winberry, 1979a); but, by 1774-75, exports from South Carolina exceeded one million pounds. Even after the Revolution, indigo was still an important export for the Carolina economy, but it disappeared after 1795. The increased production of indigo in South Carolina, Louisiana, the West Indies, Latin America, and even in India created a surplus on the world market, which lowered prices and drove many marginally profitable producers like South Carolina out of the market (Winberry, 1979b).

Population and Back Country Settlement

By 1770, North Carolina's population totaled about 180,000, almost triple what it had been in 1750, but much of this growth had occurred in the Back Country along the Piedmont. During this period, the population of Rowan County in the interior had almost quadrupled, while Onslow County had seen a doubling of its inhabitants (Merrens, 1964, 55). South Carolina's population in the early 1770s totaled about 175,000, of which about two-thirds or 110,000 were slaves (Petty, 1943, 47). Only in the Cape Fear area of North Carolina did slaves constitute a significant portion of the population (39% in 1767), and only in New Hanover and Brunswick Counties did slaves represent half or more of the inhabitants.

In 1759, the European population of the Back Country of South Carolina totaled about 7,000, but, between 1760 and 1765, it increased almost 50 percent to 10,000 and would continue growing through the early decades of the nineteenth century. The growth of the Back Country in both colonies was due to the southward migration of settlers, predominantly Scotch-Irish, who came from southeastern Pennsylvania along the Wagon Road or through the Great Valley of Virginia and the Staunton River gap into the Back Country of North and South Carolina (Leyburn, 1962, 210-223). These peoples were originally Lowland Scots, Presbyterians who had migrated to Ulster in Northern Ireland in search of economic opportunity and religious freedom in the early 1600s. They crossed the ocean in the early 1700s, settling originally in Pennsylvania and then pushing westward and southward on to the frontier. By the 1750s, the first wave of them had reached North Carolina, and they totaled around 40,000 in 1763 (Merrens, 1964, 54). By the end of the 1780s, South Carolina's European population had risen to about 100,000, and a considerable amount of that growth had occurred in the interior of the colony.

The Scotch-Irish pioneer farmers of the Back Country frontier were different from the predominantly English settlers of the Tidewater. They created a landscape of small log cabins, worm rail fences, agricultural fields carved out of the forest with dead trees and stumps scattered through them, crops of Indian corn, vegetable gardens of squash, cabbage, beans, turnips, peas, white potatoes, pumpkins, and carrots, bottles of rye whiskey, and a basically subsistent way of life (Jordan and Kaups, 1989, 100-118). The Scotch-Irish also were known for their quick temper, recklessness, individualism and disregard for authority. Charles Woodmason, the itinerant Anglican missionary priest in the Back Country of South Carolina just prior to the Revolution, who had a rather jaded view of the people in his pastoral care, described many of them as "Rude—Ignorant—Void of Manners, Education or Good Breeding . . . [who] Live in Logg Cabins like Hogs," and noted that after one service the congregation turned to drink "and most of the Company were drunk before I quitted the Spot" (Hooker, 1953, 6-7, 56). Despite the differences, the Back Country was not totally isolated; it shipped wheat, flour, and tobacco to the coastal ports through Back Country entrepots like Hillsboro, Salisbury, Halifax, Cross Creek, and Camden, or via the tobacco warehouses established in both colonies.

The Revolutionary War

The Revolutionary War had its origins in the colonies of New England and was fought largely in the Middle Atlantic. The war in the South was more a partisan conflict, and most engagements were more skirmishes involving Carolinians rather than formal battles between British troops and Continental Regulars. However, the battles of Kings Mountain, Cowpens, and Guilford Courthouse, fought in the Carolinas in 1780-1781, were critically important to the ultimate British defeat at Yorktown.

Antebellum Geography of the Carolinas

The Revolution was another of the watershed events in the settlement geography of the Carolinas as it ushered in the antebellum era. It witnessed the decline of indigo, the arrival of cotton, which would dominate the Carolina landscape for some 150 years, and the spread of the plantation into the Piedmont, which along with changing economic, social, and transportation geographies began to erase the differences between the Low Country and Back Country. Although the influence of the coastal region and especially Charleston (the name was changed from Charles Town in 1783) remained dominant, a new power base, the cotton planters, began to express itself in the Back Country, especially of South Carolina. On the other hand, many traits remained unchanged, including the persistence of slavery and the production of rice, tobacco, and naval stores.

Sectional Differences and New Capitals

The differences that separated the Low Country or Coastal Plain from the Back Country or Piedmont, as discussed earlier, were most pronounced during the colonial period. The coast had been settled largely by the English, whose commercial agriculture was based on slavery, the plantation, and rice in South Carolina and on tobacco and naval stores in North Carolina. The Back Country in both states, on the other hand, was characterized by small farms, few or no slaves, and a largely subsistence economy.

After the Revolutionary War, one issue that arose in both states out of the sectional differences was the location of the capital. The Back Country settlers felt their being in New Bern and Charleston, both on or near the coast, made them inaccessible. In 1786, the General Assembly of South Carolina debated a new location for the state capital. The original committee report recommended Camden, the Back Country trade center, but this was rejected by the entire body. Preference was expressed instead for a location more central in the state, and the Taylor property near Friday's Ferry was selected as the site for the new capital to be named Columbia (Moore, 1993, 42-44). By the end of 1787, town lots had been surveyed and a capitol building was under construction.

Similarly in North Carolina, discussion had long focused on a more central location for the state capital, and many existent towns were considered. In 1788 a convention recommended a location in Wake County, although strong support was expressed also for Fayetteville, a Back Country town similar to Camden that had grown up and absorbed the original trade center of Cross Creek. The site in Wake County was approved by the legislature in 1791, and the new capitol building was completed in 1796. Like Columbia, it was named for a personage out of history: Raleigh (Lefler and Newsome, 1973, 259-261).

Changes in the Economic Geography of South Carolina

The establishment of political independence did not significantly change the commercial relations between the United States and Britain, which soon renewed its role as America's major trading partner (Brown, 1940, 120). Rice continued as the leading export from the coastal plantations, but one major component of the colonial economy, indigo, did not survive the eighteenth century. Indigo production had revived by the 1780s, but exports never approached the amounts that characterized the colonial period. South Carolina exported only 96,000 pounds of the dye in 1797, and production virtually disappeared before the end of the century. Planters turned away from indigo because of a declining market and low prices and began planting cotton instead (Gray, 1958, 611).

Introduction and Spread of Cotton

Cotton had received some attention in the early colonial period, but it did not prove successful in British North America. It was grown in the West Indies, however, and, in the 1790s, was introduced anew into South Carolina. After about 1815, cotton became a part of North Carolina's agricultural landscape in two locations. It had spread from the Coastal Plain of Virginia into the northeastern part of the state and also had sprawled from Back Country South Carolina across the state line into North Carolina's southern Piedmont (Gray, 1958, 889-892; Lefler and Newsome, 1973, 317; Meinig, 1993, 285-287).

Two types of cotton were grown in the Carolinas: black seed and green seed (Gray, 1958, 673-675). The black seed had a larger boll and long staples (fibers that surround and are attached to the seed) up to two inches long that were relatively easy to extract with simple roller gins. The hardier green seed variety had shorter staples, less than half the length of the black seed and tightly attached to the seed, which made them much more difficult, if not impossible, to separate. The black seed and related varieties became the basis of the Sea Island cotton industry, while the green seed and later Mexican varieties, referred to as upland cotton, were planted across the rest of the two states and into the Piedmont (Figure 4.4).

Cotton production expanded quickly in South Carolina; and, between 1790 and 1800, the state's cotton exports increased from fewer than 10,000 pounds to nearly 6 1/2 million pounds (Klein, 1990, 247). In 1801, South Carolina produced 20 million pounds of cotton and saw that double to 40 million in 1811; in both years, South Carolina accounted for half of the nation's overall cotton production (Gray, 1958, 683). What contributed to this explosive growth in the production of cotton in South Carolina?

One factor was simply the disappearance of indigo, which forced even the most conservative planters to attempt a new crop. A visitor to the Sea Islands off Beaufort in 1796 reported that ". . . difficulties in processing and low prices [of indigo] . . . forced people to try to convert to cotton, begun two years ago in Georgia" and that indigo was "totally abandoned . . . where it is being replaced by cotton" (Rowland et al., 1996, 280). Second was the technological breakthrough of Eli Whitney's cotton gin in 1793 that allowed the processing of green seed cotton. Prior to it, as noted above, the separation of the fiber from the seed was too difficult to be commercially feasible. Third was the rapidly growing market for cotton created by the British textile industry, which was well into the Industrial Revolution, and the nascent American industry in New England. Fourth was the already-established trade networks based on Camden in the Back Country that added cotton to the products that it already shipped to Charleston.

The spread of cotton into the interior was reflected also in the changing population demographics. The 1790 census showed that South Carolina had a total population of 249,073, of which 108,895 (43% of the total) were African by origin. But these gross numbers for the state as a whole missed some important regional differences. Slaves made up about 75 percent or more of the population of Low Country parishes, where the rice plantation dominated, while they constituted only a small portion, well below 25 percent, of the population in the Back Country. Although slave holders (the developing Back Country planter elite) constituted almost one-fifth of the white population of the Back Country of South Carolina as early as 1768 (Klein, 1990, 19), the region still was characterized more by the small yeoman farmer, very few of whom had

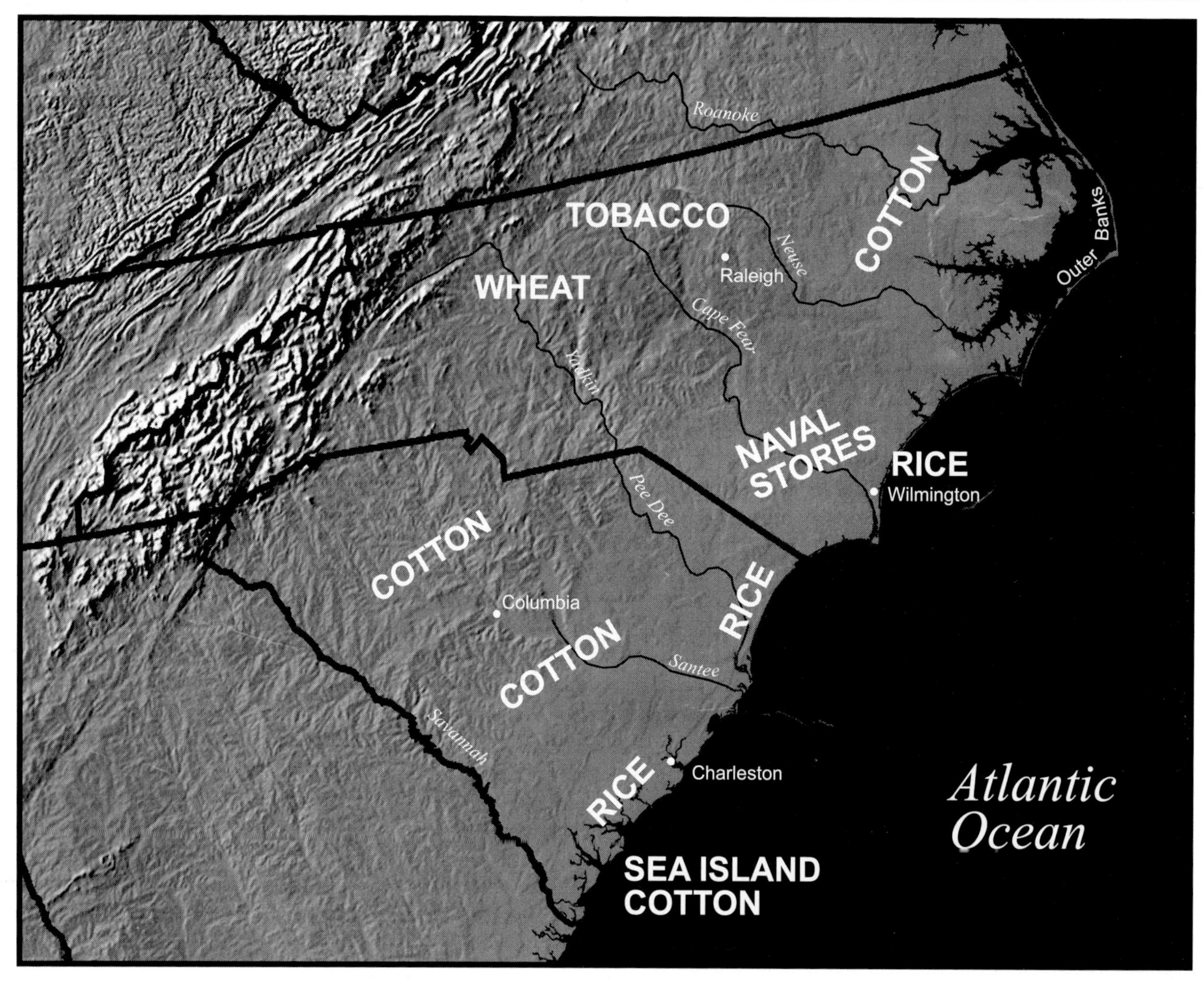

Figure 4.4. Map of Crop Regions

slaves. But the landscape was changing; even as early as 1802, John Drayton noted, "The planting of cotton increases annually both in the lower and upper country" (Klein, 1990, 247).

By the 1830 census, South Carolina's population had more than doubled to 581,185, but blacks outnumbered Europeans and accounted for almost 56 percent of the state's population. Much of this growth ironically had occurred in the Back Country with the spread of the plantation and cotton into the interior. By 1860, the population of four districts, three of them on the coast and dominated by rice or Sea Island cotton production, were more than 70 percent black, and in 16 districts across the Coastal Plain and Lower Piedmont blacks constituted more than half of the population (Hilliard, 1984, 34). These interior districts were major centers of upland cotton production in the state and the focus of another plantation system, based on green-seed cotton. In other words, the human and economic geography of South Carolina's Back Country was looking increasingly like that of its Coastal Plain.

One other factor in the changing demography of South Carolina was the out-migration of the yeoman farmers as a result of competition with the planters and because they sought new land to the west. Out-migration began in the early nineteenth century (Klein, 1990, 249-254) and continued into its second quarter as planters themselves, paying little attention to maintaining the soils, abandoned land in South Carolina and settled fresh land to the west. In 1860, and estimated 35 percent of the native-born white population, both yeoman farmer and planter, had migrated out of state, almost all westward across the South.

The Sea Island Cotton Industry

While green seed and Mexican varieties of cotton underlay the spread of the plantation and the cotton economy into the Back Country and Piedmont, black seed cotton was the basis for a successful cotton industry in the Sea Islands south of Charleston. Black seed cotton had been introduced from the Bahamas by way of the Georgia Sea Islands just after the Revolution, and the earliest recorded crop in South Carolina was harvested on Hilton Head Island in 1790. Production and export increased dramatically; in 1790, South Carolina exported only 9,840 pounds of Sea Island cotton, which included some from Georgia, but exports increased to more than eight million pounds in 1801 (Rowland et al., 1996, 281). Market demand was high for the long silky fibers of the black seed cotton, and planters worked carefully to develop improved varieties, new approaches to fertilizing the crop and soils, and better techniques for preparing the cotton for export. With successful crops and high prices, frequently double those for green seed cotton, the industry thrived as planters acquired more land to expand production and more slaves to provide the necessary labor (Kovacik and Mason, 1985, 83-85).

Sea Island cotton could be cultivated successfully only within a very limited geographic area that included the islands and adjacent mainland along the coast from Charleston southward into Georgia and Florida. The reason for this specificity is not clear, and explanations have included the soils, the sea air, the salinity, the slightly drier conditions on the immediate coast, and/or the presence of a longer growing season because of the influence of the Gulf Stream. Once planters realized these limits, they were faced with the problem of increasing production within a restricted geographic area. They introduced intensive fertilization of the soils, using salt marsh grasses to maintain production, and worked on higher quality and more productive varieties of cotton. Even though the crop did best on the higher lands of the islands, planters built dikes to reclaim marsh areas and plant cotton in them. The market for Sea Island cotton and the intensive labor of a large black work force made the great plantations of the Sea Islands successful and supported the growth of Beaufort as the commercial center for the region.

Transportation Geography

Transportation improvement was a critical component of the changing human and economic geography of the Antebellum period. While wagon roads and rivers led into the interior of the state, the former were barely adequate to handle the developing commerce and the latter were navigable only to the Fall Line and its rapids and other obstructions (Wallace, 1951, 373). Infrastructural improvements were important to the further expansion of cotton production and the plantation into the interior and the continued role of Charleston as a major port.

Three stages in the development of transportation can be identified. The first involved the completion of the Santee Canal, a 22-mile long canal built from the Santee River at about the present-day location of Lake Moultrie southeastward to the Cooper River. It allowed canal boats to navigate the Santee River and its tributaries from Columbia and Camden right to the docks of Charleston. It was completed in 1801 and operated until about 1858 (Wallace, 1951, 374-375).

The second stage began in 1818 with the state's commitment of $1 million to the construction of additional canals, improvements in navigation, and the building of roads. The canal system was completed between 1819 and 1825 and included about 25 miles of canals and granite locks that bypassed critical obstructions to navigation, primarily at the Fall Line and on the Saluda and Broad rivers above Columbia and the Catawba/Wateree River above Camden. The canals allowed navigation of the rivers almost to the state line and solidified Charleston's control

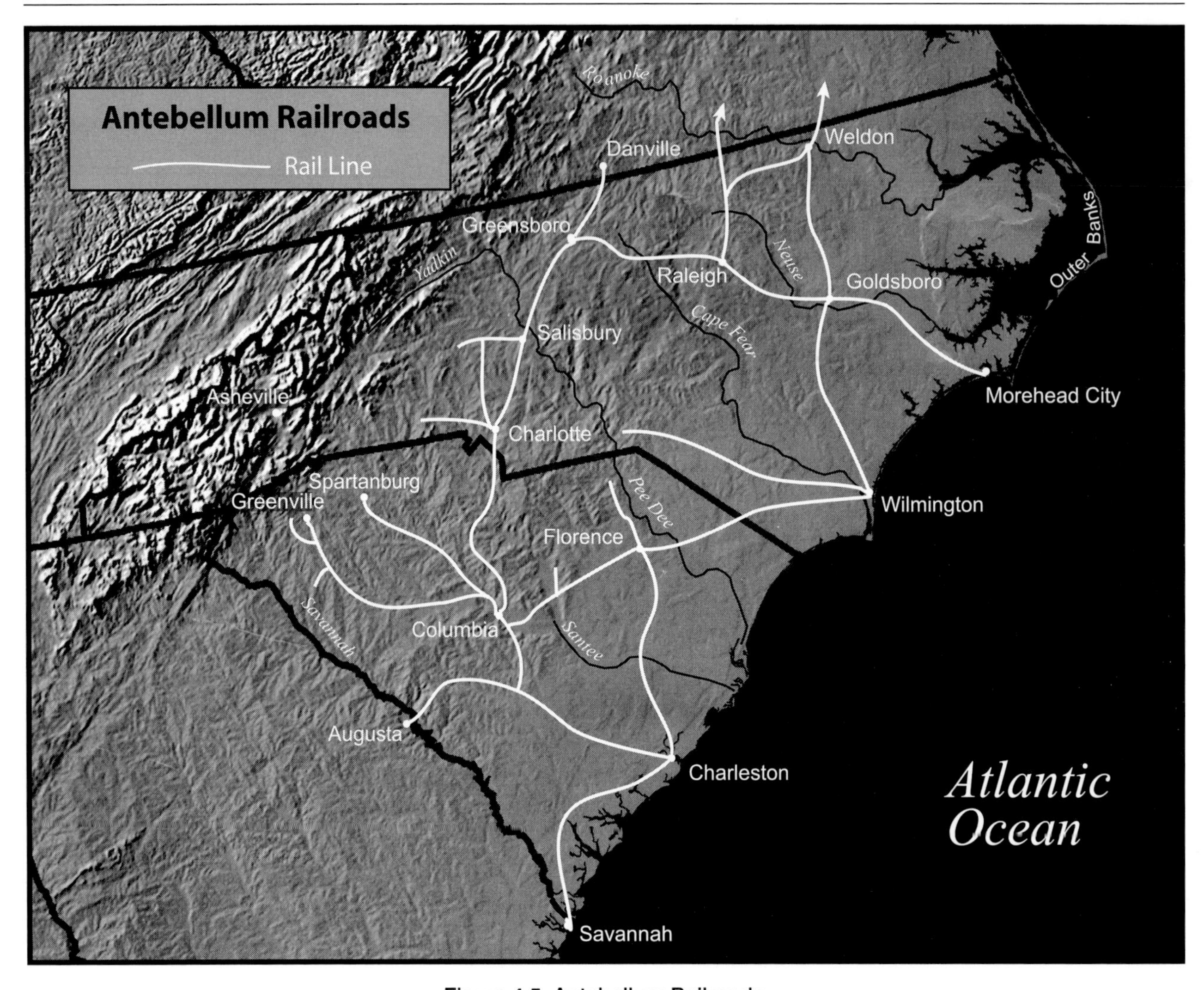

Figure 4.5. Antebellum Railroads

of its hinterland in the central part of the state. The port city relied on river steamers on the Pee Dee and Savannah to protect its hinterlands to the east and west (Hammond, 1883, 621-628). On the Savannah, however, the cotton trade of eastern Georgia and the western South Carolina Piedmont was intercepted by the port city of Savannah near the river's mouth, which was cutting into Charleston's hinterland.

The third stage of transportation improvement was foreshadowed in 1821 when Charleston merchants founded the town of Hamburg just across the river from Augusta, Georgia (Phillips, 1968, 77-81). It was intended to protect trade to Charleston instead of allowing it to be captured by Savannah. The plan was unsuccessful because it did not solve the problem of the shorter shipping route that led to Savannah (boats still had to negotiate the inland waterway from the Savannah River to Charleston). The solution, however, lay with a revolution in surface travel, the railroad.

The first railroad in the United States, the Baltimore and Ohio, began construction from its namesake port city in 1828. Charleston began work on its own South Carolina Railroad to Hamburg, which was completed in 1833 (Stover, 1997, 12-13) (Figure 4.5). But the railroad promoters in Charleston had even greater ambitions for the expansion of the port's hinterland and dreamed of a railroad across the Blue Ridge Mountains into the Ohio Valley. This dream was not realized for many decades, but, by the 1840s, railroads were extended to Columbia and

Camden and, by the 1850s, from Columbia to Abbeville, Anderson, Greenville, Spartanburg, and Florence in South Carolina and to Charlotte in North Carolina (Phillips, 1968, 335-351). The last antebellum railway line completed was to Savannah in 1860, just prior to the Civil War. The completion of these lines gave South Carolina nearly 1000 miles of railroad, brought an end to the canal era, and positioned the railroads to dominate the transportation geography of the state for almost the next century.

Changes in the Economic Geography of North Carolina

The developments that characterized South Carolina in the late 1700s and early 1800s were slow to affect North Carolina. In fact, the turn of the nineteenth century found the state little changed from the colonial period as it relied on a largely subsistent economy, was almost completely rural, and found itself in penurious straits with little chance of changing things. One legislative committee reported in 1830 that North Carolina was a State without foreign commerce for want of seaports, or a staple; without internal communication by rivers, roads, or canals; without a cash market for any article of agricultural product; without manufactures; in short without any object to which native industry and active enterprise could be directed (Lefler and Newsome, 1973, 314).

The causes of these problems were many. As noted above, the large sounds and shallow inlets through the Outer Banks constrained trade, and even the port of Wilmington was hampered by obstacles near the mouth of the Cape Fear River. The predominantly subsistent economy and orientation of much of the trade from Back Country towns toward ports in neighboring states instead of to the port cities of North Carolina also weakened the economy. But the most significant issue was the absence of transportation infrastructure—the lack of rivers, roads, and canals which the report mentioned. Not only were the roads in generally poor condition, but freight charges were high. It was not uncommon "that the profits of one-half the planter's crop were consumed in getting the other half to market" (Connor, 1929, 31).

Cotton and Tobacco

Cotton and tobacco slowly became a part of the antebellum agricultural landscape of North Carolina. Tobacco had been an important crop in the Coastal Plain during the colonial period, but the nineteenth century found it well established in the northern Piedmont. This apparently was related to the southward extension of Virginia's tobacco-growing region into North Carolina. Similarly, soon after the War of 1812, cotton production spread into northeastern North Carolina from Virginia's Coastal Plain and was well established there by 1835 (Gray, 1958, 888), replacing tobacco production and centering on an area bounded by the towns of Halifax, Goldsboro, and Washington.

Cotton also was introduced into the southern Piedmont from South Carolina where production focused on a band of southern counties from Roberson to Mecklenburg (Hilliard, 1984, 67, 76; Lefler and Newsome, 1973, 352, 392; Meinig, 1993, 287-288). The state was just beginning its involvement with cotton in 1820 when it produced about 10,000 bales for market, far short of South Carolina's crop of 170,000 bales. Even as late as 1860, though, South Carolina's cotton production more than doubled that of North Carolina on an annual basis.

Rice, another commercial crop, continued to be grown along the Cape Fear River and especially in Brunswick County, which alone accounted for 95 percent of the state's 8 million pound crop in 1860; in the same year, South Carolina rice plantations produced nearly 120 million pounds. Other crops also remained important for North Carolina, the most significant being corn. It was grown in just about every county across the state and had many uses as a food and for the making of corn whiskey. While corn production was static in the nineteenth century, wheat output more than doubled during the 1850s to nearly five million bushels (Lefler and Newsome, 1973, 392-393). It was grown primarily in the central Piedmont, with tobacco to the north and cotton to the south (Figure 4.5).

Population Geography and Demographics

The 1790 census recorded a total population in North Carolina of 388,776, of which 100,572 or about 25 percent, were African in origin (a much lower ratio than that of South Carolina). The great bulk of this black population, probably about 85 percent, was located east of the Fall Line, in the areas associated with tobacco and rice plantations and the production of naval stores. No county to the west of the Fall Line, an area dominated by yeoman farmers, had a population that was more than 20 percent black at that time (Hilliard, 1984, 28; Lefler and Newsome, 1973, 129). The arrival of cotton and tobacco in the Piedmont, however, changed those demographics.

In 1860, North Carolina had a total population approaching one million, of which about one-third were slaves. Between 1790 and 1860, the white population more than doubled, but the black population increased at a rate more than two-and-one-half times that of whites (Powell, 1977, 126). While slave owners continued to be a small minority in North Carolina (less than 28 percent in 1860), the concentration of slaves and the sizes of slave holdings had increased. Slavery also had crossed the Fall Line and was more significant in the Piedmont, especially where tobacco and cotton were grown. Here, the sizes of slaveholdings approximated those of the eastern part of the state, and the Piedmont region began to resemble increasingly the social and economic geography of the Coastal Plain. Overall, though, the population of slaves, the size of individual slave holdings, and the percentage of slave holders remained lower in the western part of the state than in the east.

One other factor affecting the demographics of North Carolina's population during the nineteenth century was out-migration. As in South Carolina, it began in the 1820s with the departure of small, white farmers. Later, it comprised planters who had given little attention to improving or restoring the soils and who sought new, fresh lands to the west. As early as 1815, it was observed that thousands of our wealthy and respectable citizens are annually moving to the West [Tennessee, Alabama, Ohio] in quest of that wealth which a rich soil and commodious transportation never fail to create in a free state, while thousands of our poorer citizens follow them . . . (Powell, 1989, 240).

The 1850 census indicated that some 31 percent of native North Carolinians were living in another state.

Transportation Geography

A basic necessity for the improvement of North Carolina was the development of a transportation infrastructure to tie together the different parts of the state and strengthen its economic geography. North Carolina early saw that the means to do this was the railroad, and, in 1827, the President of the University of North Carolina published a series of essays extolling this new means of transport. But what routes should the new railroads follow—north-south, connecting the rivers, or east-west to serve the Back Country?

In 1834, the first railway company in North Carolina was chartered: the Wilmington and Raleigh. While support for it was strong in Wilmington, it received little interest from Raleigh. As a result, the railroad promoters built the line instead to a town on the Roanoke River—Weldon—to connect with the railroad to Virginia. The 161-mile road was completed in 1840. Not to be outdone, the spurned Raleigh in 1835 received a charter of its own for the Raleigh and Gaston Railroad to build a line between those two points to connect with the Petersburg Railroad from Virginia. The 85-mile line was also completed in 1840. Both railroads also were north-south lines that served only the Coastal Plain, while the western counties continued to clamor for an east-west railroad. In 1849, the legislature chartered the North Carolina Railroad to construct what was intended to be an east-west line. It began in Goldsboro on the Wilmington and Weldon and connected the interior towns of Raleigh, Greensboro, and Salisbury in a broad arc before reaching its western terminus in Charlotte, a distance of 223 miles. It was completed in 1856. A 95-mile eastward extension of this line, the Atlantic and North Carolina, carried the road from Goldsboro to New Bern and Morehead City.

The Wilmington and Manchester, chartered in 1847, was built westward from Wilmington to the town of Manchester on the Wateree River in South Carolina. Completed in 1854, its strategic intent was to expand Wilmington's hinterland by pirating Charleston's trade with the Pee Dee, a region that was not served by a rail connection at that time. Charleston, however, soon built the Northeastern Railroad to Florence to solidify its hold on the region's trade.

While Wilmington was well connected in the Coastal Plain, and its economic hinterland extended westward into South Carolina, it needed to enhance its trade connections with the North Carolina Back Country. Under construction when the war began, the Wilmington, Charlotte, and Rutherfordton was the third railroad with its terminus in the port city and was intended to provide an important link with Charlotte and the Piedmont. It was not completed until 1886, after being reorganized as the Carolina Central Railroad. Another railroad that was incomplete at the beginning of the Civil War was the Western North Carolina Railroad from Salisbury to Asheville, eventually completed in the 1880s. Just prior to the war, North Carolina had 891 miles of rail lines (Lefler and Newsome, 1973, 380-381).

One interesting aspect of the internal improvements program unique to North Carolina in its magnitude was the construction of plank roads. Some 84 companies were chartered between 1849 and 1856 for this purpose. The idea was to provide farmers and businessmen improved roads that could be driven over at faster speeds and were passable year-round. It was anticipated but not always realized that these roads would be direct feeders to the newly completed railroads and thus enhance the network of trade across the state. Plank roads were constructed by laying stringers or "sills" end to end to form the path of the road and placing three-to-four-inch thick planks across them at right angles. Roads were eight to ten feet wide and were fairly successful financially and very popular during the 1850s. Fayetteville, perhaps because it was bypassed by early railroad building, had six plank roads going in all directions to maintain its role as an inland trade center. Roads were built across the state to serve many towns and cities, but they soon faced some basic problems. The most demanding was the constant maintenance, which worked against profit. By 1860, plank roads had all but disappeared from the landscape (Lefler and Newsome, 1973, 381-383).

Other Economic Activities

Mining and manufacturing were two other activities that had an impact on the antebellum economy of the Carolinas but to a much lesser degree. The Carolinas experienced a gold rush in the early nineteenth century, and at times mining was second only to agriculture in economic importance to North Carolina. Between 1804 and 1828, all the domestic gold used to coin money at the Philadelphia mint came from North Carolina, and, in 1835-1861, a branch mint in Charlotte made gold coins (Roberts, 1972, 11, 30-31). Gold mining began in the Carolinas in 1799, with the accidental discovery of a 17-pound gold nugget in Cabarrus County, North Carolina. (See Photo 4.2) Once its value was known, the gold rush began, and other discoveries of gold occurred in nearby Mecklenburg, Anson, and Montgomery Counties.

The gold-mining fever gradually spread to South Carolina, which recorded its first gold-mining activity in 1826 (Nitze and Wilkens, 1897, 26); in 1829, the famous Haile Mine, supposedly the most productive east of the Mississippi River, was opened in Lancaster County. Other famous mining centers in South Carolina included the

Photo 4.2. The Engine House on Upper Hill at Reed Gold Mine
(Courtesy of the North Carolina Historic Sites and the North Carolina Division of Archives and History)

Brewer Mine in Chesterfield County, opened in 1828, and the Dorn mine in present-day McCormick County, opened in 1852. The mining center of Gold Hill in Rowan County was among the most productive in North Carolina.

The two most important geologic concentrations of gold in the Carolinas were the Carolina Belt and the South Mountain Belt (Nitze and Wilkens, 1897, 45-75). The former extended northeastward-southwestward across the North Carolina Piedmont from Granville and Person counties in the north to Gaston and Mecklenburg in the south. It overlapped into South Carolina to include Chesterfield, Lancaster, and McCormick Counties. The South Mountain belt was situated in the Upper Piedmont/Blue Ridge and extended from Caldwell County, North Carolina, to Greenville and Pickens Counties in South Carolina. By 1860, North Carolina gold had been used to mint $17 million in coins, representing about half the production of gold in the southern states.

The earliest recovery of gold involved placer mining, using pans to wash the gold from streams in the mountains and Piedmont. As early as 1825, hard-rock vein mining began at the Barringer Mine in North Carolina, and the first shafts were sunk for underground mining in 1830 (Hines and Smith, 1996, 44). While small-scale placer mining persisted in much of the Carolinas, some of the larger mines employed mining engineers and up-to-date technology. Thousands of miners settled in mining centers like Gold Hill during the first half of the nineteenth century, but 1849 began the decline of the industry. The discovery of the more lucrative gold fields of California began a new gold rush that emptied many gold camps in the Carolinas during the 1850s, adding them to the mines already closed because of exhausted veins. By 1860, North Carolina had only nine operating mines reported in the census, while South Carolina recorded 58 operations (Lefler and Newsome, 1973, 395; Nitze and Wilkens, 1897, 76). Gold mining pretty much disappeared in the early 1860s but did revive after the war, only to decline again toward the end of the century.

A very different economic activity was manufacturing and especially the production of textiles. The rise and fall of textile manufacturing in the Carolinas was a function of the price of cotton: when it went down, there was strong interest in diversifying the economy; but when it recovered, those ideas lost favor as capital returned to the growing of cotton. The War of 1812 and the cutoff of the British market for Carolina cotton began the interest in manufacturing. In 1813, Michael Schenck opened the first mill near Lincolnton, North Carolina, followed by South Carolina's first mill, built by David Williams near Society Hill. The Alamance Cotton Mill's production of a diversity of cloth was unique (Powell, 1989, 315-316) as cotton mills in the Carolinas usually produced only yarn, osnaburgs, and plantation cloth of relatively low quality for the local market. A number of mill owners relied on a black labor force, but, by the 1850s, black workers were largely replaced by whites because the prices for slaves had increased so much that they were no longer competitive in an industry that was not very profitable to begin with (Lander, 1969, 91).

By 1840, after a 10-year period of expansion, the Carolinas had a total of 51 textile mills with 48,000 spindles and a capital investment of about $1 1/2 million. In contrast, two states in New England—Massachusetts and Rhode Island—had nearly 300 mills, operating 717,000 spindles, and a capital investment approaching $25 million. In 1860, the Carolinas had 57 mills operating with 68,000 spindles, still a far cry from New England in number and size. In fact, "the early mills were 'scarcely more than a step beyond the domestic manufacture with which they were so closely connected'" (Lefler and Newsome, 1973, 399). As the Carolinas approached the beginning of the Civil War, they remained predominantly agricultural economies; and the establishment of a viable textile industry lay some decades into the future.

The Civil War and its Causes

On April 12, 1861, Confederate batteries on the islands around Charleston Harbor began a 34-hour bombardment of Fort Sumter that marked the beginning of the Civil War. Events during the first half of the nineteenth century that foreshadowed the conflict involved a number of issues that led eventually to the secession of the southern states and the firing on Fort Sumter. (See Photo 4.3)

The first major issue in this debate was the basic nature of the United States and the inherent rights of the sovereign states that made it up. Essentially, it addressed the question of their right to secede, an unresolved issue put forward in the abstract in 1798 by Judge John L. Taylor of North Carolina (Powell, 1989, 333). Secession was raised a number of times before 1860 and was seriously considered in 1807, when New England states threatened to withdraw from the Union because of the Embargo Act that cut off the region's lucrative international trade (Meinig, 1993, 461-463). The limits of states' rights would be tested with South Carolina's secession in December 1860.

A second major issue was the economic differences between the two sections. The South continued to rely on cotton and other primary products in exchange for manufactured goods predominantly from Europe, the prices of which were kept down by low or non-existent tariffs. The North, on the other hand, was developing an industrial base and needed a high tariff to protect its nascent industries. The high tariffs of 1828 and 1832 directly affected the economy of the South and gave rise to the theory of "Nullification." It claimed that individual states had the right to nullify or refuse to enforce undesirable laws passed by Congress. The Compromise Tariff of 1833 lowered rates and ended the imme-

Photo 4.3. Fort Sumter

diate threat of nullification, but the issue underscored the differences between the two sections and made the South aware of the weak position it had in protecting its interests.

Despite these philosophical and economic differences, a single major issue underlay the causation of the war. As John C. Calhoun, South Carolina's leading antebellum politician, wrote in 1830 during the Nullification Crisis,

> I consider the Tariff, but the occasion, rather than the real cause, of the present unhappy state of things. The truth [could] no longer be disguised, that the peculiar domestick institutions of the Southern States [i.e., slavery] were under threat from an interfering central government (Meinig, 1986, 465).

The South relied on a voting balance in the Senate to protect the region from any legislation that threatened "the peculiar domestick institution," realizing as early as 1800 that it held a minority in the House of Representatives. The admission of Tennessee in 1795 balanced the number of free and slave states at eight each, and each new admission of states maintained that balance until the Compromise of 1850, which admitted California to the Union without a balancing slave state. Debate on the ultimate limits of slavery characterized the 1850s, which did not assuage southern concerns over the institution's future status.

The final strain leading to secession was the 1860 election of Abraham Lincoln who already had expressed his opposition to slavery and who represented to the increasingly anxious southern delegation the ultimate threat to the institution of forced bondage, which was the foundation of the southern economy. His decision to react to the secession of the southern states by calling up troops and planning the reinforcement of Union installations in the South seemed to end moderation and contribute to the unrestrained plunge into war. On May 20, 1861, North Carolina joined its sister states across the South by repealing its act of ratification of the Constitution.

When the war began in 1861, expectation on both sides was that it would be over within months. The South entered the war at a distinct disadvantage with regard to a number of measures (Powell, 1989, 350-352). First, it comprised less than half the 24 states that made up the North. Second, the population of the North was more than two and a half times that of the South (whose slave population represented about 40 percent of that total). Third, the North fielded about two and a half times as many troops as did the South during the war. Fourth, railway mileage of the North was more than twice that of the South, was better networked, and was not broken by different gauges (a common problem for southern railroads). Fifth, the value of Northern manufactures was more than 10 times that of the South, and the region produced more than twice as much corn. Sixth, the North produced 480,000 tons of iron per year, while the Confederacy made only 31,000 tons, 1/15 what the Union could turn out. Finally, the Confederate Constitution lodged the greatest power within the states, leaving a weak central government.

The strategies of the two governments were very different, that of the North committed to forcing the southern states back into the Union by invasion, while the Confederacy saw its strategic posture as one of defense. The Union had four objectives (Meinig, 1993, 508):

1. Establish a blockade of the South Atlantic and Gulf coasts to disrupt the trade on which the southern economy was based and to strangle the Confederacy into submission;

2. Capture the Confederate capital of Richmond and remove Virginia from the war;

3. Invade the South along the Mississippi River, capturing New Orleans, the Confederacy's largest port and biggest city, and separating the states of Arkansas, Louisiana, and Texas from the rest of the South; and

4. Invade the South through Tennessee, capturing the railway junctions of Chattanooga and Atlanta, marching to Savannah, and splitting the Gulf South from the South Atlantic states.

The Union was generally successful in its strategic goals by the end of 1864, but the two Carolinas had not yet felt the full impact of the war. Both states sent troops to fight for the Confederacy and had suffered heavy casualties. North Carolina dispatched an estimated 125,000 troops during the war, of which about 40,000 were killed in battle or died of disease, while South Carolina supplied about 60,000, of which at least one-fourth to one-third were lost. Both states suffered from shortages of supplies and the general deterioration of an infrastructure worn out by heavy use and lack of maintenance.

Still, fighting had arrived early on the Carolinas' shores, since part of their coasts were occupied by Union forces through much of the war. In November 1861, Union naval and military forces invaded and occupied Beaufort, Port Royal, and the adjacent Sea Islands, turning Port Royal Sound into a naval base to support the blockade. Troops periodically attacked and occupied other islands off the coast of South Carolina, tried to cut the Charleston and Savannah Railroad, and, in summer and early fall 1863, attempted unsuccessfully to capture Charleston.

Beginning in August 1861, and continuing into April of the following year, Union forces in North Carolina captured fortifications on the Outer Banks that gave them control of Albemarle and Pamlico sounds, cities such as Plymouth, New Bern, Roanoke Island, and Elizabeth City, and the coast of North Carolina to the south of Cape Hatteras. The Confederacy continued to hold Wilmington because of Fort Fisher and its protection of the Lower Cape Fear River.

By 1865, however, the two states were pulled into the war with the arrival of General William T. Sherman and his 60,000 troops, fresh from their "March to the Sea." Crossing the Savannah River into South Carolina from Georgia in January 1865, Sherman feinted moves toward Charleston and Augusta, but headed to Columbia, capturing the city on February 17. That night much of the city burned to the ground, the cause of which is still debated. Sherman continued northward and eastward, crossing into North Carolina and occupying Fayetteville; his army forded the Cape Fear River by mid-March and continued northward to Goldsboro. The end was approaching very rapidly. Fort Fisher had already fallen on January 15, 1865, and Confederate forces evacuated Charleston on February 17. The Carolinas were feeling the full brunt of the war; as one officer in Sherman's army later wrote,

> Over a region forty miles in width, stretching from Savannah and Port Royal through South Carolina, to Goldsboro, in North Carolina . . . the head, center, and rear of our column might be traced by columns of smoke by day, and the glare of fires by night (Wallace, 1951, 549).

Reconstruction and the Post-Bellum Landscape

The surrender of Robert E. Lee's Army of Northern Virginia on April 9 and Joseph E. Johnston's surrender of the Army of Tennessee on April 18 ended the Civil War after four exhausting years. Although the Carolinas had escaped much of the direct impact of the war, its last months saw Sherman's army and other Union forces crisscross the two states, destroying infrastructure, houses, and the means of production. The Carolinas also had suffered from the heavy mortality rate among their young men in battle; and the officer corps, the region's future leaders, was especially hard hit. The war marked another of the major watersheds in the history of the settlement geography of the Carolinas, and the post-bellum landscape would see changes in the economies, societies, populations, and transportation systems of the two states.

President Abraham Lincoln planned a very rapid return of the former Confederate states to the Union under Presidential Reconstruction, but that vision died along with the assassinated president in 1865. Congressional Reconstruction replaced it, and the Carolinas and other southern states were subjected to a military occupation that would last more than a decade. The Reconstruction period has long been considered among the bleakest of days by southern whites who recounted how their world had been turned upside down: poverty stalked the land, former state leaders were disfranchised from political participation, property was lost because of nonpayment of taxes, the state built up large debts, and carpetbaggers from the North and ill-educated freedmen held political power with support of the Union military. Notwithstanding such memories, it has been shown that Reconstruction was also associated with some important developments, especially the increased democratization of politics, the establishment of public education, and the construction of railroads. As well, the era found some strong leaders in the African-American community, such as Robert Smalls of Beaufort, who, earning respect from black and white, represented South Carolina in Congress from 1874 to 1886.

Agriculture and the Tenant System

Some major changes affected the agricultural landscape of the Carolinas, and this began with the end of slavery. Many of the new freedmen fled the plantations to occupy and cultivate the lands so frequently promised them by Union officials. Although some became independent landowners, the great bulk found themselves without land or employment and forced to return to the plantation. On the other hand, the former planters had recovered their old landholdings but had no labor and no crop. The solution was to bring these two groups together.

The first attempt involved contractual agreements, frequently negotiated and overseen by the military, between the freedmen and planters (Shlomowitz, 1979). These committed the freedmen to work for one year, quite frequently occupying former slave quarters and working the fields in gangs under the direction of overseers. The similarities to the days of slavery were too

Photos 4.4a & 4.4b. Tenant Shack and Farm

much, and freedmen sought a different relationship, one that allowed personal responsibility to decide success or failure. Little by little, they distanced themselves from the labor contracts in favor of a new system—tenancy.

Tenancy made an individual freedman responsible for working a specific plot of land, and it also brought change in the landscape. Instead of houses being grouped together near the big house and referred to as the "quarters" or "street," which had been the case during the plantation era, they were scattered across the countryside, on the lands that the tenants worked. (See Photo 4.4) Each tenant was responsible for the crop that he planted, cultivated and harvested. Only the initial plowing and preparation of the land was done by gangs of labor.

The tenancy system was well-established by the 1870s among planters involved with the growing of cotton and/or tobacco. During the early stages of tenancy, one could identify two types of tenant (more categories were created over time by the Census): the sharecropper and the cash tenant (Aiken, 1998, 29-35). The former had little or no capital, tools, nor work stock and relied on the planter for everything including seed and half of the fertilizer (the other half was supplied by the tenant on credit). Usually, the crop was split equally between them. The cash tenant, on the other hand, had some capital and tools and livestock and received use of the land and the house on it from the planter; he paid a set fee for the right to use the land to grow the crop. Cash tenants, therefore, were more independent than the sharecropper, thus lessening the planters' control of how his land was worked.

The increasing importance of tenancy was reflected in the agricultural statistics. In 1860, the average size of farming unit in South Carolina, dominated by the plantation, was 569 acres, while in North Carolina it was 316 acres, reflecting the greater import of the small farm. By 1880, however, that average in South Carolina had declined almost 75 percent to 143 acres and to 142 acres in North Carolina. This continued into the twentieth century, and by 1920 the average size in both states had fallen to about 65 acres. Tenants accounted for 43 percent of the farming units in South Carolina and 45 percent in North Carolina (Kovacik and Winberry, 1989, 106-107; Lefler and Newsome, 1973, 576-577; Powell, 1989, 416-417).

The tenant system did allow the post-bellum revival across the Carolinas of some important crops. In 1860, South Carolina produced more than 350,000 bales of cotton, a total that almost doubled to more than 600,000 in 1880 and almost reached 750,000 just before the turn of the century. North Carolina more than doubled its own production from 141,000 bales in 1860 to 390,000 in 1880 and 460,000 bales in 1900—cotton production overall had increased three-fold in the two states. On the other hand, it was not until 1890 that North Carolina achieved a harvest of tobacco equivalent to that before the war (33 million pounds), but, in 1900, North Carolina harvested nearly 130 million pounds of the leaf.

Other institutions that grew out of the post-war economy to complement tenancy were the "furnish" and the "crop lien." Tenants, whether sharecroppers or cash tenants, had inadequate capital to carry them through the agricultural year, but they still had to acquire seed and fertilizer for the crop and food, clothing, and other essentials for their families. To do this, they relied on either the landowner, who owned a plantation store, or country or small town merchants for the "furnish." This was the provision of basic needs through an extension of credit. The merchant sold the items to the tenant usually at an increased price and then charged interest, a total of which could raise the price of an item between 50 and 100 percent of its cash value. The merchant also needed collateral to cover his risk, but many tenants had little or nothing of value to put up beyond their share of the season's upcoming crop. Post-war laws allowed merchants to hold a "lien," or first claim on the tenant's share of the crop (Clark, 1964, 271-291). The amount of credit was therefore related directly to the anticipated size of the crop; the more cotton or tobacco that a tenant grew, the more credit that could be encumbered. As tenancy became more common and lien laws more important, the acreage devoted to cotton especially increased because it had market value. Unfortunately, one basic problem was that the more cotton grown, the lower the prices for it. The crop was valued at $0.25 a pound in 1868, but declined to less than $0.05 a pound in 1894. Many tenants thus could not pay their obligations and saw their debts to a particular merchant increase each year (Lefler and Newsome, 1973, 523-524).

Declining prices for cotton were related directly to the introduction of one crop that already had a long history in the Carolinas. Tobacco had been grown in the Piedmont of South Carolina at least by 1769, and a number of interior tobacco inspection warehouses were established in the 1770s and 1780s. Though never a major component of South Carolina's colonial economy, tobacco was grown through the Piedmont until the end of the eighteenth century when cotton largely replaced it (Kovacik and Winberry, 1989, 75-76, 112). So also was tobacco an important crop in North Carolina's Coastal Plain until it too was displaced by cotton in the early nineteenth century.

As cotton prices continued their decline in the late 1800s, farmers in eastern North Carolina and the Pee Dee region of South Carolina developed a replacement crop: bright tobacco (Hart and Chestang, 1978, 436). Encouraged by newspaper editors advocating the region's ending its reliance on cotton, they made a successful commitment to the crop. By 1900, South Carolina farmers were planting 25,000 acres and producing almost 20 million pounds of tobacco, while North Carolina's overall production was 128 million pounds. Tobacco was widely distributed across the Piedmont and Coastal Plain of North Carolina, but its production in South Carolina centered on Florence, Williamsburg, Marion, and Horry Counties.

Two important commercial crops in South Carolina that did not recover after the war were rice and Sea Island cotton. Rice, successfully cultivated since the early eighteenth century and the basis of the plantation economy of coastal South Carolina and the Cape Fear of North Carolina, could not be grown under a tenant arrangement. Rice production needed gangs of laborers to prepare dikes and trunks and plant and harvest the crop; by 1890, South Carolina barely produced 30 million pounds of rice, about 1/4 the 1860 harvest. By the turn of the century, the crop had disappeared. Sea Island cotton was grown around Beaufort, the one part of the state in which many former plantations were sold and much land transferred to African-American freedmen. They devoted their properties to a largely subsistence economy, but continued to grow some cotton. Their success was limited, however, because much of the carefully selected seed and many of the processing techniques that had made the cotton of such high quality had been lost during the war. The arrival of the boll weevil finished off the last of the Sea Island cotton in the early twentieth century.

Textiles and Other Industries

The end of the Civil War began the industrial and urban revolutions in the Carolinas. While industries involved a wide range of manufacturing, three especially would take leadership: textiles, tobacco, and furniture. As noted above, the Carolinas had a small antebellum industrial base. The two states in 1860 had, for example, 57 textile mills, 68,000 spindles, and 1655 employees, and produced goods valued at $1.75 million (Lander, 1969, 79). By comparison, Massachusetts, the leader in the textile industry, alone had 217 factories, almost 1.7 million spindles, 38,451 workers, and a product worth $38 million.

The 1870s saw the Carolinas take significant steps toward industrialization; by 1880, they had recovered the ground lost during the war and together had 67 textile mills which produced goods valued at more than $5 million. For the most part, the industry developed on southern regional capital and local leadership, and the mills produced coarse and medium-quality yarn and some low-quality cloth. The last few decades of the nineteenth century, however, saw investment coming from the North and a general improvement in the quality of textiles as the industry became more competitive (Carlton, 1982, 42-46).

Between 1880 and 1900, the number of textile mills in the Carolinas grew rapidly. While about half of South Carolina's and a large proportion of North Carolina's mills were located in the Piedmont, all of the growth took place there. By 1900, North Carolina had 177 cotton mills with more than a million spindles and produced goods valued at $28 million. It considered itself "the Massachusetts of the South [that exceeded] any Southern state in number and value of manufacturing establishments" (Lefler and Newsome, 1973, 508-509). While this may have been true overall, South Carolina was ahead in textiles, second only to Massachusetts. It had 115 mills in operation, but they held 1.9 million spindles and produced goods valued at $30 million. South Carolina's textile industry formed a belt from Anderson and Greenwood, through Greenville, Spartanburg, and Union, to Gaffney and Rock Hill in the state's Upper Piedmont. The textile belt then lapped over into North Carolina, extending through small Piedmont towns that eventually were incorporated into such cities as Charlotte, Gastonia, Concord, Greensboro, Durham, and Roanoke Rapids (Glass, 1992, 42). These Piedmont locations provided important railway connections, access to a cheap labor market, and water power. Opening of power plants in Anderson and Columbia, South Carolina, by 1894, marked the beginning of hydroelectric power in the Carolinas, which would become another important locational variable for the textile industry.

One institution found throughout the textile belt of the Carolinas was the mill village. On one hand, the mill village provided a ready and controlled supply of labor, but, at the same time, it gave the workers better housing. One mill worker stated, "I like it here. We live in a nice house, lots better'n than that cabin in the country" (Lander, 1970, 91). The mill provided low-cost housing and basic services, such as schools, churches, stores, and the delivery of wood and coal. Since many workers were just off the farm, the village had surrounding land for gardens and pasturage for some livestock. As many families provided a number of workers (father, sons, and daughters), their identity with the mills frequently carried on through generations.

A second major industry that developed in North Carolina after the war, but which had antecedents dating to the colonial period, was tobacco manufacturing. The leaf was grown in the northern Piedmont and, later, the Coastal Plain of the state. Some farmers processed a portion of the harvest on their farm for chewing or smoking tobacco. During the war, Confederate and Union troops passing through Durham became familiar with North Carolina's smoking tobacco and made it popular around the country. One postwar processor in Durham developed the trade name of "Bull Durham Smoking Tobacco," and the Duke family, headed by the father Washington, moved to Durham in 1869 to begin a company that eventually would become the American Tobacco Company.

The years immediately after the war saw a relocation of tobacco manufacturing off the farm into towns, and almost every county from Cumberland in the east to Buncombe in the west had at least one factory within its bounds (Lefler and Newsome, 1973, 510-511). In 1880, North Carolina counted 126 tobacco factories with a product worth almost $14 million, six times what it had been in 1870. In 1882, the Dukes began to make cigarettes and soon controlled the market by buying up smaller manufacturers—their American Tobacco Company, however, was broken up by the federal government in 1911. By 1890, many of the small producers had been absorbed or pushed out of business as the industry focused on three cities: Durham, Winston, and Reidsville.

Like the other industries that grew up after the war, the furniture industry was tied to particular resources—the great hardwood forests of western North Carolina—and had a specific geographic location. Begun initially by a lumber salesman who felt that it made sense to have manufacturers near the source of wood, the first furniture factory was built in High Point in 1889. North Carolina slowly established its reputation because it had no real furniture tradition, but, in time, names such as Drexel, Tomlinson, and White competed for recognition as quality products (Powell, 1989, 410-413). In 1902, North Carolina had more than 100 furniture factories; although scattered across the state, they tended to concentrate in Guilford, Davidson, and Caldwell Counties. Access to raw material and transportation were major locational factors for the industry.

Population and Urban Development

Between 1860 and 1900, the Carolinas' population grew slowly, not quite doubling during that 40-year period. The United States population, on the other hand, was 142 percent larger in 1900 than it had been in 1860 (a 50 percent faster growth rate than that of the Carolinas). This reflected two trends. First, the Carolinas received few of the European immigrants who flooded into America during the late nineteenth and early twentieth centuries. Although they were enticed to the South by specially created state agencies, such as the Commissioner of Immigration and the Department of Agriculture, Commerce, and Immigration in South Carolina, the enervating climate, low wages, and the control of land by the plantation discouraged many from staying. Second, the slower rate of growth of the Carolinas also indicated the continued out-migration of population from the two states. After the turn of the century, this would become a virtual hemorrhage and involve especially the northward trek of millions of blacks from the southern states. The last few decades of the nineteenth century, however, saw just the beginning of this movement. In 1880, African-Americans made up more than 60 percent of South Carolina's total population, but this ratio fell to a little over 58 percent in 1900, initiating a slide that would continue into the 1970s and reduce the African-American portion of the state's total population to about 30 percent. In North Carolina, black population, never as high as in South Carolina, declined from 38 percent in 1880 to 33 percent in 1900.

The Carolinas' population also remained predominantly rural. In 1880, North Carolina's population was only 4 percent urban and still less than 10 percent in 1900. The corresponding figures for South Carolina were 7.5 percent and 12.8 percent, still far behind the average for the entire United States (28% in 1880, 40% in 1900). That meant that about 85-90 percent of the Carolinas' population was rural in 1900, living in the countryside, unincorporated places (like mill villages), or small towns with fewer than 2500 inhabitants.

Although the Carolinas remained overwhelmingly rural, urban growth in both states was becoming more significant in the two decades leading up to the turn of the twentieth century, foreshadowing the rapid urbanization of the South in the middle of that century. In 1880, only one North Carolina city, Wilmington, had more than 10,000 inhabitants; and South Carolina was dominated by Charleston, whose population of about 50,000 was more than five times the size of Columbia. Between 1880 and 1900, however, the percent urban populations of the two states, while very low, almost doubled. Furthermore, by 1900, North Carolina had six towns with populations greater than 10,000: Wilmington, still the largest city in the state, Charlotte, which would become the largest in 1910, Asheville, Winston, Raleigh, and Greensboro (Powell, 1989, 414-415). South Carolina had four: Charleston, the largest city by far, Columbia, Greenville, and Spartanburg (Petty, 1943, 114-115). The distribution of these newly developing urban centers was also important because they were part of the changing pattern of the economy and reflected the importance of the new lines of transportation.

Wilmington and Charleston were two port cities, each with a rich history tied to the export of agricultural products and a maritime transportation system going back to the colonial period. Their position was maintained during the antebellum period by radiating rail lines that connected the port cities with their hinterlands. But the fastest growing cities, those that had eclipsed the 10,000-inhabitant mark by 1900, were located in the interior and related to the industrialization of the Piedmont and the development of new railroad systems oriented toward the northeast rather than to the old port cities. They began drawing trade to themselves, which then would be handled by the railroads running toward the Northeast rather than toward the old coastal ports.

NEW RAILROAD SYSTEMS

In 1860, North Carolina had 891 miles of railroad, while South Carolina had 987. By 1900, North Carolina had quadrupled its mileage to 3,656, while mileage in South Carolina had almost tripled to 2,792 (Stover, 1955, 5, 256). Instead of focusing on Charleston, Wilmington, and Virginia ports, the new railway construction connected the Carolinas to a national system, making an important contribution to the two states' economic transformation. In the last decade of the nineteenth century, the railroads of the Carolinas were largely incorporated into three major systems: the Southern Railway, that controlled the upper Piedmont but whose branch lines connected with other parts of the two states; the Seaboard Air Line, which ran through the lower Piedmont; and the Atlantic Coast Line, that served the two old port cities and ran north-south through the Coastal Plain (Black, 1998; Stover, 1955, 1997).

The core of the Southern Railway was the Richmond and Danville and especially its 48-mile extension to Greensboro, which connected with the North Carolina Railroad. Acquired by the Southern on a long-term lease, this line provided access to Charlotte and connection with the Atlanta and Charlotte Air Line, completed in 1873. By 1881 the core of the Southern Railway in the Carolinas was complete, and the parent Richmond and Danville also controlled the Western North Carolina to Asheville, the Greenville and Columbia, the South Carolina Railroad to Charleston, and the Charlotte, Columbia, and Augusta. It officially became the Southern Railway in 1894.

During the 1870s, a series of short lines in Virginia and the Carolinas, including the Wilmington and Weldon, the Wilmington, Columbia, and Augusta, and the Northeastern Railroad from Florence to Charleston, began referring to themselves as the Atlantic Coast Line. A short cut from Wilson, North Carolina, directly to Florence, South Carolina, thus bypassing Wilmington, was completed in 1893 and became the railroad's new north-south main line (Connor, 1929, II, 422). In 1902, the Atlantic Coast Line acquired the Plant System, railway lines primarily in Georgia, Alabama, and Florida, but which also included the Charleston and Savannah Railroad.

The last of the three major rail systems that dominated the Carolinas was the Seaboard Air Line. Beginning as the Seaboard and Roanoke, serving Portsmouth and Weldon, North Carolina, the company soon acquired the Raleigh and Gastonia to the state capital and a newly completed line, the Raleigh and Augusta Air Line, from there into South Carolina. By 1892, the railroad had entered Atlanta and also controlled the Carolina Central to Wilmington, which had been acquired the year before. In 1899, the Seaboard purchased the Florida Central and Peninsular Railroad, which included a branch line between Savannah and Columbia. The final gap in the system in the Carolinas was filled by the completion of a rail line from Cheraw, South Carolina, along the Fall Line to Columbia in 1900.

The completion of these major systems was related to the industrialization and urbanization of the interior of the two states. One could argue that as the nineteenth century neared its end, the glimmer of the New South—the industry, new cities, and north-south railway systems—was taking form in the Piedmont of the Carolinas.

Conclusion

The settlement geography of the Carolinas experienced what could be considered five watersheds in its history. The first occurred 12,000 years ago with the arrival of humans and the beginnings of a cultural landscape; the second, thousands of years later, was the appearance of agriculture that allowed sedentary lifestyles, the development of pottery, and the subsistence base for the Mississippian and Woodland traditions. The third watershed was the arrival of the Europeans, the introduction of commercial agriculture, the creation of the plantation, and the forced in-migration of Africans to meet its heavy labor demands. The fourth was the aftermath of the Revolution, the disappearance of indigo and the introduction of cotton, the movement of the plantation into the Piedmont, and the weakening of the sectional differences that had characterized the two states during the colonial era. The fifth was the Civil War and the post-bellum adjustments to new social and economic patterns. By 1900, the Carolinas were but a few decades away from the Depression, the New Deal, and the Second World War, a watershed that would lead ultimately to the landscape of the modern South.

References

Ackerman, R. K. 1977. South Carolina Colonial Land Policies. Columbia: University of South Carolina Press.

Aiken, C. S. 1998. The *Cotton Plantation South since the Civil War.* Baltimore: Johns-Hopkins University Press.

Anderson, D. G. 1989. The *Mississippian in South Carolina*. Studies in South

Carolina Archaeology: Essays in Honor of Robert L. Stephenson, Anthropological Studies 9. Columbia, South Carolina: Institute of Archaeology and Anthropology, 101-132.

_____.1992a. A History of Paleoindian and Early Archaic Research in the South

Carolina Area. In *Paleoindian and Early Archaic Period Research in the Lower Southeast: A South Carolina Perspective*, ed. K. E. Sassaman and C. Judge. Columbia: Council of South Carolina Professional Archaeologists, 7-18.

_____.1992b. Models of Paleoindian and Early Archaic Settlement in the Lower Southeast. In *Paleoindian and Early Archaic Period Research in the Lower Southeast: A South Carolina Perspective*, ed. K. E. Sassaman and C. Judge, Columbia: Council of South Carolina Professional Archaeologists, 28-47.

Anderson, D. G. et al. 1981. Cal Smoak: Archeological Investigations along the Edisto River in the Coastal Plain of South Carolina. Occasional Paper No. 1. Columbia: Archeological Society of South Carolina.

Black, R. C. 1998. *The Railroads of the Confederacy.* Chapel Hill: University of North Carolina Press.

Brown, R. H. 1948. Historical *Geography of the United States*. New York: Harcourt, Brace, and Company.

Carlton, D. L. 1982. *Mill and Town in South Carolina, 1880-1920.* Baton Rouge: Louisiana State University Press.

Carney, J. and R. Porcher. 1993. Geographies of the Past: Rice, Slaves and Technological Transfer in South Carolina. *Southeastern Geographer* 33: 127-147.

Clark, T. D. 1964. Pills, *Petticoats, & Plows: The Southern Country Store.* Norman: University of Oklahoma Press. First published 1944.

Clowse, C. D. 1971. *Economic Beginnings in Colonial South Carolina, 1670-1730.* Columbia: University of South Carolina Press.

Connor, R. D. W. 1929. *North Carolina: Rebuilding an Ancient Commonwealth,* Chicago: American Historical Society (reprinted 1973 by Reprint Company, Spartanburg).

Cumming, W. P. 1998. *The Southeast in Early Maps,* 3rd ed., revised and updated by L. De-Vorsey. Chapel Hill: University of North Carolina Press.

DePratter, C. B. 1989. Cofitachequi: Ethnohistorical and Archaeological Evidence. Studies in South Carolina Archaeology: Essays in Honor of Robert L. Stephenson, Anthropological Studies 9. Columbia: South Carolina Institute of Archaeology and Anthropology, 133-156.

DePratter, C. B. and S. South. 1995. Discovery at Santa Elena: Boundary Survey. Research Manuscript Series 221. Columbia: South Carolina Institute of Archaeology and Anthropology.

Enscore, S. 1996. Old Salem's Pennsylvania Heritage. In *Snapshots of the Carolinas: Landscapes and Cultures,* ed. D. G. Bennett. Washington: Association of American Geographers, 29-32.

Ernst, J. A. and H. R. Merrens. 1979. "Camden's Turrets Pierce the Skies!" The Urban Process in the Southern Colonies during the Eighteenth Century. In *Geographic Perspectives on America's Past,* ed. D. Ward. New York, Oxford University Press, 308-320 (originally published 1973).

Ferguson, L. G. 1971. South Appalachian Mississippian. Unpublished Ph.D. Dissertation, University of North Carolina.

Gibbons, A. 1998. Mother Tongues Trace Steps of Earliest Americans. *Science* 279, 1306-1307.

Glass, B. D. 1992. *The Textile Industry in North Carolina: A History.* Raleigh: North Carolina Department of Cultural Resources.

Goodyear, A. C. et al. 1989. The Earliest South Carolinians. Studies in South Carolina Archaeology: Essays in Honor of Robert L. Stephenson. Anthropological Studies 9. Columbia: Institute of Archaeology and Anthropology, 19-52.

Gray, L. C. 1958. *History of Agriculture in the United States to 1860.* Gloucester: Peter Smith, 2 vol.

Greene, J. P. 1989. *Selling a New World: Two Colonial South Carolina Promotional Pamphlets.* Columbia: University of South Carolina Press.

Hammond, Harry. 1883. South Carolina: Resources and Population, Institutions and Industries. Charleston, State Board of Agriculture.

Hammond Corporation. 1989. *United States History Atlas.* Maplewood, New Jersey: Hammond, Inc.

Hart, J. F. and E. L. Chestang. 1978. Rural Revolution in East Carolina. *Geographical Review* 68, 435-458.

Hilliard, S. B. 1978. Antebellum Tidewater Rice Culture in South Carolina and Georgia. In *European Settlement and Development in North America*: Essays in Honour and Memory of Andrew Hill Clark, ed. J. R. Gibson. Toronto: University of Toronto Press, 91-115.

_____.1984. *Atlas of Antebellum Southern Agriculture.* Baton Rouge: Louisiana State University Press.

Hines, E. and M. Smith. 1996. Gold Mining in North Carolina. In *Snapshots of the Carolinas: Landscapes and Cultures,* ed. D. G. Bennett. Washington: Association of American Geographers, 43-47.

Hoffman, P. E. 1990. *A New Andalucia and a Way to the Orient: The American Southeast During the Sixteenth Century.* Baton Rouge: Louisiana State University Press.

Hooker, R. J., Ed. 1953. *The Carolina Backcountry on the Eve of the Revolution: The Journal and Other Writings of Charles Woodmason, Anglican Itinerant.* Chapel Hill: University of North Carolina Press.

Hudson, C. 1976. *The Southeastern Indians.* Knoxville: University of Tennessee Press.

_____.1990. *The Juan Pardo Expeditions: Exploration of the Carolinas and Tennessee, 1566-1568.* Washington: Smithsonian Institution Press.

Jones, M. S. 1996. The Cherokee Reservation. In *Snapshots of the Carolinas: Landscapes and Cultures*, ed. D. G. Bennett. Washington: Association of American Geographers, 33-36.

Jordan, T. G. and M. Kaups. 1989. *The American Backwoods Frontier: An Ethnic and Ecological Interpretation.* Baltimore: Johns Hopkins University Press.

Klein, R. N. 1990. *Unification of a Slave State: The Rise of the Planter Class in the South Carolina Backcountry, 1760-1808.* Chapel Hill: University of North Carolina Press.

Kovacik, C. F. and R. E. Mason. 1985. Changes in the South Carolina Sea Island Cotton Industry. *Southeastern Geographer* 25, 77-104.

_____ and L. S. Rowland. 1973. Images of Colonial Port Royal, South Carolina. *Annals of the Association of American Geographers* 63, 331-340.

_____ and J. J. Winberry. 1989. *South Carolina: The Making of a Landscape.* Columbia: University of South Carolina Press (originally published 1987).

Lander, E. M. 1969. *The Textile Industry in Antebellum South Carolina.* Baton Rouge: Louisiana State University Press.

_____.1970. *A History of South Carolina, 1865-1960,* 2nd ed. Columbia: University of South Carolina Press.

Lee, L. 1965. *The Lower Cape Fear in Colonial Days.* Chapel Hill: University of North Carolina Press.

Lefler, H. T. and A. R. Newsome. 1973. *The History of a Southern State: North Carolina,* 3rd ed. Chapel Hill: University of North Carolina Press.

Lemon, J. T. 1987. Colonial America in the Eighteenth Century. In *North America: The Historical Geography of a Changing Continent,* ed. R. D. Mitchell and P. A. Groves. London: Hutchinson, 121-146.

Leyburn, J. G. 1962. *The Scotch-Irish: A Social History.* Chapel Hill: University of North Carolina Press.

Meinig, D. W. 1986. *The Shaping of America: A Geographical Perspective on 500 Years of History, Vol. 1, Atlantic America, 1492-1800.* New Haven: Yale University Press.

_____.1993. *The Shaping of America: A Geographical Perspective on 500 Years of History, Vol. 2, Continental America, 1800-1867.* New Haven: Yale University Press.

Meltzer, D. J. 1997. Monte Verde and the Pleistocene Peopling of the Americas. *Science* 276, 754-755.

Meriwether, R. L. 1940. *The Expansion of South Carolina, 1729-1765.* Kingsport: Southern Publishers.

Merrens, H. R. 1964. *Colonial North Carolina in the Eighteenth Century.* Chapel Hill: University of North Carolina Press.

_____.1969. The Physical Environment of Early America: Image and Image-Makers in Colonial South Carolina. *Geographical Review* 59: 530-556.

_____.1977. *The Colonial South Carolina Scene: Contemporary Views, 1697-1774.* Columbia: University of South Carolina Press.

Milling, C. J. 1969. *Red Carolinians.* Columbia: University of South Carolina Press (originally published 1940).

Mitchell, R. D. 1978. The Formation of Early American Cultural Regions: An Interpretation. In *European Settlement and Development in North America*: Essays in Honour and Memory of Andrew Hill Clark, ed. J. R. Gibson. Toronto: University of Toronto Press, 66-90.

Moore, J. H. 1993. *Columbia and Richland County: A South Carolina Community 1740-1990.* Columbia: University of South Carolina Press.

Nitze, H. B. and H.A.J. Wilkens. 1897. Gold Mining in North Carolina and Adjacent South Appalachian Regions. Bulletin No. 10. Raleigh: North Carolina Geological Survey.

Petty, J. J. 1943. The Growth and Distribution of Population in South Carolina. Bulletin No 11. Columbia: South Carolina State Planning Board (reprinted 1975 by Reprint Company, Spartanburg).

Phillips, U. B. 1968. *A History of Transportation in the Eastern Cotton Belt to 1860.* New York: Octagon Books (originally published 1908).

Powell, W. S. 1977. *North Carolina: A Bicentennial History.* New York: W. W. Norton.

_____.1989. *North Carolina Through Four Centuries.* Chapel Hill: University of North Carolina Press.

Quattlebaum, P. 1956. *The Land Called Chicora.* Gainesville: University of Florida Press.

Roberts, B. 1972. *The Carolina Gold Rush.* Charlotte: McNally and Loftin.

Robinson, W. S. 1979. *The Southern Colonial Frontier, 1607-1763.* Albuquerque: University of New Mexico Press.

Rogers, G. C. 1970. *The History of Georgetown County, South Carolina.* Columbia: University of South Carolina Press.

_____.1980. *Charleston in the Age of the Pinckneys.* Columbia: University of South Carolina Press (originally published 1969).

Rowland, L. S. et al. 1996. *The History of Beaufort, South Carolina, 1514-1861.* Vol 1. Columbia: University of South Carolina Press.

Sauer, C. O. 1971. *Sixteenth-Century North America: The Land and the People as Seen by the Europeans.* Berkeley: University of California Press.

Schulz, J. J. 1976. The Hinterland of Revolutionary Camden, South Carolina. *Southeastern Geographer,* 16: 91-97.

Shlomowitz, R. 1979. The Origins of Southern Sharecropping. *Agricultural History* 53: 557-575.

South, S. A. 1976. Indians in North Carolina. Raleigh: Division of Archives and History.

Stover, J. F. 1955. *The Railroads of the South, 1865-1900.* Chapel Hill: University of North Carolina Press.

_____.1997. *American Railroads,* 2nd ed. Chicago: University of Chicago Press.

Swanton, J. R. 1979. *The Indians of the Southeastern United States.* Washington: Smithsonian Press (reprint of Bulletin 137, Smithsonian Institution, Bureau of American Ethnology, 1946).

Wallace, D. D. 1951. *South Carolina: A Short History, 1520-1948.* Columbia: University of South Carolina.

Wesley, E. B. 1979. *Our United States: Its History In Maps.* Chicago: Denoyer: Geppert Company.

Winberry, J. J. 1979a. Reputation of Carolina Indigo. *South Carolina Historical Magazine* 80: 242-250.

_____.1979b. Indigo in South Carolina: A Historical Geography. *Southeastern Geographer,* 19: 91-102.

_____.1996. Gullah: The People, Language and Culture of the Sea Islands of South Carolina. In *Snapshots of the Carolinas: Landscapes and Cultures,* ed. D. G. Bennett. Washington: Association of American Geographers, 11-15.

Wood, P. H. 1974. *Black Majority: Negroes in Colonial South Carolina from 1670 through the Stono Rebellion.* New York, W.W. Norton.

Chapter 5

THE EVOLVING URBAN AND ECONOMIC STRUCTURE SINCE 1900

Ole Gade

Appalachian State University

Across the Piedmont curve in the Carolinas lies a swath of urban development that is gradually evolving into a megalopolitan character. Defined by the undulating I-85 and I-40, intersected by I-26, I-73, I-74, and I-77 and interconnected by land devouring suburbs, are six major metropolitan regions, ranging from 140,105 to 1,318,613 people, as of July 1, 2005 (Figure 5.1). From a nearly 300-mile stretch of these Carolina interstate highways is a welter of easily accessed and closely distributed malls, big box retail centers, and shopping strips. The image of a Carolina Main Street comes to mind. With its economic growth and spatial expansion dampened by recent decades of the closing or relocating of textile, apparel and furniture industries, this Piedmont urban region is gradually redefining its economy in a dynamic global context. Here, during the initial decade of the twenty-first century, a population surge is indicative of its economic prowess. With a current annual growth rate exceeding two percent per year, one of the highest among metro areas in the country, the estimated 2005 population of these metro areas and their intervening Micropolitan Statistical Areas reached 5,749,165. Elsewhere in the Carolinas, other urban areas have experienced more of a mix in recent growth and change, from a volatile population expansion in some coastal metros to losses and very low levels of growth in places tied more closely to textile, furniture and military bases.

It is the objective of this chapter to derive some broad generalizations concerning the evolution of urban settlement patterns in the Carolinas over the past century and to examine the degree that changing economic circumstances have impacted this urbanization process. In order to facilitate the latter, several critical industries will be assessed in greater detail in terms of their historic growth .and influence on employment and community development. Globalizing influences reign supreme in the economic landscape of the Carolinas, with varying impacts on communities and regions.

A Century of Urban Development

Patterns of urban growth and development in the Carolinas were dominated initially by several factors influencing the spread of settlement. Early city sites were confined to colonial and antebellum settlements tied to Tidewater ports and Coastal Plain agriculture. A few of these places, those capable of taking advantage of changing conditions of growth, have persisted in importance. Here are included the ports of Charleston and Wilmington, with the former, a historic seaport (then called Charlestown) and the fourth largest in the country in 1790, gradually gaining additional strength by attracting to it a naval base and shipyards. As labor intensive industries

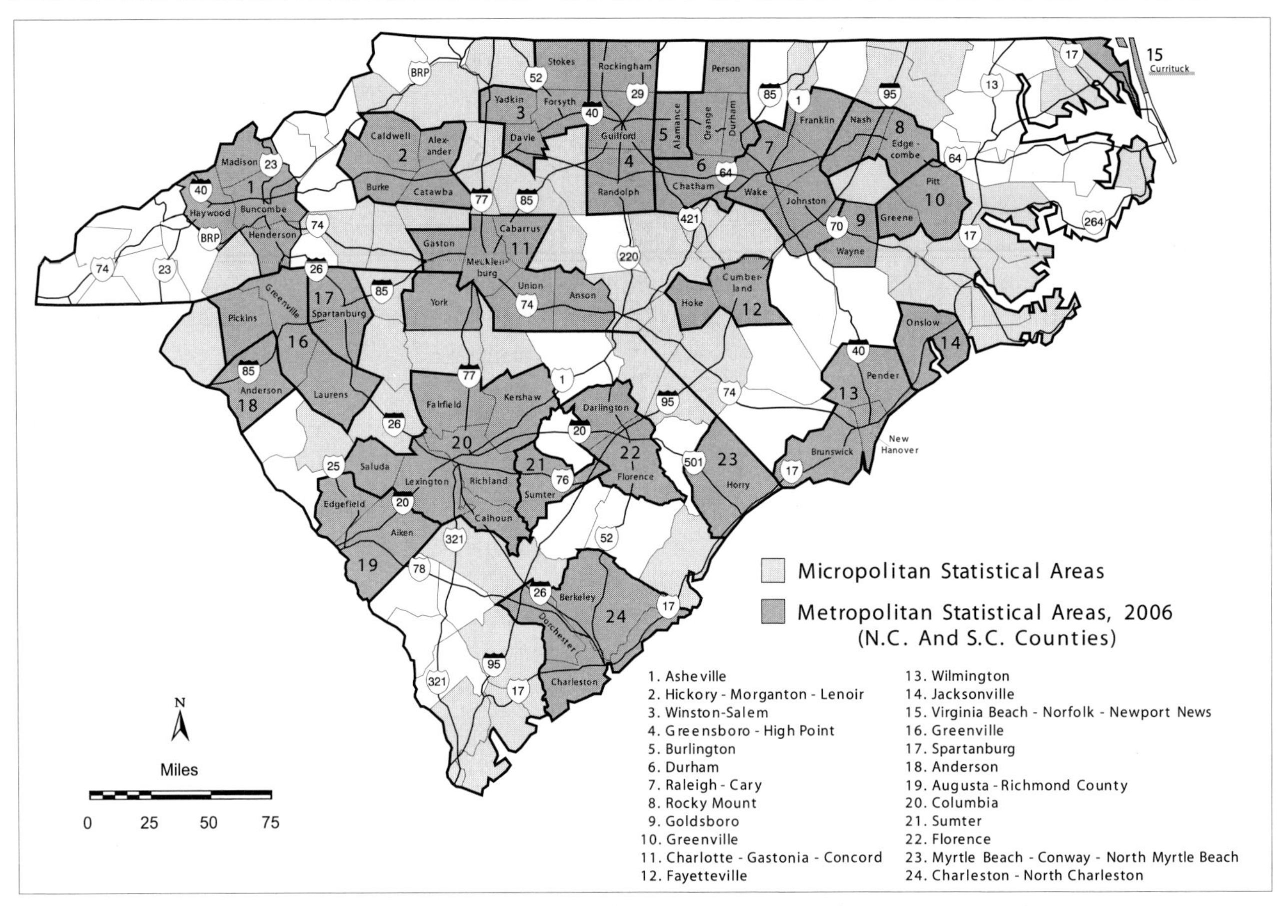

Figure 5.1. Carolina Metropolitan Statistical Areas (as of December, 2006)
Sources: United States Bureau of the Census, 2007, Mackun (2005)

began making their mark on the landscape in the late decades of the nineteenth century, most Tidewater and Coastal Plain towns gradually diminished in importance, especially where they did not emerge as centers of manufacturing.

It was railroad-spurred manufacturing that drove the initial urbanization of Carolina's Piedmont in the latter half of the nineteenth century. Early industrialists proclaimed that the Piedmont should not be raising cotton, it should be making cloth. The physical-cultural environmental mixture appeared perfect. Water from rapidly flowing streams emanating in the Appalachians to the immediate west powered early mills, hardwood forests covered square miles of nearby hills, plentiful and cheap labor came off the farms, rural land was available almost for the asking, and there were plentiful state and local incentives, including low taxes and subsidized public utilities and access. As the political and social environment persisted in favoring a union-free labor force, capital readily appeared for investment in low value-added industries (Hanham and Hanham, 2001).

Matters jelled with the defeat of the Democratic contender for the Presidency in 1880. Southern leaders hoped that here was the opportunity for the South to be seen as once again a viable option for industrial investment by northern interests, an idea that would be put in place by a Democratic administration. Their decades of attempts at establishing a secure agricultural economic base had wilted in the politics of the plantation, rural poverty and racial turmoil. City growth appeared possible only where goods were shipped out of the Carolinas. Thus, Charleston in 1880 had reached a population of 50,000, only exceeded in size among southern cities by New Orleans, Louisville, and Richmond. Wilmington figured as the second largest Carolina town with 17,350. Among the remaining towns, only Columbia passed the significant 10,000 mark, and by just 36 people (Larsen, 1990, 101). By 1890 Raleigh, Charlotte and Asheville had also passed the 10,000 mark. In 1910, according to a new Bureau of the Census volume detailing changes 1790-2000, the largest cities in the Carolinas were Charleston (58,833), Charlotte (34,019), Columbia (26,319), Wilmington (25,795), and Winston-Salem (22,700). (See Dodd, 2001.)

Before the 1880s, North Carolina's exports were pretty much confined to naval stores and tobacco. The textile industry during much of the last half of the nineteenth century, though much lauded in the literature, was largely of local importance (Carlton, 2003a, 80). But then, in response to the lack of national leadership and a fierce determination to go it alone, the "Cotton Mill Campaign" began taking hold. Hundreds of communities in the Carolinas contributed to the solicitation of entrepreneurs for building cotton mills. By the mid-1890s, their efforts were being recognized by northern entrepreneurs and especially the cotton mill owners of New England. Richard Edwards, in 1895, convincingly informed the New England Cotton Manufacturer's Association that it would be far better to encourage a southward migration of the cotton textile industry than it would be to have to face the increasing southern competition (Wood, 1986, 64). In seeming response, Cone began building his textile empire, in what was to become the national textile capital, Greensboro, North Carolina, in the late 1890s; Weldon, N.C. saw the Cameron mills being built; and J. A. Chanler, heir to the Astor fortunes, single-handedly created Roanoke Rapids in Halifax County on the Virginia border (Wood, 1986, 65). And they were followed by many other investors. In his *History of Mecklenburg County*, D. A. Thompkins wrote: "Within a radius of 100 miles of Charlotte there were 300 cotton mills" (as reported in B. Stuart, 1979, 13). Subsequent writers have indicated that this constituted about fifty per cent of the country's total. To be sure, many of these were powered not by naturally flowing water but by coal fired steam engines. This was in 1903, 23 years following the beginning of the Cotton Mill Campaign. Relocation of the cotton mill industry from New England during these decades was also fueled by the

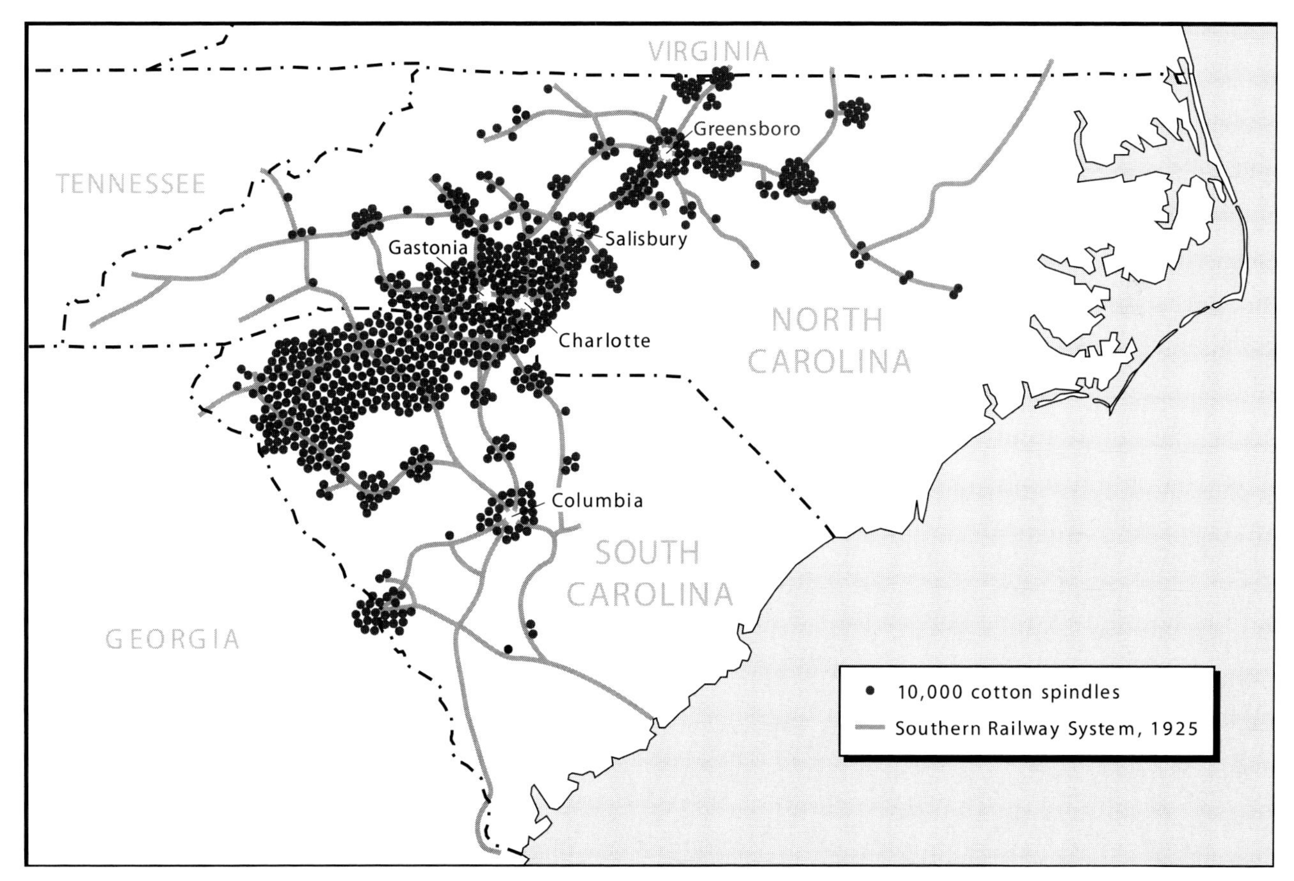

Figure 5.2. Cotton Spindles, 1925. Source: Reprinted from Gade, et al (2002), p. 441.

problems perceived to exist in the Northeast, but readily ameliorated in the South. As noted by Wood (1986, 59):

> For the ailing New England cotton textile industry in the depressed conditions of the 1920s, relocation permitted a significant increase in the production of absolute and relative surplus value through increases in the number of daily hours worked, the expanded exploitation of women and children, reduced wage levels, and increases in the intensity of work.

The results by 1925 are demonstrated in Figure 5.2, which incidentally also shows the initiating of the urban geography of Carolina Main Street. An interesting aspect of this cotton spindle distribution is that it was confined to the Piedmont. The mills were mostly quite small, 10,000 spindles on the average per mill as compared to the norm of 40,000 in New England mills, and they were quite dispersed in the context of that day's settlement density. Remember that the goal of the "Campaign" was to bring the jobs to where people live! Every small town in the Piedmont, it seemed, had one or more cotton mills. Though the mills were mostly confined to the smaller towns, their raison d'etre, the larger cities grew in wealth by providing the machinery, tools, funding resources, and with their distribution system facilitating marketing and sales of cotton goods across the country. Charlotte became the largest center of textile equipment manufacturing in the nation.

Also, in the latter decades of the nineteenth century, Winston-Salem and Durham benefited from the invention of the cigarette manufacturing machines. There were family owned furniture plants flourishing in small towns and rural communities in the northwestern part of the North Carolina Piedmont, and textile and apparel plants were abundantly distributed elsewhere through the Carolina Piedmont. In addition there was the continuing emphasis on export agricultural products, such as tobacco in the Coastal Plain. Indeed, the stage was set in the Carolinas on a rural and small town dominated political objective of fostering a geographically all-embracing prosperity, focusing on balanced growth and development. The notion of the Carolinas as essentially rural prevailed to the late 1970s for South Carolina and to the 1990 census for North Carolina, and in some quarters still holds sway. In the context of national urban development Carolina cities, particularly those in the Piedmont, were medium-sized. No "Atlantas" are in sight.

Low capital, labor intensive, single industry "company or mill" towns, with their largely paternalistic plant ownership environment, dotted the Piedmont by the 1920s, the nation's Progressive Era. It should be noted here that 12 counties of the South Carolina Upper Piedmont, termed the Upper Piedmont Manufacturing Region (UPMR) by Kovacik and Winberry (1989, 212), had emerged as the economic core region in the state by the 1920s with a focus on cotton mills. Here, the arrival of Southern Railway and the Seaboard Air Line Railway in the early part of the century had encouraged the further growth of a cluster of mill villages. Major textile companies were headquartered within or close to the UPMR, such as Delta Woodside in Greenville, acclaimed as the largest mill in the world, when built in 1902, and Springs Industries in Fort Mill (Kennedy, 1998). Even more so than its northern neighbor, South Carolina had evolved a clear core-periphery economy, in part because of the obvious advantages of water power, but also due to the abundance of white yeoman farmers dominating the rural excess labor pool, as well as to the natural overspill of development from the North Carolina Piedmont where the textile industrialization had been initiated years earlier. As a core region the UPMR also gained the additional advantage of much earlier paved road development and the diffusion of electrification than was the case for the remainder of South Carolina (Carlton, 2003b, 142-143). While 11 Southern cities had grown to beyond 100,000 in population by 1930, none of these were in the Carolinas.

By this time, in the 1930s, the federal "New Deal" had appeared and, with the strength of Southern congressional legislators, benefited southern manufacturers disproportionately essentially by "reap(ing) the financial and political benefits of federal efforts at economic stimulation and regulation . . ." (Wood, 1986, 155). One result was a groundswell of opposition to the evolution of a narrowly controlled export economy, dependent on a few low wage manufacturing industries like cotton textiles, furniture, and lumber. Academics, such as Harriet Herring (1940), Howard Odum (1936), and Rupert Vance (1935), led the attack.

More recently, locational advantages have shifted toward capital intensive manufacturing and service industries, such as robotics, and other labor-saving devices and approaches have been releasing excess human labor. Equally, the global labor market has been attracting out of the Carolinas and out of the country those industries still wedded to low cost, low skill labor. Following the shift in employment characteristics, which at the turn of the twentieth century saw close to 70 percent of the Carolinas labor force still engaged in primary activities, especially agriculture, the region by mid-century led the nation in the percentage employed in secondary activities, manufacturing, with its attendant urbanizing process.

This, however, was still the Carolinas. Ideas of equally distributed economic opportunities continued to influence the legislative agendas, and governors won elections under the banner of, "good roads and educational opportunities for everyone" with a subsequent concomitant healthy effort at rural industrialization through a state policy of "balanced growth." The idea of disarming urban growth tendencies by influencing the spread of small manufacturing industries across the state was central to Democratic Governor Robert Scott's administration in the early 1970s. It became a major plank in the campaign of the first Republican Governor in North Carolina, Jim Holshouser (1972-1976). A state planning policy promising ". . . uncongested, livable cities and a more prosperous countryside dotted with industries," (B. Stuart, 1979) came very close to implementation during the early years of Democratic Governor Jim Hunt (1976-1984), whose early years coincided with President Carter's administration. Governor Hunt was able to gain initial federal approval for diverting to localities billions of federal dollars flowing through the EPA, HUD, EDA, and FmHA directly, and without state oversight. Perceived federal excesses directed toward the urban renewal programs of the 1960s had become the political rallying point. Having defined a set of "growth centers," rural located places deemed capable of self-sustained growth given initial state financial encouragement, the Hunt administration succeeded initially to divert the annual $500 million allocated to the state through the Farmers Home Administration, to a Governor-created committee. Here, decisions on allocation could be made within the framework of state directed 'balanced growth' policy. President Jimmy Carter came to Asheville airport on September 29, 1977 to announce this new direction in federal-state cooperation. He further indicated that the FmHA (Department of Agriculture) agreement was to be followed by similar agreements with the Department of Housing through its HUD programs, and the Department of Labor. A magnitude of $1.2 billion per year was the initial objective. Consider that with an annual state budget of just $4 billion, already carefully targeted, the Governor would suddenly have incredible power for directing funds toward a great variety of federally sponsored rural and small town development programs. And all of this was to be achieved without legislative advice and consent! In South Carolina, the incoming Democratic Governor Richard Riley (1977-1985), first two-term governor of the state, was sympathetic to the balanced growth idea, but his administrations focused emphatically on education as its lead objective. Not that this would have mattered very much. Federal administrations have a habit of changing political party, hence major objectives. Carter was followed by President Ronald Reagan, and support for federal-state extraordinary agreements was immediately deeply buried in the sands of time, and the prospects for targeted subsidized rural development with them. Small scale manufacturing plants provided anchors for small towns and rural areas for another decade or so. Meanwhile, major economic changes clearly favoring urban concentrations of people and markets were taking place.

Employment in service activities began this major shift in economic activities. For the decades of the 1970s and 1980s, this amounted to a 179 percent increase in services employment for North Carolina, and a 222 percent increase for South Carolina, while employment in manufacturing industries increased only 21.5 percent and 14.4 percent, respectively. Into the late 1980s and early 1990s, with the easing of international trade restrictions, manufacturers were increasingly finding it profitable to close local plants and transfer production processes to countries where easily trainable, much less expensive labor was readily available; a seeming repeat of the process which a full century earlier contributed to cotton mills opening in the Carolinas as those in New England closed. This process was greatly aided by the gradual disappearance of the family-owned textile and apparel plants through buy-outs by international conglomerates, themselves without concern for the fate of labor and local communities. This was not the 1930s when plant owners shared their relative largesse with workers and communities during hard times.

One of the more notable buy-outs was that of the Cannon Mills, a family business that by the early 1980s had nine mills distributed throughout their company-owned town of Kannapolis where more than 10,000 of their roughly 17,000 workers lived in company owned housing. The mills were owned briefly by David Murdoch (California) who bought the company in 1982 for $413 million. With $200 million in immediate investment he sought to turn the production of towels and bed sheets into the world's most famous linens through expensive and splashy advertising. The reader may recall the racy advertised image of Hollywood stars lying amid Cannon bed sheets, or seductively wrapped in a Cannon towel in the bathroom, with the notation "Two of the most famous names in America sleep (bathe) together" (Mecia, 2003). But this was to no avail. The expansion of world trade in textiles brought in low cost competition, three plants were closed and 3,000 workers were laid off. In 1986, Murdoch sold the Kannapolis holdings for $312 million to Fieldcrest, at that time headquartered with major plants in Eden, North Carolina. Fieldcrest Cannon, Inc. gained 12 plants, 14 sales offices, and 12,900 employees in the sale. Eventually, having overextended itself, Fieldcrest Cannon was bought out by Pillowtex, a Dallas textile combine for $700 million in 1997. (See Gade, et al, 2002, 166-169.) A few years later, Pillowtex succumbed to globalizing competition and declared bankruptcy. Perhaps, this is the end of the story for several thousand mill families; but not so for the town of Kannapolis, whose new benefactor, yes, the same David Murdoch, now CEO of Dole Foods, has committed to investing $1.2 billion in a new bio-science/tech research campus focusing on agricultural production. Dole Foods is simultaneously building new capacity for vegetable and fruit processing in western Carolina Piedmont counties, an interesting new form of vertical clustering industrial enterprise. It bears noting here that North Carolina agriculture, in the throes of moving away from tobacco, has become the third leading producer of greenhouse and nursery products (nearing $1 billion in 2005 sales) in the country after California and Florida. Building the North Carolina Research Campus is well under way. Abandoned plants and their hallmark Cannon smokestacks, have been demolished, and new office buildings and laboratories are making their appearance. North Carolina is aiding the process through its tax-increment finance law, which allows local governments to issue bonds without voter approval; and the state universities are playing an increasingly active role. Cabarrus County, however, is smarting from the increasing expectations from Murdoch in the size of the bond package, now (January, 2007) pegged at $150 million, having escalated from an initial request some months ago of $7 million in public money. The bond issue is to be equally shared by the Town of Kannapolis, and the county (Cherrie, 2007).

Graniteville, near Aiken, South Carolina, provides another example of the shift in economic activities. William Gregg founded Graniteville Manufacturing Company in 1845; the company was gradually expanded with the building of additional plants or through the purchase of others in the vicinity. To provide for their workers, the company constructed and maintained homes, schools, recreation centers, stores, a post office, a hotel, and a cemetery. When being bought out by Avondale Mills in 1995, during the year of its 150 year anniversary, the company owned 1,500 acres in the heart of town in addition to manufacturing plants covering two million square feet. The Graniteville plants already had been contributing to the economic shifts occurring in this part of the state. Workers trained in the textile mills found themselves competitive for openings in the nearby and rapidly expanding Savannah River Site, as well as with higher paying Bridgestone-Firestone tire manufacturers and other plants in Aiken. Apparently, the decades of gradual modernizing life conditions removed the burden of the lower caste self-image and isolation traditionally preventing mill workers from entering employment beyond the mill (Wood, 1986, 42). Their replacements in the textile plants were people who commuted in from smaller towns and rural communities in the hinterland. Thus, the loss of 1,600 workers and five plants was perhaps not as great as it might have been for Graniteville itself, which, incidentally, also is now within the urban shadow of Aiken and impacted by advancing suburbs. (See coverage provided by the Columbia newspaper, *The State*, July 21, 23, 24, and 27, 2006.) Of considerable interest to job seekers in the Aiken metro area are the current efforts on the part of Duke Energy in mixed-oxide fuel fabrication for its nuclear plant in York County in order to have these materials available nearby, such as at the Savannah River Site. Here, U.S. Department of Energy workers had already broken ground for a plutonium-conversion plant, when a budget conscious federal administration threatened to cut off funding in 2006. As Congressional Representative John Spratt (York County) noted, "Without the MOX fuel-fabrication, South Carolina will be stuck with tons of weapons grade plutonium with no clear path for dispersal" (Rosen, 2006, A4). From an employment perspective, this would involve several thousand construction workers, plus several hundred high salaried permanent employees once the plant is operating. Meanwhile, Duke Energy, the Charlotte based provider of electrification for the early mills and mill towns of the Upper Piedmont, will no longer have to import MOX from Europe.

Different versions of these scenarios have played out for most of the Carolinas textile and apparel industries, except only a very few have had options for joining the "New Economy." Meat production, especially chicken and turkey plants, and the tobacco plants equally suffered from closings as a result of changing labor and market conditions, and the consolidation fever. The gradual closing of square miles of tobacco plants in the hearts of Winston-Salem, Durham, Reidsville, and Wilson, has left an enduring mark on these cities. And by the turn of the twenty-first century, the furniture industry began showing very similar tendencies. The dot.com bust of the early years of this century brought its own labor uncertainties; in this case, to electronics and high-tech industries. Where call centers already were being relocated to English speaking countries, initially to Ireland and, then, with a vengeance it might seem, to India, their positive results for the international corporations led to a massive outsourcing of knowledge industries.

The Collapsing Manufacturing Economy

How have the Carolinas fared in this incredible economic shift? For two states that had become the leading producers of textiles, apparel, tobacco and furniture, not so good! Figure 5.3 illustrates the story of manufacturing plants closed and their related employment losses. For the six-year period (2000-2006), the Carolinas saw the closing of 2,071 industrial plants and the loss of 238,327 manufacturing jobs, with two-thirds in North Carolina and one-third in South Carolina. An initial thought about the distribution of job losses is the remarkable resemblance it has to the location of cotton spindles in 1925 (Figure 5.2). Disaggregating the data for North Carolina provides a reasonable reflection for the entire Carolinas. Of this state's 158,501 job losses, half were in textiles and apparel, a tenth in electrical equipment, and a twelfth each in furniture and machinery. The North Carolina Department of Commerce has been submitting this data in annual reports in response to a federal edict. The Department freely admits to it being collected from chambers of commerce and newspaper reports. (It is available online at: <http://eslmi23.esc.state.nc.us/masslayoff/MLSFrame.asp?contentsFrame=5>. Note in this listing the extraordinary job loses in the textile and apparel industries. South Carolina job loss data specific to mass firings and plant closings, though not defined by NAICS Code, is available from: <http://www.sscommerce.com/wia/annual2000.htm>.)

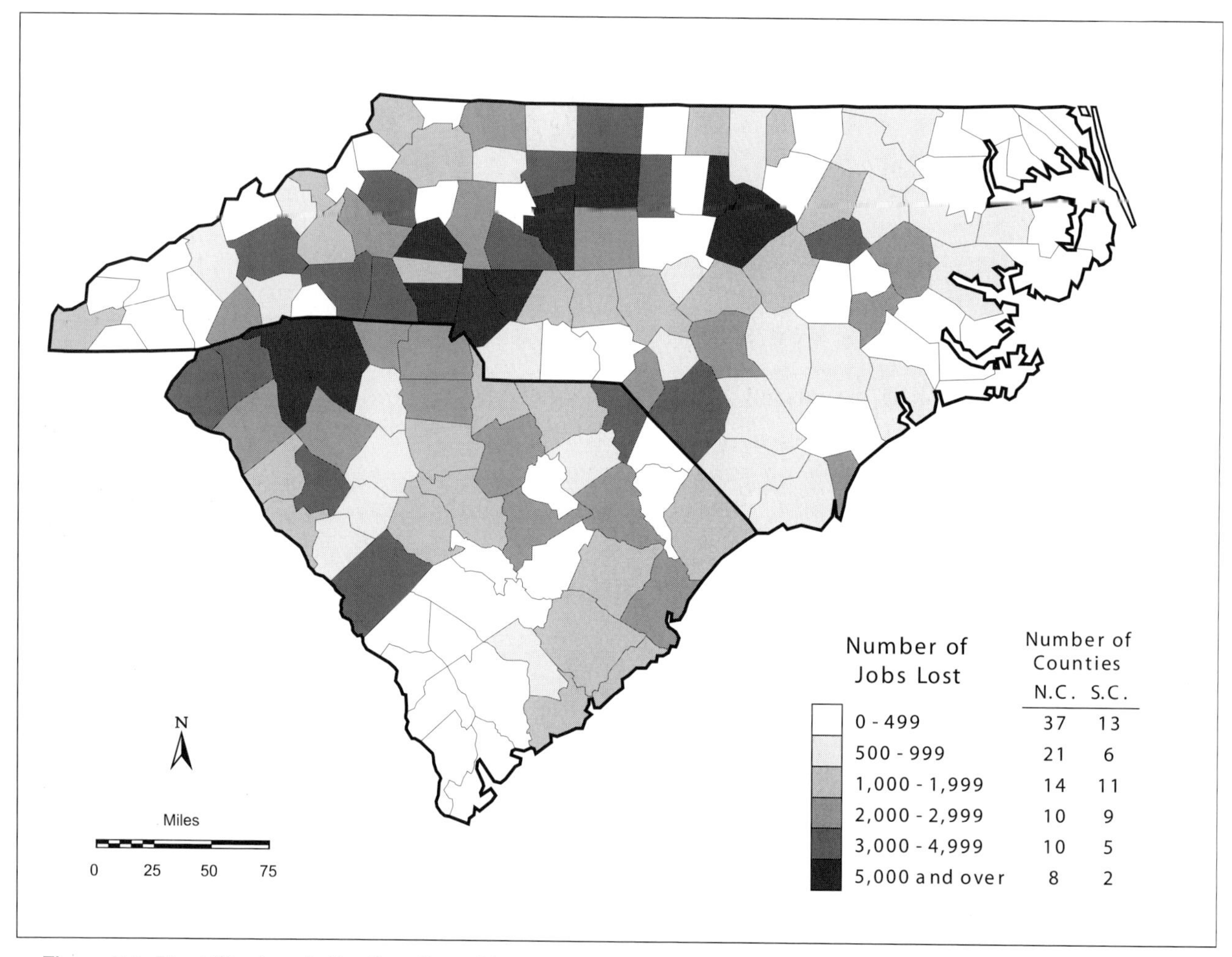

Figure 5.3. Plant Closings in the Carolinas, 2000-2006. Source: Derived by author from web sources listed above.

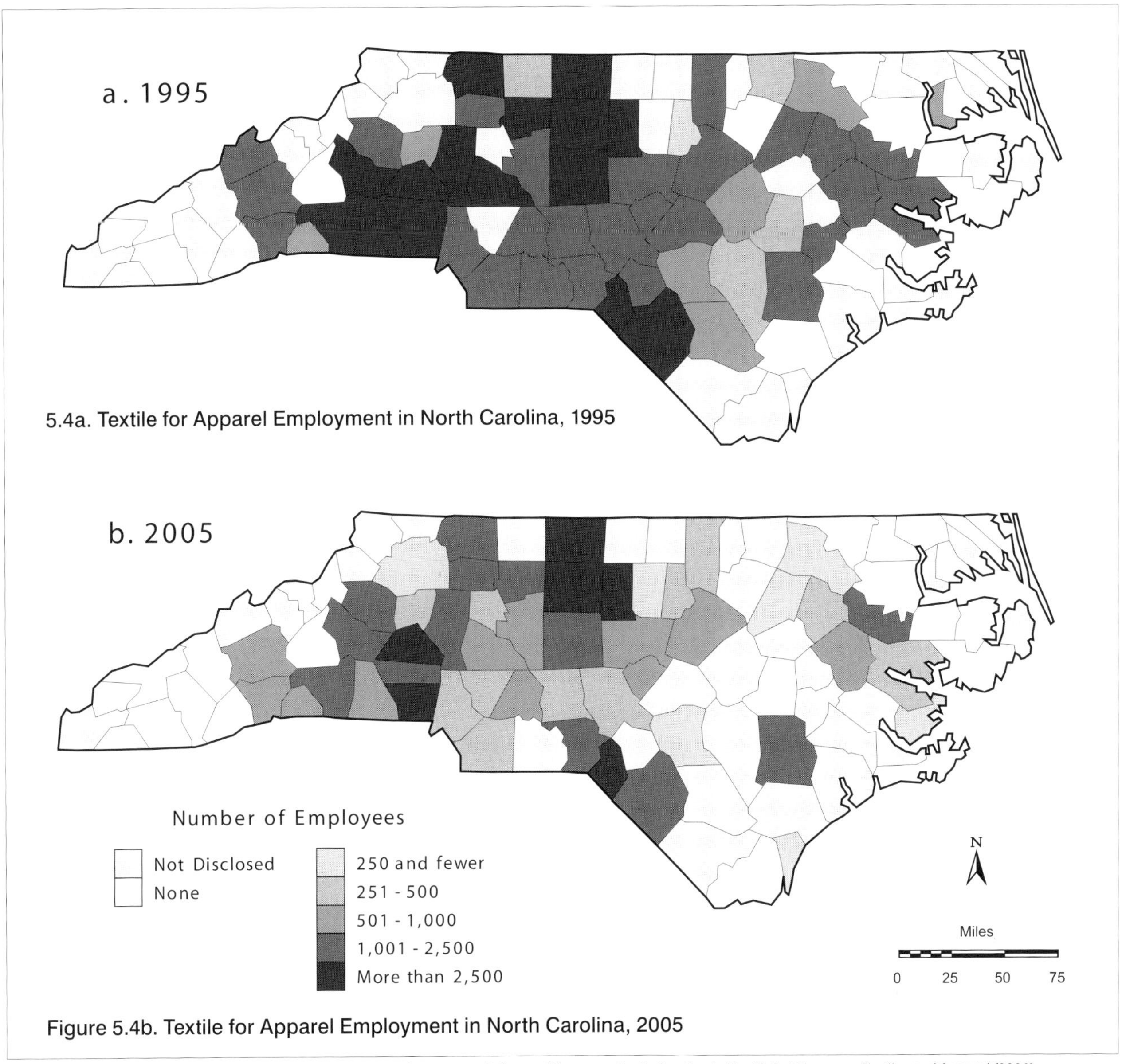

5.4a. Textile for Apparel Employment in North Carolina, 1995

Figure 5.4b. Textile for Apparel Employment in North Carolina, 2005

Source: Duke University Center of Globalization, Governance & Competitiveness, *North Carolina in the Global Economy: Textiles and Apparel (2006)*

A cursory glance at Figure 5.4 a and b will reveal the dramatic changes in the textile production landscape. This shows the change in employment in plants that manufacture textiles for use in the production of apparel, such as mens-, womens- and childrens-wear. These include "down stream" producers only, mostly smaller plants generously distributed throughout the small-town and rural community landscape. Though there have been large scale job reductions throughout the state, it is the eastern and central Piedmont, largely the more rural counties, that are bearing the brunt of the changes.

Perhaps, a more holistic way to dramatize the decline in manufacturing industry employment is shown in Figure 5.5a and b. The year 1980 is chosen as exemplifying probably the greatest importance of manufacturing as a source of employment in both the Carolinas. This is before the industry achieved its largest employment in the early to mid-1990s, but also at the very beginning of the service employment revolution, an inherent feature of the "boomers" generation. For many counties, manufacturing plants were the most significant economic contributor, though in only 11 Piedmont counties did it comprise over 50 percent of total civilian employment. With the exception of the seven most urbanized counties of Carolina Main Street, over 35 percent of its labor force worked in manufacturing

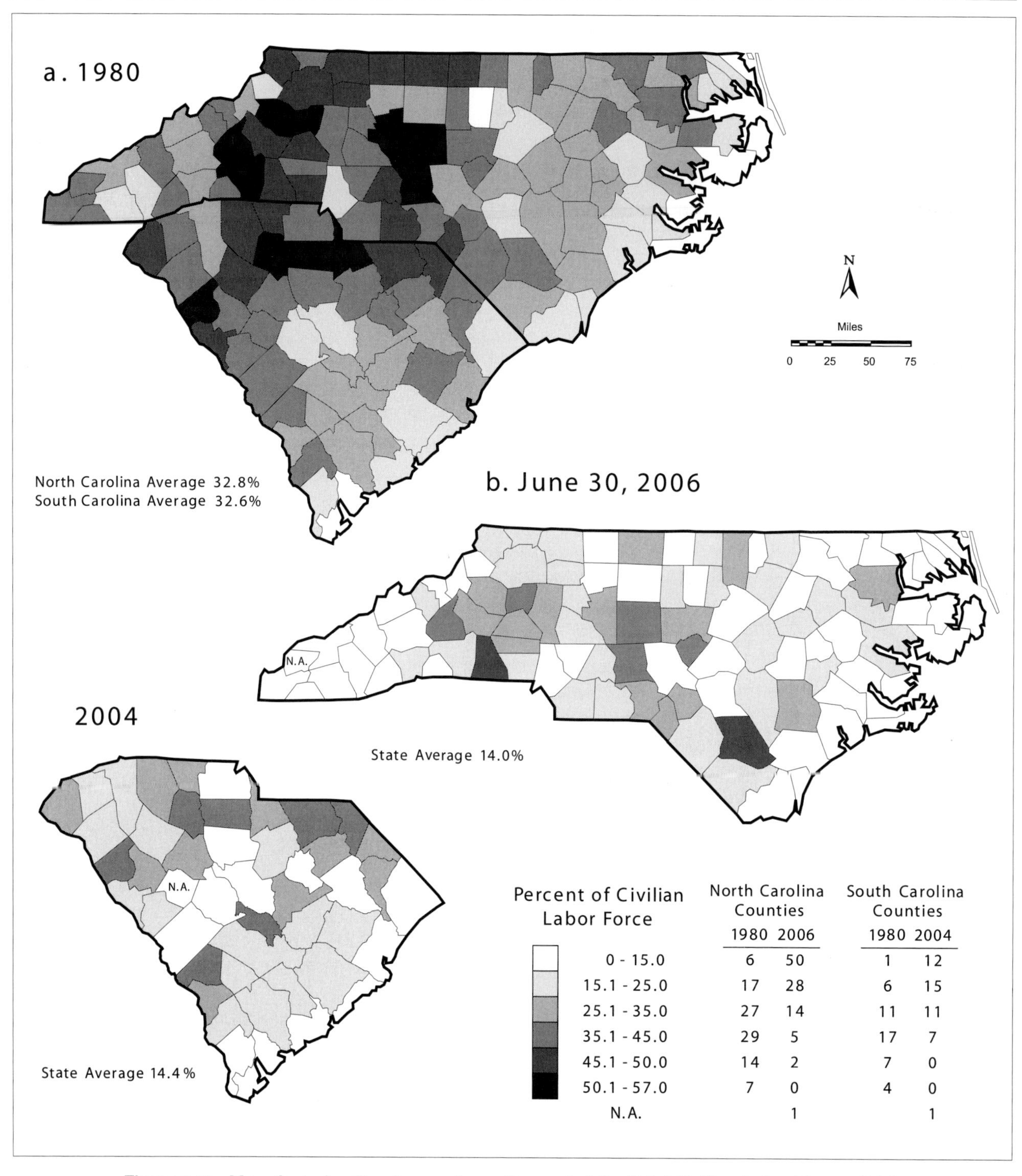

Figure 5.5a. Manufacturing Employment as a Percent of the Total Civilian Labor Force, 1980.
Figure 5.5b. Manufacturing Employment as a Percent of the Total Civilian Labor Force, 2004-2006.
Source: United States Bureau of the Census.

plants, making the Piedmont clearly the heart of the industrial economy of the Carolinas. Plants spill generously into adjacent Coastal Plain and Mountain counties, but their much smaller populations compete with agriculture and agricultural service center type employment.

The incredible impacts of the New Economy with its globalizing shifts and new emphases on service and knowledge activities are written in the manufacturing landscapes of 2006 (for North Carolina) and 2004 (for South Carolina). For both states, the decrease in manufacturing employment as a percent of total civilian employment decreased during the last nearly quarter century from about 33 percent to about 14 per cent in various counties, though the vast majority of this decrease occurred since 1995. The decrease was experienced throughout the Carolinas, especially along Carolina Main Street and the more urban counties. Remnant manufacturing strength is found in a few counties located on the periphery of metro areas in the Piedmont, as well as in a couple of counties where food processing is still critical, or is gaining in importance.

With the ascendancy of hog farming in North Carolina's Coastal Plain, meat slaughter and packing plants have assumed an extraordinary scale, as evidenced by the 5,000-plus employee Smithfield Foods meat plant in Tar Heel (Bladen County). For the rural counties, the loss of their anchor manufacturing plants, without jobs, other than in low-wage retail and food service to replace those lost, conditions verge on economic depression. Complicating the picture further has been the massive influx of foreign labor, most notably Hispanic, seeking employment at low wages and zero benefits, primarily in large scale agricultural enterprises and food processing plants, into many of the same counties. In urban areas, Hispanics dominate construction, building maintenance, restaurants, and landscaping jobs. During the 1990s, North Carolina led the country in foreign-born population increase with 278 percent and South Carolina came in tenth with a 147 percent increase; while the United States, as a whole, recorded an increase of 61 percent. For the 2000-2005 period, South Carolina led the nation with an estimated 46 percent increase in its foreign-born population (AP, 2006). For these two states, which through most of the twentieth century had been at the very bottom in the percentage of foreign-born population, this was

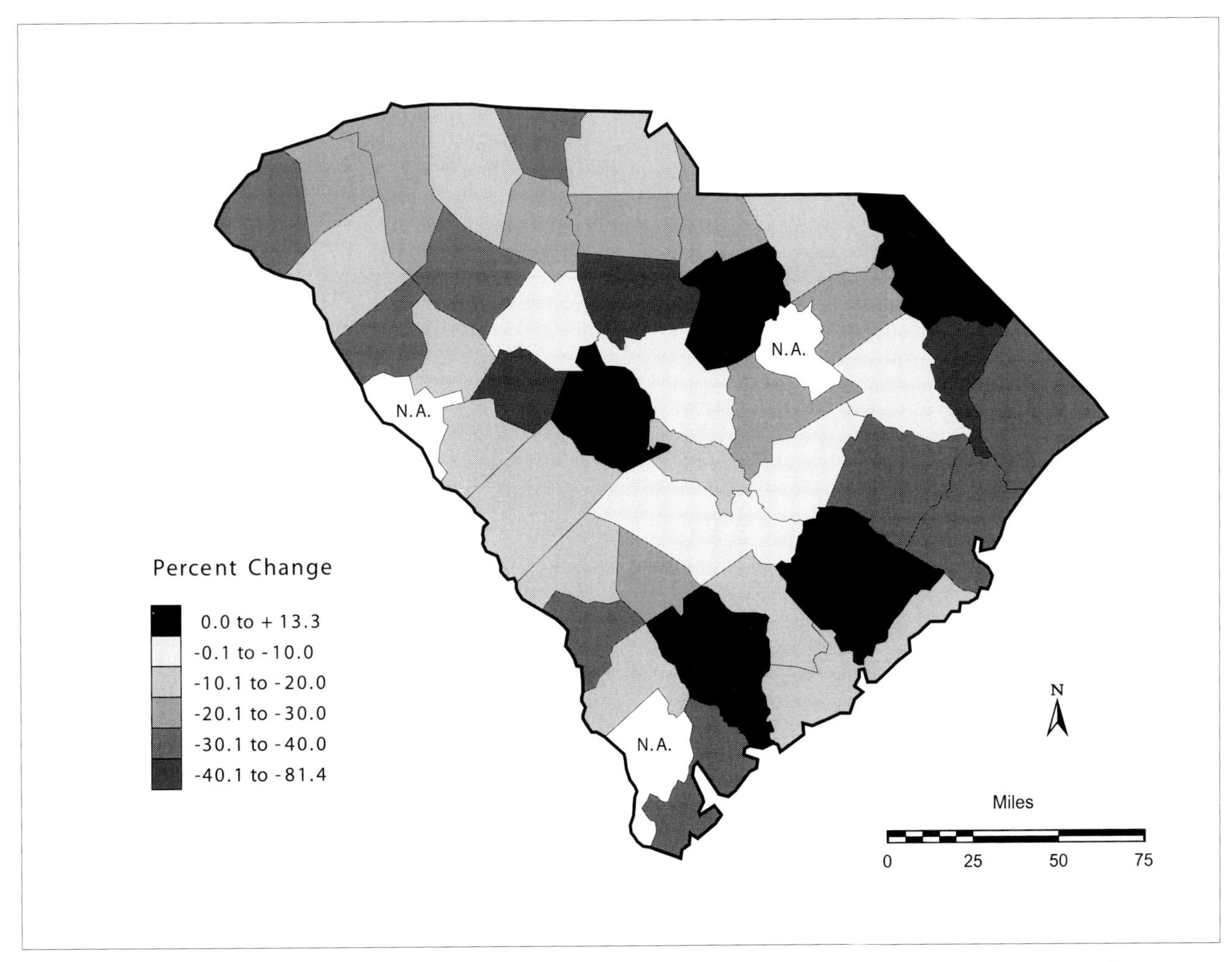

Figure 5.6. Total Manufacturing Employment Change, South Carolina, 2000-2004. Source: United States Bureau of the Census.

revolutionary. In both cases, the largest single element of the increase was the Hispanic. Of course, this estimate does not include the illegals. Jim Johnson, population geographer with UNC Chapel-Hill's Kenan Institute has been quoted as suggesting this figure is in the vicinity of 48 percent of the total Hispanic population in 2006 of 601,000 (Phillips, 2006; Kasarda and Johnson, 2006). At any rate, the Carolina landscape is blossoming with newfound energy. In the smallest of Coastal Plain and Piedmont towns, former Baptist churches have been converted to Spanish-speaking Catholic ones, and what remains of their downtown business districts have a healthy helping of tavernas and tiendas, Hispanic Mom & Pop general stores.

Another approach to demonstrating the severity of South Carolina's manufacturing decline is comparative data on manufacturing employment change for the state during 2000-2004 (Figure 5.6). County Business Patterns comparisons for 2004 shows not only the overall major impact of the manufacturing decline, but more especially it demonstrates the incredible relative decreases experienced by the more rural counties (U.S. Census Bureau). (Compare to Figures 5.6 and 5.3.) Some of the counties experiencing very large losses are on the urban periphery, such as Fairfield (-61.3%), Saluda with an astonishing decrease of 81.1 percent (actually accounting for about half of its total nonagricultural wage and salary employment), and Laurens, Oconee, Georgetown, and Beaufort hovering around a one-third loss in manufacturing employment; and all of these counties are showing these changes over a period of only four years. They, with the likely exception of Saluda, are apt to recover faster, due to their urban shadow location, and/or their situation in environments with other growth possibilities, i.e. recreation and leisure activities. A number of otherwise endangered mostly rural counties are blessed with a significant and reasonably stable food processing industry, such as Dillon with Perdue Farms (1,002 workers), Newberry with Louis Rich (1,236), Cherokee with Nestle USA (1,215), Sumter with Gold Kist (2,015), and even Saluda appears here, with Arrick Farms (1,250). Other counties with significant manufacturing job losses are some distance removed from the metros. These include Marion (-42.1%), Abbeville (-35.4%), Allendale (-33.1%), and Williamsburg (-35.9%). In these very rural counties, there are few prospects for economic investment and job opportunities to replace those lost with the closing plants. In our globalizing world, this represents the most abject example of one major result, the increasing spatial differentiation in economic opportunity and welfare.

Metropolitan Carolina

In recent decades, it has been the complex processes characterizing globalizing economies that have been dominating economic change and the localizing of investments which fuel urban development. With its emphatic Piedmont focus, one can now begin to recognize in the Carolinas a "new" Main Street, a transportation thruway linking the major cities of the Carolina Piedmont, from Anderson to Raleigh, as an agglomerated distinctly elongated centroid of the eastern seaboard region of the United States. Carolina Main Street finds itself in a critical location, about midway between Boston and Dallas, between New York and Miami, and between Washington D.C. and Atlanta.

The discussion of the character of urbanization in the Carolinas will first consider the larger impression provided by the newly defined (December, 2000 and revised annually) Office of Budget and Management Metropolitan Statistical Areas (MSAs) and their smaller cousins, the Micropolitan Statistical Areas (MiSAs). Collectively, the MSAs and MiSAs are known as Core Based Statistical Areas (CBSAs). (For a Census Bureau discussion of this new formulation, see <www.census.gov/www/ua/ua_2k.html>). There are 25 MSAs in the Carolinas. Fifteen of them include 40 of North Carolina's 100 counties, and ten incorporate 21 of South Carolina's 46 counties. Thus, the metropolitan share as defined by inclusion in an MSA is well over one-third of the total number of counties in each state. Table 5.1 tabulates the recent population changes in the Carolina MSAs for the two states. Complementing this table is Figure 5.1 which shows the location of each of the Carolina MSAs. While two MSAs include counties in an adjacent state, only the population of those counties within each respective state is included. These two MSAs are (1) the Norfolk/ Virginia Beach MSA, with its Currituck tongue extending southward into North Carolina; and (2) the Augusta-Richmond County MSA, which extends from the larger central place of Augusta, Georgia into two counties in South Carolina.

The Carolinas clearly outpaced population growth in the United States during the past several decades. During the 1990s, the United States saw a population expansion of about 11 percent, while the Carolinas increased by 19.2 percent. The initial half decade of the twenty-first century continues this pattern at a much reduced rate, and that in spite of the extraordinary economic shifts occurring in the two states, though the U.S. increased by 6.0 percent while the Carolinas showed a 6.4 percent advance. During this period, North Carolina MSAs increased their share of state population from 68.1 to 69.1 percent, while South Carolina's MSAs had a gain in their share from 74.8 to 75.5 percent.

In general terms, the geographic distribution of Carolina MSAs form three west-east sequences, separated by sections of designated micropolitan and non-CBSA designated counties. The more populated sequence, by far, have been designated as Carolina Main Street. From Anderson, Greenville, Spartanburg and York in South Carolina's Upper Piedmont, Carolina Main Street further extends across North Carolina from Charlotte to beyond Raleigh (Figure 5.2). With nine proximal MSAs, Carolina Main Street also includes five Micropolitan Statistical Areas, which were among the fastest growing MiSAs in the country during the 2000-2003 period (Figure 5.1). All together, this interconnected urbanized region included 27 counties and tallied 5,192,165 people as of July 1, 2005. Notable, perhaps, this is a population larger than that of the designated Atlanta-Sandy Springs-Marietta GA MSA with its 28 counties and 4,917,717, for the comparable time period.

During the 1990s, the Carolina Main Street MSAs as a whole exceeded the overall United States MSA growth rate of 14.0 percent. The Raleigh-Cary MSA exceeded the national average by more than three times, and the Charlotte MSA by more than double. During the initial years of the twenty-first century, this rapid growth condition continued, with the Charlotte MSA staying at more than double the national rate and the Raleigh-Cary MSA ranking as the eleventh fastest growing metro area in the nation, with a 3.7 percent annual rate of growth (Mackun, 2005, 12). The micropolitan areas of Thomasville, Lexington, Salisbury, and Statesville-Mooresville were ranked 9th, 16th, and 17th among the nation's fastest growing MiSAs during 2000-2003 (Mackun, 2005, 15). For the South Carolina portion of Carolina Main Street, the population expansion has been less vigorous, declining to below the national average for the early years of this century, with the notable exception of York County, which continues to reflect its inclusion in the Charlotte MSA.

Carolina MSAs not located on Carolina Main Street have experienced a notable variation in population growth. The central South Carolina tier of MSAs, overlapping the Midlands, Upper Coastal Plains, and Pee Dee geographic regions, have seen their 2000-2005 growth rates vary from 1.2 percent, in the case of Sumter, to 6.7 percent, in the case of Columbia. In North Carolina, it is the MSAs with their economy tied to textiles, furniture and military bases that have fared least well. Coastal MSAs have done much better, with Myrtle Beach seeing an 11.8 percent increase, Wilmington a 12.2 percent rise and the Currituck appendage an incredible 23.6 percent gain from 2000-2005. The exception in the Coastal belt of MSAs is the Jacksonville MSA, which actually experienced a slight decrease in its population.

Carolina Cities

No powerhouse urban center like Atlanta directing the fortunes of Georgia exists in the Carolinas, though Charlotte appears to be poised

Table 5.1. Carolinas Metropolitan Statistical Areas, 1990-2005 (Dec. 2005 defnition)

	MSA	# Counties (out of state)	2000 Pop. Census	%Change 1990-2000	2005 Pop. (estimate)	% Change 2000-2005
1.	Asheville	4	369,171	19.9	391,007	5.9
2.	Burlington*	1	130,800	20.9	140,105	7.1
3.	Charlotte-Gastonia-Con.*	5 (6)	1,165,834	30.6	1,318,613	13.1
4.	Durham*	4	426,493	23.7	458,673	7.5
5.	Fayetteville	2	336,609	13.1	348,352	3.5
6.	Goldsboro	1	113,329	8.3	113,461	0.1
7.	Greensboro-High Point*	3	643,430	19.1	673,134	4.2
8.	Greenville	2	152,772	23.0	162,183	6.2
9.	Hickory-Lenoir-Morgan.	4	341-851	14.5	356,039	4.2
10.	Jacksonville	1	150,355	0.3	149,754	-0.4
11.	Raleigh-Cary*	3	797,071	46.5	939,136	17.8
12.	Rocky Mount	2	143,026	7.0	145,779	1.9
13.	Currituck/Virginia Beach	1 (7)	18,190	32.4	22,484	23.6
14.	Wilmington	3	274,532	37.2	307,944	12.2
15.	Winston-Salem*	4	421,961	16.7	447,369	6.0
A.	North Carolina MSAs	40	5,485,424	24.2	5,974,033	8.9
	State Total	***100***	***8,049,313***	***21.4***	***8,642,524***	***7.4***
1.	Aiken-Edgefield/Augusta	2 (6)	167,147	20.0	174,629	4.5
2.	Anderson*	1	165,740	14.1	175,558	5.9
3.	Charleston-North Charl.	3	549,033	8.3	586,872	6.9
4.	Columbia	6	647,158	18.0	690,422	6.7
5.	Florence	2	193,155	10.0	198,594	2.8
6.	Greenville*	3	559,940	17.0	588,729	5.1
7.	Myrtle Beach	1	196,629	36.5	219,791	11.8
8.	Spartanburg*	1	253,791	11.9	264,032	4.0
9.	Sumter	1	104,646	2.0	106,517	1.2
10.	York*/Charlotte	1 (6)	164,614	25.2	186,816	13.5
B.	South Carolina MSAs	21	3,001,853	15.9	3,192,960	6.0
	State Total	***46***	***4,012,012***	***15.1***	***4,230,879***	***5.5***
	Main Street (*) MSAs	20	4,729,674		5,192,165	9.8
	Micropolitan Counties included in Main Street	5	516,834		557,287	7.8
	Carolina Main Street	25	5,246,508		5,749,452	9.6
C.	Carolinas MSAs	61	8,487,277		9,166,993	8.0
	Carolinas Total	***146***	***12,061,325***	***19.2***	***12,873,403***	***6.4***

Sources: North Carolina State Demographics (www.http://demog.state.nc.us/demog/home.html; South Carolina Statistical Abstracts (www.http://www.ors2.state.sc.us/abstract/index.asp.; Rand McNally Commercial Atlas, 2006.

and waiting in the wings for its desired anointment as "Queen City of the Carolinas," much as she has been self-promoted for some time. In a report authored by Neil Pierce and C. Johnson (1995), and bankrolled by the Charlotte Chamber of Commerce, Pierce reiterated his earlier stated opinion that the large Carolina cities have evolved much like city states. A city state is "(a) region consisting of one or more central cities surrounded by cities and towns, which have a shared identification, function as a single zone for trade, commerce and communication, and are characterized by social, economic and environmental interdependence" (cited in Gade, et al., 2002, 442).

Certainly, the data in Table 5.2 can be seen in support of this contention. Charlotte is a very rapidly growing central city, by far the largest in the Carolinas and buttressed by even more rapidly growing nearby centers feeding housing and amenities to her burgeoning employment. These neighbors include Rock Hill, Monroe, and Huntersville, with Concord now also beginning to feel the hurt of providing access, health, education, and public safety to invading subdivisions. Bedroom communities spreading within the limits of other smaller neighboring towns are encouraging these to expand at an even faster rate, as exemplified by Cornelius, Indian Trail, Mint Hill, and Stallings. These are all in excess of 10,000 people and grew by 30 to 200 percent during the 2000-2005 period. Also note Raleigh and the Research Triangle Park bedroom community of Cary. Perhaps, the most extraordinary example is that of Charleston. Hemmed in by water and swamplands, Charleston has given birth to North Charleston, which due to its expansion room is likely to eventually exceed its parent in population size. Charleston has also contributed by its overflow of humanity to the growth of Mt. Pleasant and Goose Creek. Even Summerville, some 30 miles away on I-26, is seeing a crowding of residential subdivisions with their attendant retail baggage. But proximity to "THE" central place in a metropolitan area is only part of the answer for understanding population growth or decline. Being positioned along the Carolina Main Street seems to carry considerable weight in influencing growth, as does being located in a coastal or mountain, or educational institution, environment. However, from a "string of pearls" of city states strung along the necklace provided by the interstate freeways have evolved a massive urban diffusion developing the gradual coalescence of expanding city states reaching out and embracing formerly independent towns on the periphery (Morrill, 2006). Carolina Main Street stands sovereign, not only as the dominant population feature in the Carolina landscape, but also as an increasingly significant megalopolis in the Southeast and in the country at large.

There are some striking differences in city population change between North and South Carolina. A scan of Table 5.2 will reveal extraordinary differences in the rates of population growth. And should one expand the inquiry to all incorporated places in the Carolinas, about 806 in all, then one finds that 26 percent of North Carolina's 538 municipalities lost population during the 2000-2005 period, while 47 percent of South Carolina's 268 municipalities did as well. Given the overall population growth in the Carolinas this is a remarkable statistic.

Remember, this was the post-dot.com bust period, as well as being the vanguard of the textile/apparel and tobacco production decline and the beginning of the furniture plant closings. Incorporated places across the Carolinas are more apt to be in the lost column if they are smaller, if they are company towns with plant closing(s), or if they are in rural areas beyond reasonable commuting distance to metro areas. Most North Carolina city population declines were largely tied to the impact of these economic shifts, as was true for South Carolina's, but here only in part. A very notable difference between the two states for municipality population change is the variance in state ordained annexation policy. For North Carolina, it is considered quite permissive, while, for South Carolina, many consider it truly archaic. So, here is the opinion of the Keynote Speaker at the Annual Meeting of the Municipal Association of South Carolina held at Hilton Head July 16, 2004 (Rusk, 2004; Tyer, 1995):

> . . .in contrast with Texas or North Carolina, South Carolina has the most unwork-

Table 5.2. Carolina Incorporated Places of 30,000 and Over in 2005 (Estimates)

City	Population Census 2000	Population 7/1/2005-est.	%Change 1980-90	%Change 1990-2000	%Change 2000-05
1. Charlotte	540,167	640,270	20.9	36.4	18.5
2. Raleigh	276,094	338,357	32.0	30.2	22.6
3. Greensboro	223,891	237,316	7.8	21.8	6.0
4. Durham	187,035	209,123	26.0	36.8	11.8
5. Winston-Salem	185,776	198,593	0.4	29.5	6.9
6. Fayetteville	121,015	130,646	16.0	59.5	8.0
7. Columbia	116,273	117,088	-3.1	18.9	0.7
8. Cary	94,536	115,967	96.5	112.9	22.7
9. Charleston	96,650	106,712	9.0	20.2	10.4
10. Wilmington	75,838	97,135	6.2	36.6	28.1
11. High Point	85,839	92,491	4.2	23.6	7.7
12. North Charleston	80,537	86,313	NA	NA	7.2
13. Asheville	68,889	73,189	4.7	11.4	6.2
14. Jacksonville	66,715	73,121	28.2	99.8	9.6
15. Gastonia	66,355	70,243	7.0	21.3	5.9
16. Greenville/NC	61,209	68,852	24.6	32.2	12.5
17. Concord	55,977	63,429	11.6	104.6	13.3
18. Rock Hill	49,816	59,544	14.3	19.6	19.5
19. Mount Pleasant	47,656	57,932	NA	NA	21.6
20. Greenville/SC	56,052	56,676	0.1	-3.8	1.1
21. Rocky Mount	55,977	56,291	11.1	12.0	0.6
22. Chapel Hill	46,019	52,397	17.6	18.9	13.9
23. Wilson	44,405	47,815	7.3	20.2	7.7
24. Burlington	44,917	47,295	6.0	13.7	5.3
25. Kannapolis	36,910	40,139	-2.0	24.2	8.7
26. Sumter	40,073	39,679	51.7	-4.5	-1.0
27. Hickory	37,222	39,018	13.4	30.7	4.8
28. Spartanburg	39,803	38,379	-0.8	-8.4	-3.6
29. Goldsboro	39,147	38,186	14.7	-3.8	-2.5
30. Summerville	27,819	37,714	NA	NA	35.6
31. Hilton Head	33,862	34,497	110.8	42.9	1.9
32. Goose Creek	29,271	32,516	NA	NA	11.1
33. Monroe	26,228	32,454	NA	NA	23.7
34. Huntersville	24,960	31,646	NA	NA	26.7
35. Florence	30,211	31,269	-0.1	1.3	3.5

Sources: North Carolina Department of Commerce, State Data Office, 2006, 2000; South Carolina Statistical Abstract, 2005; United States Bureau of the Census, 2000.

able, unreasonable annexation laws in the USA. Of the 40 states where annexation is even possible, South Carolina's annexation laws make it harder to bring new subdivisions, shopping centers and regional malls, office and industrial parks into the city limits than any other state. As a result, in the Age of Urban Sprawl, as a percentage of their metropolitan areas' populations, South Carolina has the USA's fifth smallest central cities, following only Rhode Island, Massachusetts, Connecticut, and Michigan.

In a document entitled "Local Government and Home Rule in South Carolina: A Citizens Guide," the authors identify the specifics "as urban growth occurs . . ., municipalities have the powers to grow through annexation. Adjacent property owners can request annexation and be accepted by ordinance. Larger areas can be annexed by petition if 75 percent of the landowners representing 75 percent of the property to be annexed agree to annexation" (Ulbrich and Steirer, 2004). Of course, the latter follows a public hearing. Where in the rare case there is a majority of 75 percent of land owners ready for city takeover, even this may be doomed by one landowner owning more than 25 percent of the value of the land slated for annexation. It is truly fascinating to view the data showing the decreasing significance of incorporated areas in South Carolina, particularly as compared to the state and the MSA increasing populations. In 1970, MSAs held 47.3 percent of the total population, while incorporated cities/towns/villages held 44.8 percent, most of their population, of course, held within the MSAs (Tyer, 1995). Since then, there has been a downward slide in the importance of municipalities. As derived from Table 5.1, 25 years later, with massive ongoing urbanization, South Carolina's eight MSAs had 75.4 percent of the state's population, meanwhile the incorporated places had dropped to 35.1 percent (North Carolina = 53.2%). While the population of the state has increased over the past quarter of a century by 63.1 percent, the population of incorporated centers has increased by a mere 28.9 percent. Indeed, cities, as legal entities, are not faring all that well in the South Carolina political environment. The implications for urban governance and community well-being are all too clear. Since no land use planning or taxing authority is easily extended to rapidly growing areas in the urban shadow, infrastucture development will continue to be hampered, as will the global competitiveness of South Carolina.

This aids in understanding the lesser impacts made by South Carolina cities as suggested in Table 5.2. Part of the growth spurts experienced by a number of Carolina central cities over the past quarter century is due not only to economic prosperity, but also the ability in North Carolina to readily annex adjacent urbanizing communities. This municipal administration land acquisition tool almost became a fad in North Carolina in the 1980s, an otherwise slow growth decade; it then extended into the 1990s. The state had empowered the cities to extend their territory, albeit within a set of well-defined land use regulations, but without the expectation of advise and consent of those to be annexed. Notice in Figure 5.7 the steeply sloping curve in the 1980-1990 urban expansion experience of Raleigh, Greens-

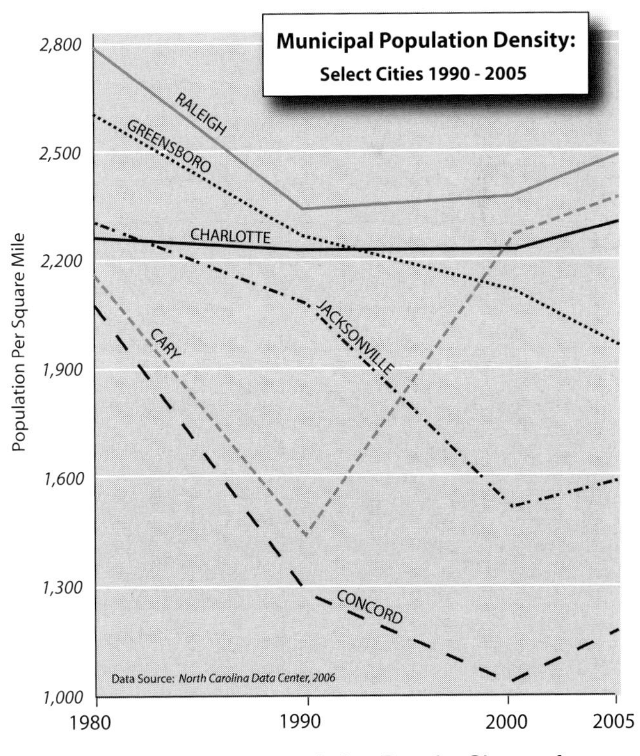

Figure 5.7. Municipal Population Density Change for a Selection of Cities, 2000-2005

boro, and Jacksonville, and especially for Cary and Concord. While these cities were gaining population internally, many more people were added by territorial annexation, as evidenced by the decreasing population density. For Cary, the subsequent incredible growth spurt of the 1990s was finally slowed at the turn of the century due in part to a considerable increase in residential construction impact fees, assessed for new home owners for them to share more proportionately in the new schools and improved road systems they require. Impact fees went from $315 in 1998 to $2,535 in 2001, per new house built. Due to an antagonistic construction and real estate environment, Mayor Lang, who was influential in the impact fee escalation, lost his job. Cary has now lowered the impact fee with builders and home buyers again flocking to Cary (Coleman, 2006). Charlotte, already pretty much surrounded by smaller incorporated places, had very little room to maneuver, and was otherwise growing rapidly enough to fill in vacant spaces absorbed through annexation. During the 1990s, for example Charlotte expanded its municipal boundary by 66 square miles absorbing 83,979 people in the process. Its internal growth, in the meantime, was 60,254. Greensboro, on the other hand, has rather limited room for expansion in most directions, as a result of several incorporations around it during the 1990s. During the 1990s, as well, Jacksonville's tenuous dependence on the Camp Lejeune Marine Corps Base caused the town to have a decline of 112 persons. This, however, was offset by the town's annexation of all of the base personnel living quarter areas, while expanding the municipal territory by 200 percent. And who had planned for the development of an urban area the size of Cary to be squeezed in between Raleigh and the Research Triangle Park? Cary was farsighted in land acquisition during the 1980s, but has run out of room and is clearly feeling the pressures on continued population growth. Not so for Concord, a traditional textile, and now tobacco, town in the urban shadow of Charlotte. With the latter's vigorous expansion, Concord has rushed to annex in order to control the overflow from Mecklenburg County. Concord is growing rapidly, but will have considerable space to accommodate the newcomers for the near future.

Though municipalities in North Carolina continue to be intimidated by the power of the land development interests, they do have the ability to assert their prerogatives relating to taxation and land use control over their urbanizing regions. Not so for their kin to the south. Being saddled with an outmoded and idealistic perspective on local governance and home rule will continue to stand in South Carolina's path of urban livability and progress.

Carolina Main Street and Spersopolis

In the Carolinas Main Street region is found ample support for Hart and Morgan's (1995) notion of Spersopolis (dispersed city), where urban areas are intermixed with a disproportionately high incidence of rural nonfarm population, of intercounty commuters, of persons employed in manufacturing industry, and of mobile homes. See especially Hart and Morgan's map of Spersopolis (Hart and Morgan, 1995, 107), which in its Carolina coverage correlates well with Carolina Main Street. In terms of land usage, the "Main Street" image is especially operative. Between Greenville, S.C. and Raleigh, inter-digital industrial, research and office parks, shopping malls, and residential subdivisions are competing for land with ready access to one another, as well as to employment and amenities offered by the central cities all along that great lifeline, I-85/I-40.

The notion of "Main Street" as a meaningful moniker providing an image for the urbanizing Carolina Piedmont has its origin in the transportation system that over the years has linked contributing cities. Most important has been the development of the federal interstate highway system. Though politically inspired and financially driven by legislative support for building a system of "defense highways," the overriding context was a united mind of linking effectively and efficiently the far-flung communities of the United States. A report commissioned by the American Automobile Association in the 1960s indicated that the major objectives of the Interstate Highway System were to (1) improve accessibility and make the central business district more attractive; (2) stimulate the planned development of metropolitan areas; and (3) influence the redevelopment of the older central city. As Ziegler (1992) points out, more likely the opposite has been achieved. Undoubtedly, the initial objective was reached with major public and corporate investments in the central city expanding service sector jobs. But the urban renewal process replaced existing neighborhoods with asphalt and office buildings, and improved access further encouraged a white flight to the suburbs. Left behind in most Carolinas Main Street central cities are deeply segregated residential neighborhoods, and faltering public schools. The second objective of planned development continues to befuddle decision makers, primarily due to the absence of broad-based voter support for regional and metropolitan planning, combined with developers' general view that land use planning and regulations robs them of potential profits as they speculate in land use changes on the urban periphery. Thus, state departments of transportation are left to their own devices and employment centers, residential subdivisions, and retail development follow the Interstates, focusing on critical road intersections and building along new beltways. Sinewy extensions of exurbia submerge existing rurality to a pale reflection of its former sovereignty. Spersopolis has arrived (!), though now guilded by the occasional one to two acre lots with McMansion homes spread across the landscape safe and secure within their gated community, in a sea of swelling mobile home parks. Concerning the third objective of influencing redevelopment in older central cities, continuing large scale corporate investments in the heart of the city is especially impacting the larger centers, most particularly Charlotte. This has clearly stimulated other kinds of redevelopment, notably high rise condo and apartment buildings, the reuse of old textile plants (Ingalls and Moore, 2001), as well as continuing gentrification of long established center city neighborhoods. Mostly, the larger central cities in the Carolinas are still populated by a daytime working population that leaves their downtown areas almost devoid of all but the homeless from 5 p.m. to 8 a.m. each workday, and all day on weekends and holidays. In essence, the hum of activity along Carolina Main Street, and in other Carolina MSAs, masks the nagging problem of a continuing "demographic balkanization" (Manson and Groop, 2000).

Figure 5.8 provides better visual support for understanding the locational continuity of urban places inhabiting Carolina Main Street, as well as the urbanizing areas elsewhere in the Carolinas. This map is derived from the so-called Ranally definition of urban development, where central cities are depicted by size category, with their urbanized area delineated around them, as well as areas beyond included where they contain over 70 persons per square mile. Clearly, a megalopolis is spreading across the Carolinas, anchored firmly to the I-85 "corridor" identified earlier by Hartshorn (1997). *Business Week* in 1993 referred to the I-85 passage as the "boom belt", though the magazine clearly had in mind Atlanta and Richmond as the polar termini. "From the air, enormous swaths of economically successful regions like the area stretching from Atlanta to Raleigh-Durham along the (P)iedmont region of the Carolinas . . . . appear as a series of large boxes surrounded by parking lots and connected by highways" (Bruegmann, 2005, 70). State newspapers are also picking up on the relevance of the freeways in evolving an inter-digitated urban system. Labeling the phenomenon The Carolina Crescent, Price noted that "The Interstate I-85 corridor from South Carolina (Greenville-Spartanburg) to the Research Triangle is experiencing some of the fastest job growth in the nation" (1998), and in an article on the 50th anniversary of the Interstate Highway System, Coleman and Bracken discussed the interconnective role of I-40 in linking "cities, (and su)burbs to region's economic engine" (2006). This article also reflected on Charles Kuralt's disclaiming comment, upon hearing of the finishing of I-40, "thanks to the Interstate Highway System it is now possible to travel from coast to coast without seeing anything!" Others have hit upon names like "Charleighboro" to capture an evolving Charlotte to Raleigh megalopolis whose life is dominated by exit ramps (Tuttle, 1994).

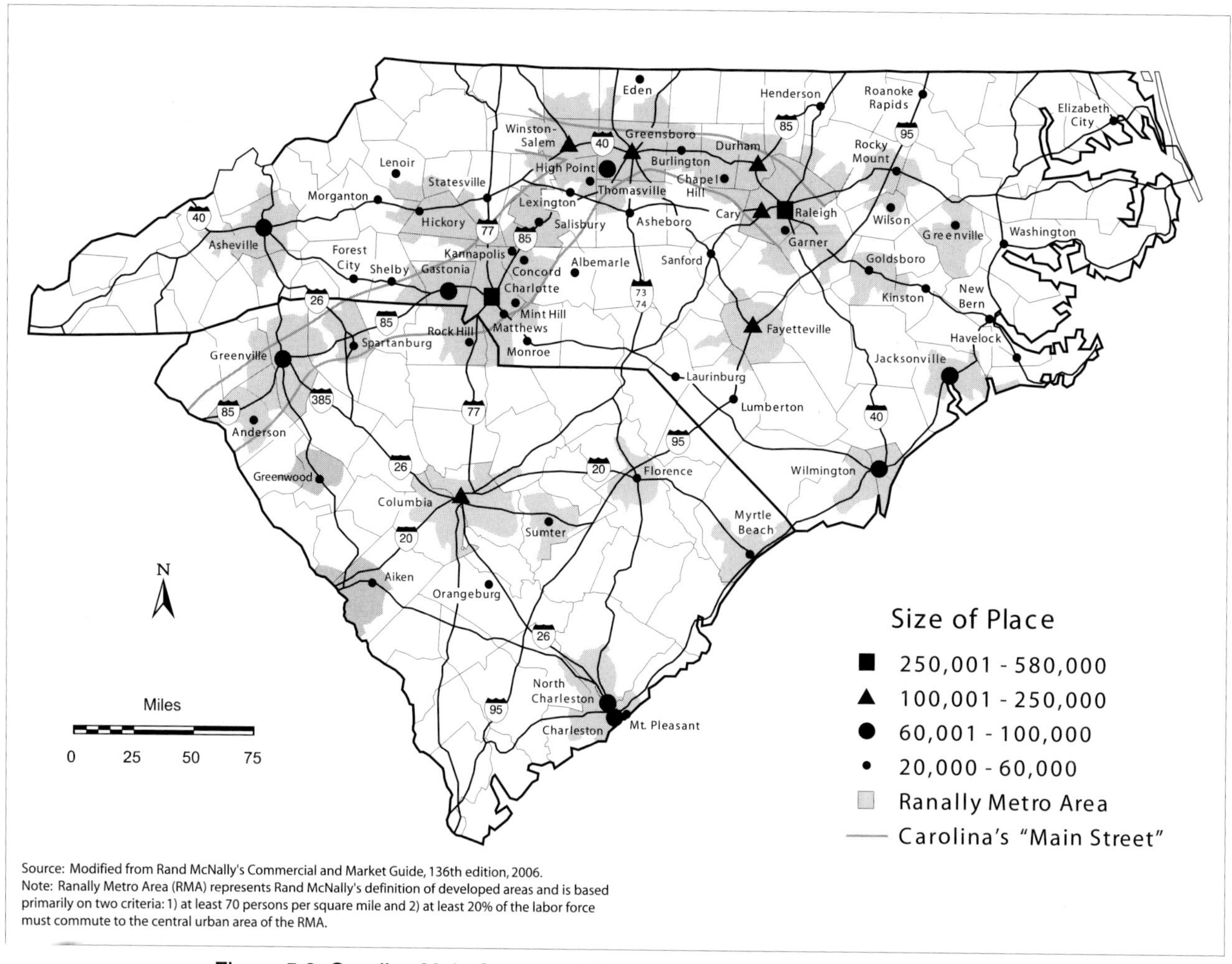

Figure 5.8. Carolina Main Street and Other Urbanized Carolina Regions, 2006. Source: North Carolina State Data Center, 2006

The Commuting Reach of Carolina Main Street

One of the major indicators of the dynamics of change affecting urban areas is the shifting pattern of commutation, the daily journey to work. As Morrill (2006) has suggested, metropolitan dominance is reflected in the intense commuter labor market of central cities. Initially, the commuting patterns for the Carolinas will be examined as a whole for 2000. As shown in Figure 5.9, these readily identify the crescent shape of Carolina Main Street, with six counties exceeding 40,000 commuters daily and another five ranging from 16,000 to 40,000 daily roundtrips to work from outside the county indicated. Mecklenburg stands out with its 146,211 commuters, while Wake (Raleigh), Durham (Research Triangle Park), and Guilford (Greensboro) are nearly tied at 87,735, 81,795, and 81,786, respectively, followed by Greenville, S.C. with 59,715 and Forsyth (Winston-Salem) with 50,978. Beyond Carolina Main Street, Richland (Columbia) with 69,043 and Charleston with 55,374 exceed 50,000 daily commuters. A UNC-Charlotte study is quoted as stating that almost two-thirds of travel in North Carolina crosses county borders (Whitacre and Mellnik, 2003).

The dynamics of recent change in contrasting employment and residential locations of households is well captured when comparing the changes in commuting habits for Mecklenburg workers over a decade, 1990-2000. Figure 5.10a illustrates this very well. Stuart (1996) has shown that in 1960 a total of 15,100 workers daily came into Mecklenburg County. In 1990, this had increased to 102,576. About 85 percent of this flow was generated in the counties displayed. Stuart

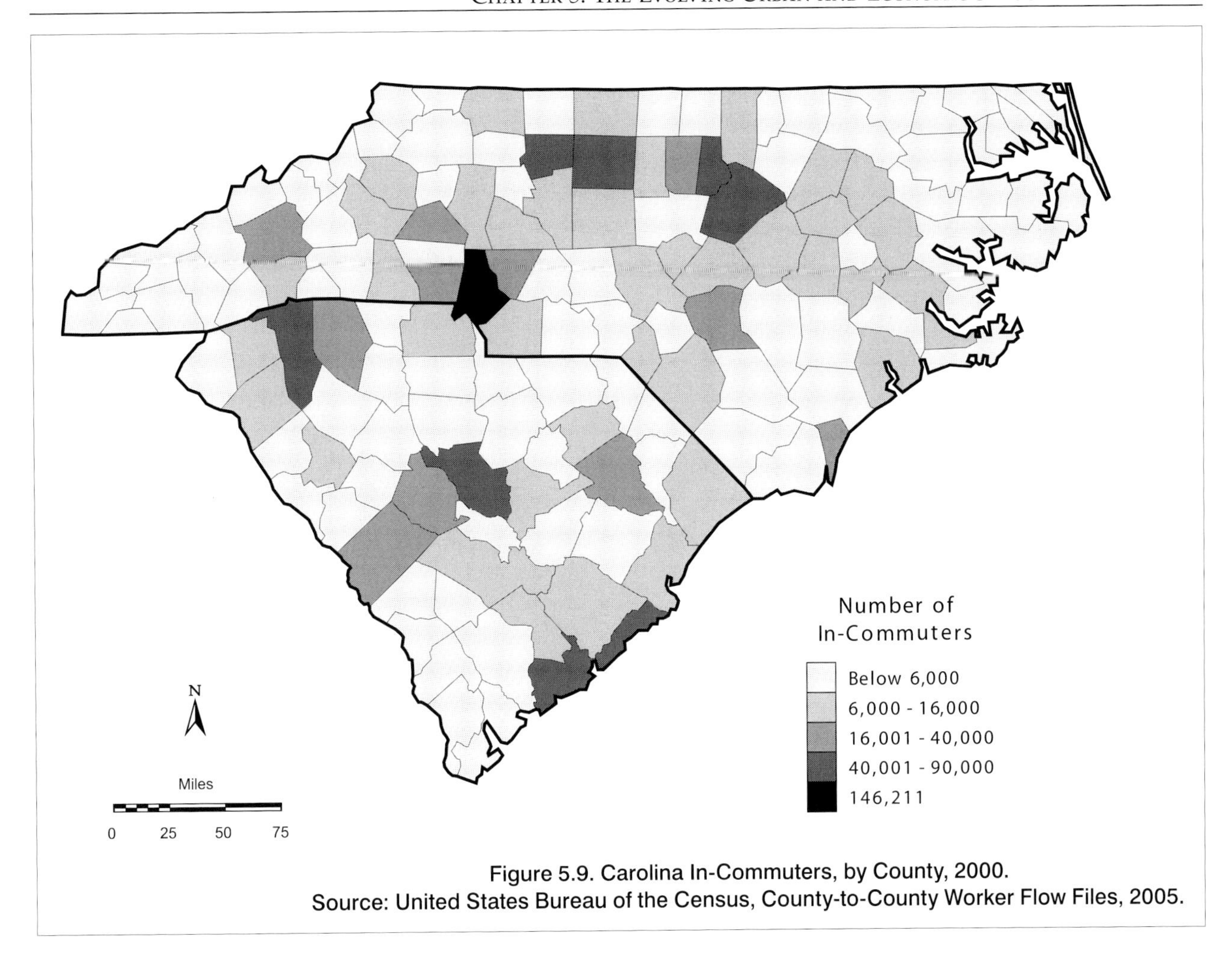

Figure 5.9. Carolina In-Commuters, by County, 2000.
Source: United States Bureau of the Census, County-to-County Worker Flow Files, 2005.

further points out that though all of the counties were engaged in the exchange of commuters, for six of these counties commuters constituted over 30 percent of the employed. The trend continued over the next decade (Figure 5.10b) with the four leading counties of Union, York, Gaston and Cabarrus nearly tied for the lead in workers flowing into Mecklenburg County, each of them exceeding 22,000 daily, and commuting totals reaching 128,471 for the illustrated nearby counties. Increasingly, of course, commuters are finding employment in job centers accessible by the Outerbelt freeway, such as University City, Arrowood Industrial Park, and Ballantyne office parks and commercial areas. As Harrison Campbell, UNC-Charlotte geographer noted, "As job centers have diffused toward the county boundary, suddenly those jobs are in striking distance of workers who live farther away" (quoted in Whitacre and Mellnik, 2003). Of course, this demonstrates not only that residential subdivisions are being built an increasing distance from Charlotte's major sources of employment, but also that living conditions are less expensive in some adjacent county areas and that their rules on mobile home parks are less restrictive. There is also the unhappy fact that local job replacements for those persons discarded from the closing textile/apparel, furniture and tobacco plants are difficult to come by. All along Carolina Main Street there is a relatively greater employment concentration, as is true for all of the Carolinas MSAs. Thus, highway traffic is becoming more and more onerous, with increasing cost of fuel competing with the costs of expanding said highway system. The *News and Observer* reported that road building costs have grown by 46 percent from 2003 to 2006, driven by higher prices for steel, concrete and other construction materials. One must not forget the increasing air pollution of a vastly expanded traffic volume, itself moving at a much slower and hesitant pace.

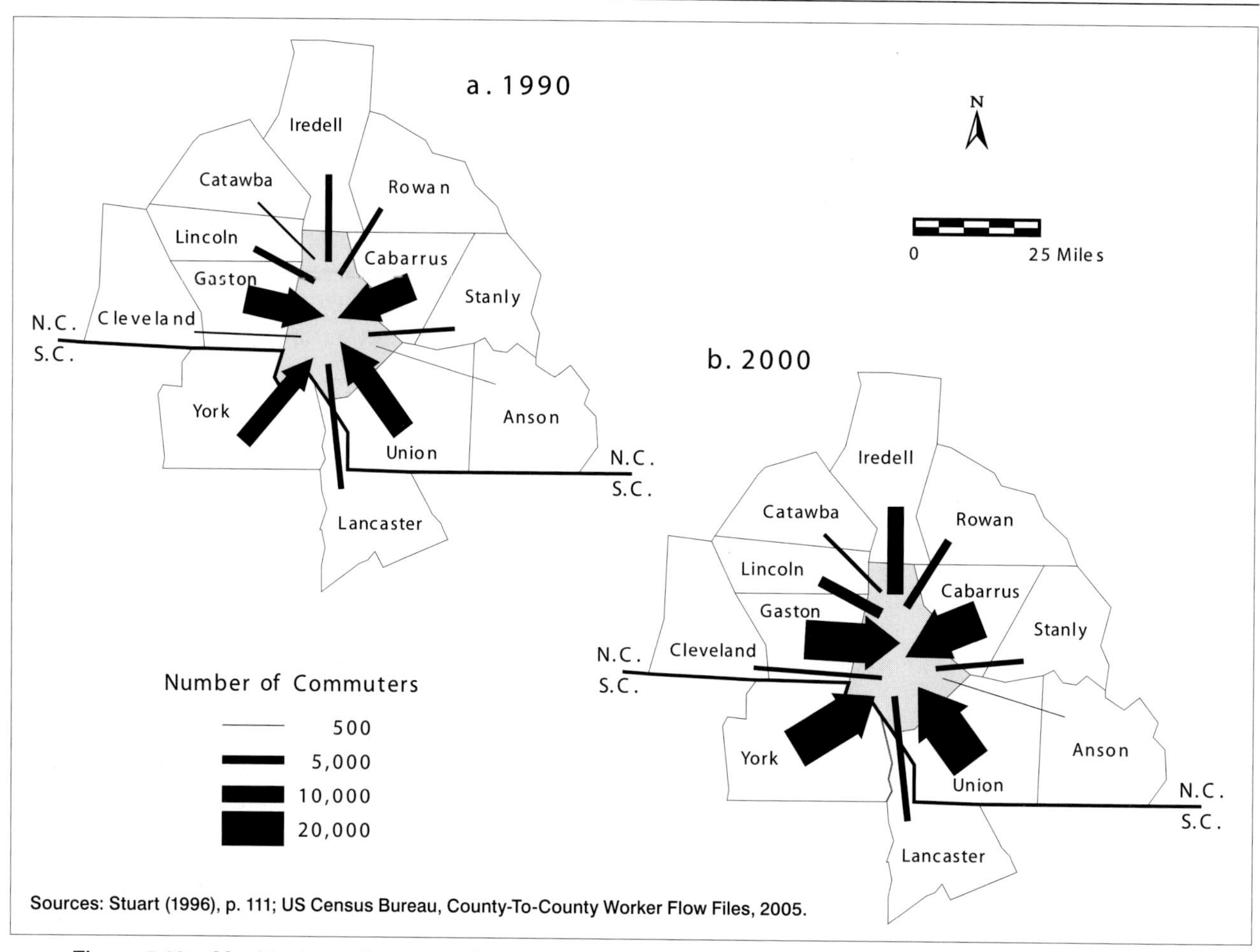

Sources: Stuart (1996), p. 111; US Census Bureau, County-To-County Worker Flow Files, 2005.

Figure 5.10a. Mecklenburg County In-Commuters, 1990 5.10b. Mecklenburg County In-Commuters, 2000

Surely, the preceding identifies a complexity of problems that will be reflected in a discussion of the problem of the near total absence of mass transit along Carolina Main Street. Commuter rail in the Charlotte region is now finally being built and slated to open its initial service in 2008, but Raleigh-Durham's Triangle Transit Authority seems persistently stymied in attempts over the past decade to get its rapid rail system started. Meanwhile, interregional bus transportation in the Midlands of South Carolina, in the Columbia MSA region, appears perpetually in the process of being abandoned. Thus, one can expect to see commuters continue to travel, twice daily, mostly singly in their autos, spewing pollutants while they negotiate some of the nation's heaviest highway traffic.

There is ample evidence that the world-at-large is well aware of the potential in locating along Carolina Main Street. IKEA, the Swedish furniture and household goods giant considers a minimum customer base of two million residents critical for locating one of its two-story 345,000 square foot stores. In January, 2007, it announced its plans to locate on the north side of Charlotte with easy access to I-485 and I-77, for as noted by the IKEA real estate manager, this will readily access the 75 percent of the total customer base, which is expected to arrive from outside of Mecklenburg County (N. M. Bell, 2007). A Google hub will be arriving in Lenoir by late 2007; and a FedEx regional hub is landing in Greensboro's Piedmont Triad Airport by 2009. These are among the many large scale ventures on line; Carolinas Main Street is fairly humming with activity.

It bears noting that South Carolina cities are not as tied to the beneficial developments occurring along the Carolina Main Street simply because this urbanizing belt is dominating the broader sweep of the North Carolina Piedmont with 15 counties, while only five of its counties are in South Carolina. Similarly, 17, or about half, of the largest cities in the Carolinas (Table 5.2), are hubs along Carolina Main Street, yet only three of these 17 (Greenville, Rock Hill, and Spartanburg) are in South Carolina.

Air Transport and Urban Development

Nodal magnetism is also reflected in the increasing volume of air passenger traffic in cities along Carolina Main Street. First of all, these sizable increases over 40-plus years reflect the substantial inroads that air transportation has made in recent decades. It should be noted that Charlotte's singular importance with now 56.8 percent of the Carolina passenger traffic is determined by its growing importance as the lead hub of USAirways.

Debbage (1996) makes a number of points relevant to changing times in the air passenger industry, and in the attractiveness of Carolina Main Street. These can be readily gauged from Table 5.3. First, dramatic shifts in the competitiveness and locational advantage of specific places have followed the late 1980s deregulation of the air transport industry in 1978. Note the boom in traffic for Charlotte and Raleigh/Durham. Second, economic restructuring in the Carolinas away from the singular importance of labor intensive industries to a service-dominated economy, plus the generally booming sunbelt, has escalated passenger traffic for the ten largest airports across the board, except for Fayetteville. Finally, the introduction of the "hub and spoke" system favored just two of the airports, which then moved from a 38.5 percent share of the market for the ten airports to 80.9 percent share in 1992.

Some of these conditions have changed more recently. What reigns supreme is uncertainty. Charlotte alone has evolved as a "fortress" airport; in this case, as the leading hub of USAirways, the fifth largest employer in the Charlotte region with 5,700 workers. That airline so dominates the traffic that noncompetitive ticket prices are known to drive Charlotte area air travelers to Piedmont Triad International (Greensboro) to, for

Table 5.3. Air Passenger Volume at Larger Carolina Airports, 1973-2005

Airport Location	Total Number of Emplaned Passenger Boardings (1,000)							
	1973	%	1983	%	1992	%	2005	%
1. Charlotte	1,105	31.4	3,764	52.3	8,220	52.8	14,010	56.8
2. Raleigh/Durham	603	17.1	1,123	15.6	4,376	28.1	4,724	19.2
3. Greensboro/Winston-Salem	482	13.7	727	10.1	849	5.4	1,310	5.3
4. Charleston	326	9.2	407	5.7	592	3.8	1,072	4.3
5. Greenville/Spartanburg/Greer	234	6.6	300	4.2	424	2,7	893	3.6
6. Myrtle Beach	38	1.1	112	1.5	212	1.4	775	3.1
7. Columbia	351	10.0	377	5.2	474	3.0	726	2.9
8. Wilmington	76	2.2	107	1.5	124	0.8	340	1.4
9. Asheville	148	4.2	138	1.9	164	1.1	316	1.3
10. Fayetteville	154	4.4	140	1.9	134	0.8	154	0.6
Total	***3,518***	***100***	***7,194***	***100***	***15,570***	***100***	***24,320***	***100***

Note: Other Carolina airports with less that 100,000 enplanements in 2005 include Kinston, Hickory, Florence, and New Bern. Sources: Debbage (1996), p. 199; Federal Aviation Administration (2006).

example, catch a USAirways flight to Los Angeles that will have a change of planes in Charlotte. Merger mania has been hovering on the sidelines for this airline for decades. Now, following a recent exit from bankruptcy proceedings in 2006, USAirways made an audacious buy-out offer of larger Delta Airlines, also in financial difficulties at present. With Delta Airlines debtor group showing a lack of interest, the offer has, for the time, been dropped. Should such a merger occur, consequences for Charlotte can only be negative, since Delta Airlines is securely hubbed in Atlanta. At this time, though, Charlotte reigns supreme in the Carolinas, as well as being the tenth busiest airport (takeoffs and landings) nationally, and ranking 21st in passenger enplanements, with 575 flights daily and with nonstop flights to 168 cities.

Raleigh/Durham (RDU) remains in much more insecure waters following the airport's maturing in the 1980s. First, American Airlines failed in its several-year attempt to establish a "mini hub." By 1995, American had discontinued all commuter flights and reduced their daily jet departures from a high of 100 daily flights in the early 1990s to 35. Then, into the vacuum stepped Midway Airlines. Founded with its initial and unsuccessful hub in Chicago, Midway moved its headquarters to Morrisville and its hub to RDU. At its height, the airline offered 200 flights daily out of RDU to 25 east coast cities, then was hit hard by the dot.com bust of 2001-2002, and died shortly after following the terrorist disaster of 9/11. And now, Raleigh/Durham Airport is served by no less than eleven airlines and is likely to be offering the most competitive prices in the Carolinas, but its loss of a hub character has curtailed its potential for any kind of commanding presence.

About ten years ago, the Greensboro/Winston-Salem Airport (Piedmont Triad International PTI) went from boom to bust as the national model testing the CALite experiment. Continental Airlines had in mind trying out the low-fare, quick turnaround, short-hop, service which had made Southwest Airlines consistently profitable. It did not work! Passenger traffic zoomed to four million in 1994, only to fall back to an estimated 2.2 million in 1997. Surely an unnerving time for PTI management. Served by six major airlines, largely through their commuter service associates, Piedmont Triad is continuing losing passenger traffic with 29 percent less in June of 2006 than recorded for the same month in 2005 (Barron, 2006a).

While improving highway transportation appears to hamper PTIs competitive efforts vis a vis Charlotte and Raleigh airports, it is PTI's central location on Carolina Main Street that encouraged FedEx in a 1998 decision to locate its Eastern Seaboard hub here. Awaiting the finishing of an additional runway, necessitating buying out over 100 homes, while horrifying west-Greensboro residents with the prospects of an extensive 65 decibel noise umbrella functioning at night time, which is the dominant activity period for FedEx operations, the cargo airline is now looking at 2009 for a first flight out of its new PTI facilities. Meanwhile, Honda decided to produce jets at PTI! Having spent several years in testing their new private luxury jet at PTI, Honda has now, in February of 2007, decided that PTI will be the operational, as well as the manufacturing headquarters for the new Honda Jet. Orders are being taken for a production of up to 200 per year. Actually, the competition for this kind of product is fierce, including companies either building or developed light jets, including Eclipse Aviation, Cessna, Adam Aircraft, Diamond Aircraft, Aviation Technology Group, Spectrum Aeronautical, and Excel-Jet, none of which, unhappily, are located in the Carolinas (Barron, 2006b). Most Carolina airports are seeing notable increases in air cargo transport not only by FedEx, but also by United Parcel Service (UPS) and many smaller air cargo carriers. In addition, airport landing and takeoff data is heavily influenced by general aviation comprising personal and corporate aircraft.

Greenville-Spartanburg International (GSI) is a typical hub-served airport. Passengers connect with the major hubs of six airlines, but for a vastly enhanced opportunity at direct flights to almost anywhere, they are but a few hours driving time from Atlanta Hartsfield Airport. GSI has recently received an important upgrade in response to the need for the new BMW automobile complex to airship parts by Boeing 747 cargo planes (Kennedy, 1998, 5). Clearly, this magnetism of the anchor cities is replicated with varying intensity all along Carolina Main Street, ensuring a gradual expansion of its reach toward the hinterland, along the way capturing market strength from outlying cities. One of these is Columbia, which in spite of the continuing population expansion of its MSA, has had its comparative enplanement numbers decline, with its position among Carolina airports slipping from fourth to seventh over the four decades. Charlotte International is just too close. A continuing advantage for the operation of the Columbia Airport, though, is the location here of the United Parcel Service (UPS), a major cargo carrier.

The three coastal airports have done much better. Between 1992 and 2005, Myrtle Beach Airport has more than tripled its enplanements, while those for Charleston and Wilmington have more than doubled. Myrtle Beach's seasonal swings in passenger traffic has slowed year around permanent service by the major airlines. For example, there is a considerable inflow of Canadian vacationers in early spring encouraging customized air service. Then, there was Hooter Air, born and bred in Myrtle Beach in 2003, but sadly dead three years after arrival—so much for merriment with snug T-shirts and orange short shorts en (air) route to "fun in the sun." Charleston International Airport (CIA) is served by six airlines. Its increasing significance as a vacation and business destination airport has caused a doubling of passenger traffic over the past decade or so. This airport is significantly supported by the recent entry of the Vought-Alenia complex in North Charleston, which will be assembling the rear fuselages for Boeing's new 787 Dreamliner, a mid-sized passenger jet. With state incentives amounting to $116 million in the support of a corporate investment of $560 million, this is the largest South Carolina state support since the 1992 arrival of BMW. An expected 775 jobs, paying $40,000 to 50,000 per year, will be created in the process, with production having started in January, 2007 (DuPlessis, 2006).

Growing rapidly as well, given their position in the Carolina enplanement listings, are Wilmington and Asheville, both in part due to their favorable vacation retreat locations, beach and mountain. Fayetteville, meanwhile, is continuing to find that its major mobile population, the members of the armed forces, continue to find seats in frequent flights of cargo planes out of Fort Bragg fields or jets out of Seymour Johnson Air Force Base in Goldsboro. For others in the Fayetteville MSA, they are an hour and a half by interstates to Raleigh/Durham International.

Globalization and the New Economy

Tumultuous is a word appropriate to describe the character and impact of economic shifts in the Carolinas over the past several decades. Traditional manufacturing, the hallmark of the Carolina economy for most of the twentieth century is disappearing from the landscape, the victim of multinational coalescence and outsourcing. In the early years of the twenty-first century, the negative evidence was ubiquitous, from villages and towns with abandoned industrial plants, empty homes, and a wrecked local tax base, to the lines of former workers at the employment security offices. Were the Carolinas well prepared to meet this transition? Emergency meetings at all levels of public and private management, and workshops attracting socio/economic development experts, were rampant. Frequently was echoed the voice of Edward Fesser of UNC Chapel Hill, who at a workshop on "Globalization and North Carolina Industries" in May, 2002, opined that "North Carolina and its regions are not prepared for the changes accompanying increasingly open markets." (Nwagbara, et al, 2002). And, yet, the Carolinas also have the highest rates of in-migrating Americans and foreigners among the 50 states, as well as having some of the fastest growing metropolitan areas. How is this possible? The several hundred thousand migrants arriving over the past few years are not only low skill and low wage Hispanics, but also well educated and highly skilled people seeking high wage opportunities in the booming knowledge and high tech industries. These are banking and finance, pharmaceuticals, high tech electronics, and automotive assembly and parts manufacturing, and many others. A look at the emergence and promise of these economic activities will suggest not only which metro areas are more likely to see the greater success, but also the likelihood of a further hardening of geographic disparities in income and welfare across the two states. Initially, though, it is important to investigate what has been happening to the traditional industries, and what is being done to salvage what remains.

Changing Nature of Traditional Manufacturing Industries

Textiles and Apparel Globalize

The combined MSAs of Greensboro-High Point and Winston-Salem are commonly referred to as the Piedmont Triad. This is the North Carolina portion of Carolina Main Street that championed the industrial revolution in the state and the region that has been the slowest in moving toward industrial diversification. Even so, over the past 15 years the Piedmont Triad is seeing its population growing at a rate faster than that of the country as a whole (Table 5.1). Greensboro is the textile capital of North Carolina, if not, for a period, of the world. Headquartered here in the 1990s were Burlington Industries (11,000 Carolina workers in 1997), Cone Mills (6,160), Galey and Lord (2,300), Guilford Mills (3,025), and Unifi Inc. (6,500). Of course, these workers were not in single plants in Greensboro, rather they were distributed in dozens of plants throughout the Carolinas. And this listing does not include other important textile/apparel producers, such as Sara Lee (formerly Hanes), which, while headquartered in Chicago, had 21,700 workers in the Carolinas producing L'eggs hosery, Hanes underwear, and other clothing in several dozen Carolina apparel plants. Trade and investment advantages deriving from the North American Trade Agreement (NAFTA) made offshore investments particularly attractive for this industry. Galey and Lord quickly moved to buy out five Mexican sewing plants to allow the company improved vertical control over its production process, while remaining internationally competitive with its pricing of the finished products. Guilford Mills participated in a joint venture involving the development of the first of a planned series of "textile cities" in Morales, Mexico. For a while, this swift move toward gaining control of Latin American upstream production caused imported Asiatic apparel imports to drop from 66 percent of the U.S. total in 1993 to 35 percent in 1997, while increasing the share of Mexican/Caribbean imports (Hopkins, 1998). The project ultimately failed, and more serious changes were afoot.

While companies were closing, outsourcing, divesting some of their activities, and generally firing workers, others were joining to form stronger globalizing entities. Such was the case of Burlington and Cone Mills, which, at the behest of financier W. L. Ross in 2002, merged into a private company labeled the International Textile Group (ITG), but with its majority ownership vested in WL Ross & Co. LLC. Combining the operations of the two firms allowed a strategic focus on six discrete textile businesses units. First, *Cone Denim* is the world's largest producer of denim fabric with plants in the US, Mexico, Guatemala, Turkey and India. Second, *Burlington Worldwide Apparel* is a leading producer of materials for upstream producers of menswear, womenswear, uniforms, and other apparel. This business is partnered with firms in Hong Kong and recently announced its intent to build a textile manufacturing complex in Viet Nam. Third, *Burlington House Interior Fabrics* produces interior fabrics under the labels of Burlington House and Cone Jacquards, with sourcing offices in China, Pakistan, Australia, Turkey and Lebanon. Fourth, *Carlisle Finishing* is a dying, finishing, and printing operation serving home decorative markets. Fifth, *Safety Components International, Inc.* produces automotive safety and specialty niche engineering fabrics; headquartered in South Carolina SCI has operations in Africa, China and Europe, as well as North America. Sixth, *Nano-Tex, LLC* is an affiliated company whose nanotechnology-based textile treatments are developed to improve the performance of everyday fabrics, with its headquarters and R&D laboratories in California, while regional offices are in Greensboro, Milan, Istanbul, Hong Kong, and Osaka. Plans are afoot to expand operations of ITG into China and Japan. In August of 2006, the CEO of ITG reiterated the corporate strategy, "(w)e formed the ITG in 2004 to consolidate the businesses of leading textile and fabric manufacturers and to take forward the strategic vision of repositioning the US textile industry by leveraging its marketing and textile know-how on a global basis" (ITG web, 2006). ITG now has about 9,000 employees the world over and is much different by any measure from what used to characterize the Carolina textile industry. Certainly, it exemplifies what is commonly thought of as a globalizing industry.

Meanwhile, Sara Lee is now a food, beverage and household product corporation, which in 2006, divested its branded apparel division, headquartered in Winston-Salem, in 2006 (Sara-Lee web, 2006). Hanesbrand Inc, which emerged as a $2.5 billion indebted company out of the SaraLee spin-off, has moved quickly to adopt the kind of cost cutting currently found acceptable in the New Economy, closing more plants, cutting off medical support benefits for retirees, and reducing the number of yearly sick days. Even so, Hanesbrands, Inc. employs 50,000 workers worldwide and is the leading national producer of T-shirts, fleece, socks, men's underwear, sheer hosiery, and children's underwear (Hanesbrands web, 2006).

Springs Global offers a final example of globalizing textiles. Spring Mills Inc. was founded in Ft. Mill, South Carolina in 1887. By 1966, the firm was operating 19 plants, most located in the Upper Piedmont Manufacturing Region, with 18,000 employees, plus a newly built 21-story office building in midtown Manhattan, from where to attract department store buyers. In 2001, a long term alliance was formed with Brazilian textile giant Coteminas, an alliance that by 2006 led to a joint venture in the creation of Springs Global, the world's largest manufacturer and suppliers of home furnishing textile products. At this point, Springs Global announced the closing of plants idling 1,100 workers in South Carolina, though others continue operating. The variations on the globalizing theme staggers the imagination of the average citizen, but these are examples of textile businesses that continue to pay the workers, fewer though there be, and contribute by their presence to a smaller share of the Carolina economy.

The Tobacco Industry Restructures

A number of cities along Carolina Main Street within North Carolina have benefitted from tobacco production, especially the Triad cities of Winston-Salem, Reidsville, and Greensboro. The continuing saga of tobacco has become another important story of the impacts of restructuring and corporate globalization on a manufacturing industry, its employees, and their communities. It is abstracted here from a recent overview in Gade, Rex and Young (2002, 175-177) and is supplemented by more recent data and findings.

In a matter of just two recent decades, tobacco, which was the state's leading economic tiger, an anchor of stability and high-wage employment firmly tied to particular urban centers, fell from its lofty position. The six main tobacco players, American Tobacco Co., RJR (Reynolds), Philip Morris, Brown and Williamson, Liggett-Myers, and Lorillard, have shifted in their commitment to manufacturing in the state or been absorbed by other interests and, herewith, has come employment trauma and hardship for some communities and a healthy economy for just a few others. American Tobacco Company, the leading cigarette maker in the 1940s and 1950s (especially due to Lucky Strike and Pall Mall), was established in Reidsville, immediately north of Greensboro, in 1911, with the antitrust breakup of the Duke family tobacco monopoly. It brought to its new hometown identity, prestige, and good wages by building here its headquarters and its flagship plant. American Tobacco (AT) was bought out by American Brands, a national collector and seller of manufacturing industries, which then divested the company as an independent operating subsidiary in 1986. At this time, AT announced the closing of its 100 year old facilities in Durham, with the loss of 1,000 jobs, 260 of which were shifted to Reidsville. In 1995, AT was purchased by B. A. T. Industry, a British conglomerate. The result was a dramatic closing of all interests in Reidsville and a relocation of manufacturing to Macon, Georgia! About 1,000 jobs in a town of 12,000 were directly affected, with the rippling effect continuing to the present.

It is not only American Tobacco that abandoned Durham, Liggett-Myers, the manufacturer who broke the dam in tobacco litigation in 1996 by agreeing to help repay Medicaid bills for the treatment of habitual smokers, was essentially dispossessed by the corporate strategies of its parent company, the conglomerate Brooke Group. This left L-M without the necessary upgrading of equipment and facilities, and eventually unable to continue. In Durham, tobacco plants and warehouses have been converted to a transportation center, Brightleaf restaurant/shopping/office complex, and condominium housing units, or razed to become part of a new in-town subdivision, possibly the most successful conversion of traditional manufacturing facilities on a large scale in the Carolinas (Zimmer, 2006).

In recent decades, it has become difficult to ferret out precisely who owns or controls what in the tobacco industry, except it is increasingly clear that decisions affecting the lives of employees and the economic security of their communities are no longer made locally. In 1875, the first tobacco plant was built in what grew to become a sprawling complex of four to five story red brick manufacturing and warehouse buildings which gradually grew to define the urban structure and growth of Winston-Salem. R. J. Reynolds Tobacco Company (RJR) also bought into the prevailing 1970s corporate strategy of product diversification, by joining with Nabisco Foods. Already on the way was the closing down of all of the production facilities and warehouses in downtown Winston-Salem with the 1987 opening of a super modern cigarette factory in Tobaccoville, about 15 miles northeast of town. Also, in 1987, F. Ross Johnson, the chief executive of RJRNabisco, donated the recently constructed ten acre office headquarters building to Wake Forest University, an institution which, through the efforts of the Reynolds family in 1950, had relocated to Winston-Salem from Wake Forest north of Raleigh. Ross had decided to relocate the RJRNabisco headquarters functions and personnel to Atlanta. Two years later, in 1989, RJRNabisco was bought out for $30 billion by a New York investment firm, which promptly relocated RJRs world headquarters to Switzerland. Impacts of economic restructuring and globalization were written large. In 1990, the last of the original downtown plants closed with some workers transferred to the still operating Shorefair or Whitaker plants on the outskirts of Winston-Salem, or to Tobaccoville. Coincidentally, RJR owned or leased no less than twenty tobacco plants from China to Poland. Thus, the late 1980s brought a major shift in tobacco's commitment to Winston-Salem. Among the changes was the initial mid-1980s reduction of 4,000 RJR workers, as a result of new and more efficient plants replacing the old ones. Few white collar jobs, which totaled about 5,300 in the late 1980s, remain and, finally, another 1,500 workers were cut in the mid-1990s. In spite of the traumatic impacts that RJR corporate convulsions have made in Winston-Salem, the company, now divested from Nabisco and its world alliances, remains headquartered in the state's stately first skyscraper, the 1929 20-story Reynolds Building. It has about 7,000 employees working mostly at the Tobaccoville or Whitaker Park plants, and it is the second largest producer of tobacco products in the country (R. J. Reynolds web, 2006). Undoubtedly, some of the negative impacts (e.g., the loss of the North Carolina Dance Theatre), might have occurred naturally with the repositioning of the company as it modernized and decreased the number of workers, but the major impacts of the streamlining and office relocating resulted from a sequence of ownership changes where profit maximization reflected less traditional productivity and market conditions, and more the machinations of distant corporate buy-out strategies and whims of executive decision makers.

Likewise, Philip Morris (PM), headquartered in New York City and the largest American tobacco company controlling close to 50 percent of the domestic cigarette market, became the largest consumer products firm in the world by gradually acquiring Kraft, General Foods, Miller Breweries, Jacob's Sutured, Nabisco (oh, yes, how the world turns), Gavalia Coffee, DiGiorno, and many others. PM has now itself become one of these many operating companies, within a new parent company, the Altria Group (Philip Morris web, 2006). The subsuming of the tobacco business within a larger parent company has apparently allowed PM to avoid the divesting route in the Carolinas. It built the very large and highly efficient cigarette factory in Concord in Cabarrus County, just a few miles north of Charlotte in 1983. Here, on 1,200 rolling acres, it recently completed a $400 million addition allowing an increase in the daily production of cigarettes from 135 billion to 165 billion. In 2004, PM announced another $200 million improvement in site facilities (Philip Morris web, 2006). To place these numbers in a larger context, note that Philip Morris in the late 1990s sold annually 230 billion cigarettes in the U. S. and 660 billion overseas, especially to Asian countries. The economic impact of the plant is considerable. Comparatively high wages for some 2,400 workers who commute to the plant from 16 Piedmont counties in North and South Carolina makes the impact decidedly regional in nature.

In the meantime, Lorillard, the nation's fourth largest tobacco products producer made an unexpected move of its headquarters from New York to Greensboro, where it has its only tobacco plant and employs over 1,500 workers. Thus, in spite of the complexities of change resulting mostly in massive job losses and devastated communities, tobacco is still important in the Carolinas, and Winston-Salem is still the nation's leading tobacco city.

World Furniture Center (and a Fiber Optics Addendum)

The northwestern Piedmont was the manufacturer par excellence in yet another major product, household furniture. Here, it is High Point that became the center, not only being the focus of the largest concentration of furniture plants in the country, but also becoming the nation's leading furniture market. Producers and buyers of furniture have for several decades been arriving in High Point twice a year for an intensive week of trading at the International Home Furnishings Market (IHFM). With the market attracting over 70,000 participants from all over the world, regional airports are crowded and there is no room at hotels within 50 miles of High Point. Local families have been known to take these weeks off for either a traveling vacation or for staying with other family members, while they rent out their own homes for $5,000 plus per week. Only folks clearly associated with the design, manufacturing, marketing or buying of furniture, plus media representatives, are permitted at the market; no general public, please! The IHFM competes strongly with the annual marts in Tokyo, Cologne and a few other cities where the international marketing of furniture hold sway.

In 2006, the market functioned as in prior years, April attendance was estimated at the highest ever with 100,000 participants. One negative note was the qualification that the state subsidy for 2007 was being seriously reduced. In addition, local governments were more closely evaluating economic impacts in determining their future support (Binker, 2006). Also, an unease had been slowly creeping into the show rooms as the knowledge that a very competitive, and perhaps continuously, i.e. open year around, furniture market was being planned for Las Vegas. More ominous was the knowledge that an increasing percentage of the new bedroom and living room suites, occasional pieces, and even cocktail bars were being manufactured in China or elsewhere, but carrying the names of the most prestigious American brands. Consider Broyhill Furniture's "Humphrey Bogart" line, carefully crafted wooden furniture commanding as much as $2,000 for a bowed-out, straight legged bedroom dresser with its "Made in China" label; or that high quality, equally expensive, Bogart liquor cabinet, advertised as Broyhill, but made in the Philippines (Morse, 2004). Bogie would turn in his grave! Of course, the market was used to seeing North Carolina producers competing with high quality Scandinavian and Italian furniture and gradually lower quality (and cost) East European, Brazilian and Asian furniture; but high quality Chinese and Filipino furniture?

The writing was actually on the wall when the massive buy-outs of the leading manufacturers in North Carolina began in the 1980s. There was Ladd Furniture capturing American Drew, Lea Furniture and several other, Masco amassed Henredon, Drexel-Heritage, Lexington, Lineage, Hickory Craft and Frederick-Edward, all family-owned companies whose hundreds of furniture plants were crowded into cities and towns, or distributed along the highways from High Point, through Lexington, Thomasville, Newton, Hickory and Lenoir (Gade, et al., 2002). But it was Furniture Brands International (FBI) that took the initiative on China. FBI, a division of St. Louis-based Interco, had done its own harvesting of North Carolina furniture companies, including Broyhill Furniture Industries in 1980. In 2001, China became a member of the World Trade Organization and FBI began investing in that country's resources of low cost and easily trained labor, while appreciating the relatively weak health and environmental standards, considering that furniture manufacturing in most of its phases is a massively polluting industry, both the interior work place and outside atmosphere (Gade, 1996).

In 1999, Broyhill employed 7,000 in the Lenoir-Hickory region with its headquarters in Lenoir, its home since 1905. The last of its wooden furniture plants closed in November of 2006, and now only two upholstering plants remain. Almost all of the consumer product carrying the name Broyhill is now made in China. And so it is for most of the of the household furniture sold in the United States, the high lines still sold under the traditional brand names. Most striking is the fact hat the exact same lines being produced in North Carolina factories are being produced in the same firm's plants in Asia, same design, same materials, and the Asiatic version is equal or better quality and much lower in price on the American market, even given transport costs (Morse, 2004). As FBIs 2005 annual report stated, "(these daughter companies) are to be true marketing and sourcing operations that will design and import furniture to sell under their (own) well-known names" (Aronoff, 2006). Indeed, from Blowing Rock, North Carolina to Ottawa, Canada, Broyhill Furniture Industries have been building magnificent appearing showhouses for their customers, nary a piece displayed at the Blowing Rock two story showplace carries a North Carolina plant label, and the store is but 45 minutes from Lenoir. Meanwhile, Caldwell County (Lenoir) has been carrying one of the largest unemployment burdens in the state for several years, resulting from the extraordinary furniture employment losses. (Compare Figures 5.5a and 5.5b.)

Thus, it was with great joy and beginning hopes for an improved future that Lenoir learned at Christmas time 2006, Google had decided to build in the town one of its server farms at an expected construction cost of $600 million and with an eventual employment base of 210 (Mitchell and Cox, 2007). Perhaps, this seems quite slight by comparison with the thousands of jobs lost over the past six years, but a Google job average salary is close to $50,000 per year (not including a substantial benefits package), nearly twice the salary of the average furniture job, and their availability will certainly bring in a working population with different lifestyles and expectations than currently is the case. In the Catawba Valley, the Hickory-Lenoir-Morganton area, high school graduation rates and the percent who have completed college is significantly below the state averages, in part because these were not the expectations of families in a region diffused for a century with furniture and textile plants. Even so, it may surprise some that knowledge industries are to be had for even the most blue collar of Carolina counties, and once in place are they not apt to stimulate cognate service and industries to locate? Certainly, this is the expectation of local and state governments who provided an estimated $265 million package to entice the location decision (Cox, 2007). It is an interesting case of industrial location somewhat off Carolina Main Street, which suggests that the knowledge industry may be more fluid in its location requirements than generally considered in the literature. (See Glasmeier and Leichenko, 2000, 561-563.) In this particular case, the main location attraction is cheap electricity for a library of enormous computer memory and turnaround capacity with which to serve an ever increasing search engine demand. Google is in the process of locating these ubiquitously across the land, and the world, in response to the increasing case load of individual centers.

For Lenoir, Caldwell County, the Hickory MSA, and the state of North Carolina, this is an example of the application of Michael Porter's new model of regional growth through a collaborative location process (2003b). This goes beyond the earlier idea of government driving economic development through policy decisions and incentives. There is still a lot of this in the new model, but the complexities have grown considerably. In Google's case, first, in 2006, the General Assembly indirectly agreed to forego taxes for a period of years for industries like Google, by approving a measure eliminating sales taxes on electricity and certain purchases specific to data centers. Second, a bundle of state ordained tax credits tied to job generation; for Google this means a Job Development Grant that could yield $4.7 million should it create 168 jobs and invest $480 million by 2011. Third, additional tax credits are on line should employment reach 210. Fourth, local governments, city and county, is waiving 100 percent of business property taxes and 80 percent of real estate taxes for 30 years. Fifth, Caldwell Community College is gearing up to train people for specified positions with Google.

When considering the economic shifts occurring in the Catawba Valley, one might note the experience of the fiber industry. Hickory and Catawba County had, by 2000, evolved as one of the premier clusters in the country for fiber optics. CommScope, a fiber and copper cable company, was founded in Hickory in 1977 and is still headquartered there (web site: <http://www.commscope.com>). The company built three manufacturing plants in Catawba County and one nearby in Statesville and has recently built or acquired additional manufacturing capacity in Belgium, Australia, Ireland, Brazil, and China. Alcatel-Lucent, at that time with world headquarters in Paris, had its U.S. headquarters in Hickory with a nearby plant in Claremont. Corning Cable Systems also favored the Hickory area, in this case with five plants plus a Hickory headquarters facility. Unemployment in the Hickory metro area in 1999 was at an all time low of 1.9 percent. However, the dot.com bust of 2001-2002 had a dramatic effect on this industrial cluster. Alcatel closed its operations completely; Corning, having just completed a new but never occupied plant in 2001, closed its headquarters offices as well as three plants, and reduced employees in the remaining two. CommScope closed its Hickory headquarters and vastly reduced its manufacturing work force. Hit by the "perfect storm," a combination of major job losses in textiles, furniture, and fiber optics, the three manufacturing cornerstones of this blue collar region, the Hickory MSA unemployment rate soared beyond 10 percent by mid-2002 (Choe and Howard, 2004). However, the telecom equipment industries are once again riding high, but, as noted by Carr, now that the worldwide "backbone" cable has been largely completed with enormous capacity and a long life span, optic cable manufacturers that remain are focusing on "the last mile" and "cable to the premises" to homes and businesses with wiring cabinets, connecting details, etc. (2007).

A Boom in Automotive Assembly and Parts Manufacturing

The largest individual manufacturing sector in the United States is motor vehicles, body, trailer and parts. This comprised about 8.7 percent of the total U.S. manufacturing sector in 2003. Traditionally tied to the Midwest, the industry is gradually relocating to the Southeast because of an array of more positive locational attractions, in essence the same factors that seem to have ruled southern location strategies for a century. Recently, these factors were discussed in a Southern Legislative Conference Report, *The Drive to Move South*, published in November, 2003. As reported by Haag (2005, 12), the several reasons for the southern gravitational pull include, worker efficiency and productivity, readily available sites prepared by local government and the state road department, favorable work environment, that includes relatively low wages ($10 per hour less than those in Ohio/Michigan), Right to Work laws, highly attractive business incentives including tax relief and training assistance, proximity to an efficient and comprehensive multi-modal transportation system, and close-by availability of colleges and technical institutes. Not really much is different from southern attractions of 1900-1920.

A recent flurry of reports from the Departments of Commerce of the two states has emphasized a developing major industrial cluster. The South Carolina report refers to this as the New Southeast Automotive Cluster; though confined in this report to South Carolina, the notion is readily expandable to include all of Carolina Main Street, and some places beyond (Opportunities for Auto..., 2006). It estimates that in South Carolina's transportation equipment sector some 13,000 new jobs have been created since 2000. The star witness is the continued growth of the only BMW assembly plant in the country with 4,500 employees, located in Spartanburg County near Greer, equidistant between Greenville and Spartanburg. In addition to the exclusive world production here of BMWs Z3 Roadster and Sports Utility Vehicles, with a new version of the X5 SUV initiated fall of 2006, will soon be added a crossover vehicle and a small SUV, the X3. The United States is by far BMWs largest overseas market.

Nearby is the Clemson University International Center for Automotive Research, a public/private endeavor focusing on research and development and supported by BMW, Michelin, IBM, Microsoft, and others. Michelin's North American Headquarters is in Greenville (S.C.) where also is located its major radial tire factory with 1,468 workers. With its six more plants in South Carolina, Michelin had 7,628 workers in the state in 2006, including administration and sales. The automotive cluster, much energized by the BMW location, had an additional 91 automotive parts and supply facilities with 26,101 workers along the South Carolina portion of Carolina Main Street. Indeed, South Carolina has acquired a major contribution to its economy through its Automotive cluster, with Honda ATV production (250 employees) located in Florence, and American LeFrance, which is relocating its emergency and fire vehicles manufacturing as well as its headquarters to Berkeley County near Charleston. The latter city, incidentally, provides an important cornerstone in this cluster. Here is the South Carolina Port Authority's nationally fourth largest container port, and its Charleston Port is critical in the delivery of raw materials for the automotive cluster, as well as shipment of new BMWs to worldwide recipients. Michael Porter in his comprehensive assessment of South Carolina's Competitiveness Initiative (2003a) identifies the Automotive cluster as having 30,922 employees, and indicates that it functions as the near ideal of an industrial cluster (elaborated in Porter, 2000, 254). In his assessment, Porter emphasizes the critical importance of having in South Carolina the Savannah River Site, plus the research facilities at the University of South Carolina in Columbia, each of these capable of contributing unique fuel-cell development opportunities, as well as the comprehensive, and throughout the state easily accessible, technical college system.

When considering the portion of the Southeastern Automotive Cluster (SAC) extending into North Carolina, its size becomes truly massive. It is broadened in its manufacturing foci to heavy and light trucks, bus and heavy transport equipment, as well as to sports car design and manufacturing. Porter indicated that North Carolina had 36,663 workers in the automotive industry in 2001, while James Haag, a senior analyst with the N.C. Department of Commerce estimates that no less than 141,000 employees in 2005 derive their income from the automotive cluster, though not all are involved directly in making vehicles (Haag, 2006). The cluster is referred to as motor vehicle and heavy equipment manufacturing, with over two thirds of its capacity located within Carolina Main Street. For the reader, now introduced to the idea of cluster manufacturing, it might be instructive to show what is meant in terms of inclusive categories of manufacturing industry directly related to the cluster as opposed to those more indirectly involved (Table 5.4). Note that the primary sites include only a little over one-third of the total assigned to the cluster for accounting purposes. There are significant overlaps in the way the contribution is being assessed. Obviously, employees are tallied on several overlapping activity arenas. Regardless, the table includes some 1,000 plus North Carolina manufacturing entities. Of these, 434 companies are located in the Charlotte region, 212 are in the Piedmont Triad region, with an additional 118 concentrated in the Research Triangle area and 110 in the Asheville Basin.

The North Carolina automotive cluster ranks tenth in the nation, and second in the South, following Tennessee. In comparison to the earlier distribution of apparel and textile plants, automotive related firms form a much tighter network within the Carolina Main Street. Notable is

Table 5.4. North Carolina Motor Vehicle and Heavy Equipment Cluster by Industry, 2005

A. Motor vehicles and heavy equipment manufacturing	# Industries	# Workers	% Workers
MV and passenger car bodies (SIC 3711)	47	6,177	4.4
Truck and bus bodies (SIC 3713)	46	9,017	6.4
MV parts and accessories (SIC 3417)	206	27,564	19.5
Heavy equipment (SIC 3523, 3531, 3537)	24	5,772	4.1
Other motor vehicles (SIC 3715-16, 3751, 3792, 3799)	35	536	0.4
Subtotal	358	49,066	34.7
B. Related Manufacturing	# Industries	# Workers	% Workers
Indust. & commercial machinery and Comp. equip. (SIC 3500)	156	14,864	10.5
Primary and fabricated metals (SIC 3300-3400)	128	13,404	9.5
Rubber and micellaneous plastics (SIC 3000)	91	19,124	13.5
Textiles, fabrics and leather (SIC 2200, 2300, 3100)	75	11,553	8.2
Electronic Equipment. except computers (SIC 3600)	79	20,964	14.8
Chemicals and petrolium (Sic 2800, 2900)	59	5,548	3.9
Measuring, analyzing, controlling instruments, etc. (SIC 3800)	20	1,717	1.2
Other (SIC 1742, 2500, 2600, 3200, 3728, 3769, 3900)	28	5,074	3.6
Subtotal	644	92,248	65.3
Total	***1,002***	***141,314***	***100.0%***

Source: NCDOT, Division of Policy, Research and Strategic Planning, August 15, 2005

Daimler-Chrysler's Freightliner heavy-truck division with its plants in Cleveland, Mount Holly and Gastonia. Their work force of over 7,100 is soon to be reduced by an anticipated 1,960 because of overproduction and additional costs in the manufacturing process caused by tightened emission requirements (A. Bell, 2007). Other large firms within this cluster include Thomas Built Buses (also part of Daimler-Chrysler) in High Point; Consolidated Diesel (an engine manufacturer) and Douglas Battery in Winston-Salem, and Marconi Commerce Systems (gas station pumps and equipment) and Volvo Trucks North American headquarters, both in Greensboro. Each of these has between 1,000 and 2,000 employees. All of these larger cluster members are international firms or a part of same. Marconi, with its headquarters in Greensboro, has plant operations in Germany, Italy, the United Kingdom, Argentina, Brazil, Argentina, and China. Success again builds on these companies' global integration, markets, and profits.

Moonshiners being chased by "revenooers" is the part-mythology and part-factual initiation of the stock car racing industry. From humble beginnings on dirt tracks throughout the South, though concentrated in the Appalachian hill and adjacent Piedmont country, this sport has grown into a significant cluster of race car design, production and entertainment, with an estimated 70 to 80 percent of the racing teams located within a hundred miles of Charlotte (Alderman et al., 2003). Whereas rural dirt tracks and the paved surfaces of larger tracks used to dominate the public racing aspects of stock car racing, it, like most other twenty-first century human activities, is increasingly consolidating and concentration in the center of larger metros. Eight major stock car tracks have been discontinued from Winston Cup racing over the past decade or so, while larger tracks are becoming popular in metros across the country. Only three remain in the Carolinas. They are Rockingham, Darlington, and Charlotte, with the latter having achieved capacity and being renown to the point that the recent competition for the locating of the new NASCAR museum and activity center was won by Charlotte. Aiding this location effort was the earlier built NASCAR Technical Institute at Mooresville, where the training of automotive technicians and designers contributed a valuable knowledge industry component. This rather major emphasis on auto racing, in all of its manifestations, denotes its character as a major anchor in the Carolinas automotive clusters.

Military Cities

The Carolinas have benefited enormously from the buildup of military facilities in the 1930s and 1940s. This included a major naval base and shipyards in Charleston; Parris Island Marine Corps Training Camp at Beaufort, S.C.; a huge Fleet Marine Force Base at Camp Lejeune (Jacksonville) with its nearby ancillary Marine Air Stations at New River and Cherry Point; a major Army Training Center at Fort Jackson near Columbia with its supporting Air Force base at Sumter; the even larger Army base of Ft. Bragg in Fayetteville, the latter supported by Pope (Fayetteville) and Seymour Johnson Air Force (Goldsboro) bases; the Coast Guard Air Station at Elizabeth City N.C., and the Sunny Point Military Ocean Terminal at Southport, N.C. These were major investments by the federal government that were primarily focused on empty spaces in the Carolinas. The net effect was to strengthen the cities and communities away from Carolina Main Street. Military personnel, civilian workers at military installations, and retired military people, plus actual military procurements in the localities, are lifelines of support for many outlier cities. A persistently effective group of Carolina legislators to the U.S. Congress has been able to stem the tide of military downsizing for the Carolinas, though Charleston lost much of its navy installations and related jobs in the mid-1990s. The Charleston Navy Yard at the time of its announced downsizing had a total military, civilian, and contractual employment of 35,654 persons. In 1996, left behind by the Navy was 176 contaminated sites, following unregulated decades of spoil banks with accumulating mercury, acids, asbestos, pesticides, arsenic, PBCs, and much else (Bower, 1996, 10).

The Service Industry Explosion

By any measure, the revolution in economic activities the country over, as is certainly the case for the Carolinas, has been the growth of employment in the service sectors. This includes the growth of government in its various sectors, but dominated by hospitals and medical clinics, and by education, both K-12 and post high school institution, as well as the explosion in the food, hospitality and leisure industries. This certainly also includes that very special industry that totally and unpredictably (some 30 years ago) is capturing the employment base of the Charlotte Metropolitan region and certain others, the financial sector. Striking evidence of the importance of the service industry to the Carolinas is shown in a listing of the largest private firms: by their number of employees. With comparable information not available for South Carolina, Table 5.5 illustrates the North Carolina listing for October, 2005.

Retail sales firms are especially dominant, with Wal-Mart, Food Lion, Lowe's, and others dominating the top individual employers. Medical research and services also ranks quite high as shown by Duke University's medical complex and by several regional hospitals. Financial institutions, as well, rank high with Wachovia, Bank of America, and BB&T in the top 15. Transportation of goods and people enters the top ranks as well, with U.S. Parcel Service and U.S. Airways. Large scale manufacturers no longer rule the roost. Only IBM and Sara Lee appear on a list that fifteen years earlier included manufacturers as comprising over one-third of the total employment (Gade, et al., 2002). Should this list be extended then business and professional service companies, especially employment service, leisure and hospitality firms, many more retailers, and a more generous helping of manufacturers, will be included. For regional development, the critical aspect of these concentrations is that they, in most cases, contribute to the swelling of the metro and urban concentrations.

The Example of Banking and Financial Services

One of the most extraordinary economic developments over recent decades has been the emergence of the financial stronghold of Caro-

Table 5.5. North Carolina's Largest Private Employer in Order of Employment Size

Rank	Company	Industry
1	Wal-Mart Associates Inc.	Trade, Transportation, and Utilities
2	Duke University	Education and Health Services
3.	Food Lion, Inc.	Trade, Transportation, and Utilities
4.	Wachovia Bank	Financial Activities
5.	International Business Machines	Manufacturing
6.	Lowe's Home Centers, Inc.	Trade, Transportation, and Utilities
7.	Bank of America NA	Financial Activities
8.	Harris Teeter, Inc.	Trade, Transportation, and Utilities
9.	Sara Lee Corporation	Manufacturing
10.	Branch Banking and Trust	Financial Activities
11.	United Parcel Service Inc.	Trade, Transportation, and Utilities
12.	U. S. Airways, Inc.	Trade, Transportation, and Utilities
13.	North Carolina Baptist Hospital	Education and Health Services
14.	Lowe's Food Stores, Inc.	Trade, Transportation, and Utilities
15.	Moses H. Cone Mem. Hospital	Education and Health Services

Source: North Carolina Department of Commerce, 2006.

lina Main Street. Since national and regional banking laws began loosening their stranglehold on out-of-state bank acquisitions, North Carolina banks have been busy foraging the countryside and buying out financial institutions as if they were on a fire sale where they were the only buyers. State laws have for some time been permissive for statewide branch banking. This led to the emergence of a few highly competitive banking powerhouses, NationsBank and First Union headquartered in Charlotte, and Wachovia, initially headquartered in Winston-Salem. When North Carolina then permitted interstate banking in 1981, none of these banks had accumulated as much as $8 billion in assets; the largest, NationsBank, ranked 26th in the country. Lord (1987) noted the conditions of change, "(t)he geography of interstate banking acquisitions and the relocation of economic control points (corporate headquarters) have been influenced by the patterns, form, and timing of interstate legislation, by market and financial institution characteristics and by bank management attitudes. The magnitude of the interstate and interurban relocation of corporate control is best measured in terms of the volume of assets exchanged," (p. 11). By 1986, North Carolina banks were leading the country in new acquisitions, but the concentrations of holdings were still overwhelmingly in New York City, with assets of $596 billion, while San Francisco had $160 billion, and Los Angeles $112 billion. But among the 15 largest metropolitan areas, Charlotte had in a matter of two years moved from 14th to 9th place, while Atlanta had progressed to 12th place.

By the late 1990s, the fight was almost over. Metrolina, Queen City, newly minted "Wall Street" of the South (Newman, 1995), far exceeded Atlanta as a center of finance and banking. Branches of the two banks, NationsBank and First Union covered the country like kudzu from New Jersey and Pennsylvania south to Florida and west to Texas and Oklahoma. In 1997, Charlotte had emerged as the second leading financial center, as measured by bank assets, in the country. In 1998, NationsBank merged with BankAmerica in a $60 billion deal, to become the nation's largest bank, and amazing to the unsuspecting, with its headquarters in Charlotte. Shortly thereafter, First Union and Wachovia merged and with headquarters in Charlotte under the Wachovia name. Subsequently, there was a flurry of acquisitions of financial institutions and mergers across the country from Florida, to Boston, to Los Angeles, and elsewhere, with the Charlotte banks as senior partners. At the end of 2006, their position is unchallenged as two of the top five banks in the country, depending on just how such leadership is defined (Table 5.6). Critical to understanding why Charlotte became a banking center is not the fact that it was the focus of the nation's first gold rush, and early in its history became a regional U.S. Mint. Rather, it is the influence of entrepreneurial bankers who have determinedly carried their competitive world view of the financial industry to the point to where their respective banks, Bank of America and Wachovia Bank, are industry leaders. Graves (2001) additionally suggests that the regulatory advantage provided North Carolina

Table 5.6. Largest Banks in the United States, December 2006

Bank	Headquarters	Assets (in $billions)	U.S. Deposits (in $billions)
Citigroup	New York	1,700.0	224.6
Bank of America	Charlotte	1,500.0	590.6
JPMorgan Chase	New York	1,300.0	462.3
Wachovia	Charlotte	688.7	370.0
Wells Fargo	San Francisco	483.0	309.0

Source: R. Rothacker, "Big Banks, Big Cities, Big Fight," Charlotte Observer, December 27, 2006.

banks is a critical aspect of the applied nature of cumulative causation theory, which also aids in understanding, why Charlotte? It needs to be noted that having such an overwhelming concentration of financial interests in Charlotte has been a gift to the city itself through the urban redevelopment efforts in the city's uptown four wards (Smith and Graves, 2003).

Community impacts of bank mergers are decidedly mixed. Whereas the Carolinas can smell success in the many jobs provided and in gleaming new skylines, especially in Charlotte and Winston-Salem, states where banks are bought out are apt to lose considerably in closed branches and in consolidated or relocated operations. For example, the estimates of jobs lost with the purchase of the Barnett Bank of Jacksonville in 1997 was 6,000, of CoreStates Financial Corporation of Philadelphia was 4,400, of the NationsBank-BankAmerica merger, 8,000 positions. Charlotte banks are not much beloved for their aggressive behavior outside the Carolinas. But the other side of this coin is the massive increase in jobs in the Charlotte metro area, and other places along Carolina Main Street. Wachovia has built a customer service center in University Park north of Charlotte staffed currently by 7,000 employees. Additionally, this bank has 6,000 or so employed uptown. Bank of America has in excess of 10,000 workers in uptown Charlotte and has opened a service center near High Point with another 3,000 workers. And these figures do not reflect the rather massive increase in professional and business services enticed by the explosion in the finance industry.

Mid-sized banks may not in the end survive the assertiveness of their much larger cousins. Even so, they have also been busy consolidating and expanding. While Wachovia, Bank of America, and Branch Bank and Trust have focused on out-of-state acquisitions, the smaller banks tend the local turf. These consolidations have harmed the smaller communities and helped, once again, Carolina Main Street. In 1995, BB&T Financial Corporation of Wilson and Southern National of Lumberton combined with 8,000 employees and $18.8 billion in assets. They needed a more imposing location for their business and moved headquarters to Winston-Salem. Subsequently, BB&T picked up United Banc Shares of Whiteville and is now the lead bank in Carolina deposits, though barely surpassing Wachovia. In 2007, Royal Bank of Canada/Centura is building a $100 million edifice in downtown Raleigh and will relocate their Centura headquarters from Rocky Mount. Certainly, these changes, symptomatic of regionally consolidating economies within the globalization umbrella, favors Carolina Main Street at the direct expense of outlying cities and rural communities where banking operations and jobs lose out. To be sure, the previous large losses suffered by many smaller towns through the closing of manufacturing plants have contributed as well to the vulnerability of local banks.

Yet, there are still a large number of smaller banks serving largely local needs in the Carolinas (Table 5.7). And a growing number of much smaller community startup banks dot the Carolina landscape. The shear scale of the mammoth banks is intimidating, or thought of as impersonal, by many people (Werner, 2006). Thus, in towns as small as Boone, North Carolina, with 16,000 people and branch offices of the top five banks in the state, there has been the opening of three community banks over the past decade. From Table 5.7, it appears that some of the larger cities in the Carolinas are so diffused with regional offices or branches of the larger banks that these cities are no longer capable of supporting a significant sized home grown institution. Note the absence from this list of larger banks in cities like Charleston, Myrtle Beach, Wilmington, Asheville, Spartanburg, and others.

The Knowledge Industry and the Research Triangle

Carolina Main Street's easternmost metro areas were built initially by tobacco, universities and public administration. They all three came together in what became a developmental driving force not only in the Piedmont Triangle of Raleigh, Durham and Chapel Hill, but also for the state as a whole. Innovative entrepreneurs, university leaders, and public officials divined

Table 5.7. Leading Carolina Financial Institutions by Consolidated Assets, 2005

National Rank	Bank	Assets (in $million)
1	Bank of America, Charlotte	1,160,260
4	Wachovia Bank, Charlotte	504,270
14	Branch Bank & TC, Winston-Salem	85,215
51	Royal Bank of Canada/Centura Bank, Rocky Mount	20,772
62	First Citizens Bank and Trust Co., Raleigh	13,362
89	Carolina First Bank, Greenville, SC	8,970
100	Branch Bank and Trust Co., Greenville SC	7,549
124	First Citizen Bank, Columbia	5,431
145	First Charter Bank, Charlotte	4,316
188	National Bank of South Carolina, Sumter	3,930
280	South Carolina Bank and Trust, Orangeburg	1,828
353	Capital Bank, Raleigh	1,364
363	Fidelity Bank, Fuqua-Varina	1,332
406	Bank of Granite Falls	1,147
416	Palmetto Bank, Laurens	1,116
428	Yadkin Valley Bank and Trust, Elkin	1,080
434	Southern Bank, Mount Olive	1,069
438	First National Bank and Trust Co., Asheboro	1,063
439	Gateway Bannk and Trust, Elizabeth City	1,061
446	United Community Bank, Murphy	1,047
448	First National Bank, Southeast, Reidsville	1,035
463	Lexington State Bank, Lexington NC	986

Source: Federal Reserve Bank, Statistical Release, June 30, 2006.

the Research Triangle Park (RTP) in the late 1950s, mostly to create economic opportunities in an economy so dominated by low wage manufacturing industry. RTP gradually grew to cover some 7,000 acres in agricultural and forested lands between the centers encasing the tri-universities of North Carolina State University at Raleigh, the University of North Carolina at Chapel Hill, and Duke University in Durham (Havlick and Kirsch, 2004). Making a coup in capturing a major processing plant of IBM computers in the late 1960s, the RTP was able to gradually fill the space allocated to it and even expand this considerably in the 1990s. By 2000, the Park claimed a total of 41,297 employees with the following approximate distribution: microelectronics and software development—36.5 percent; telecommunications and internetworking—28.8 percent; pharmaceuticals and healthcare—11.8 percent; environmental science—9.5 percent; biotechnology and biopharmaceutical—4.7 percent; other research and development—7.2 percent; and chemicals—1.5 percent. High-tech and knowledge industries were arriving in the Carolinas in large measure through the portal provided by the Triangle Research Park. In parenthesis, such has also been the case through the opportunities provided for the chemicals, plastics, nuclear, and machine metals industries by federal government investments in South Carolina, such as the Savannah River Site (with 12,000 employees) and the Naval Research Facility in Charleston.

A lesson was to be learned about this mix of industries in the face of economic reversal on a national scale. The dot.com bust of 2001-2002, combined with a further increase in outsourcing, dealt the computer manufacturing and the telecommunications industry a severe blow, from which it has not recovered. IBM and others reduced their work force at RTP by several thousand, and Nortel, Lucent, and Cisco cut back severely, as well. In 2005, RTP had about 37,000 employees, with pharmaceuticals, biotechnology, nanotechnology-electronics, and environmental sciences the expanding industries. But this is hardly the whole story. The RTP has had an extraordinary impact on its surroundings, with ancillary industries, especially supporting business and professional services, in leisure and hospitality, businesses in retail strips and shopping malls, but most particularly in construction and housing. RTP literally built Cary, the fastest growing city for a decade, in North Carolina.

The Twenty-first Century: A Global Game

As indicated in large measure on previous pages, the Carolinas are deeply intertwined with the processes of economic globalization. From the wholesale restructuring of the textile industry, the heavy investments of foreign pharmaceuticals (such as the Swiss Novartis/Sandoz, the Danish Novo-Nordisk, the British Glaxo/Smith/Kline, the German Bayer, or the Japanese Eisai), to the textile giants outsourcing to Central/South America and Asia, or to German automakers, the Carolinas continue to be firmly committed to playing the global game of industrial recruitment, outsourcing and foreign investment. From 1986 to 2004, foreign company generated jobs in the Carolinas has increased from 154,100 to 319,700. In 2004, the Carolinas consistently ranked among the top ten states in the country in direct foreign investment over the past decade. On the other hand, 12,819 Carolina companies were involved in some degree of exporting raw materials or finished products overseas in 2004 (North Carolina, 2006; South Carolina, 2006). Though in a very different context, a century later the Carolinas remain wedded to an export economy!

Given what can only be termed traumatic shifts in Carolinas economic activities over the past decade, what then appear to be the major outcomes? Initially, it is important to admit that what is happening here is in a significant degree a mirror of national socioeconomic change brought about by the evolution of a globalizing world affecting all aspects of our life. But having said that, it is important to note what appear to be more specialized changes that even as they might have been focused in the South were centered in the Carolinas. From the perspective of urban development in the context of the early years of the twenty-first century, one should be aware of a few critical indicators, regional unemployment-related commuting patterns and population change. Employment change that is strongly locational in nature is apt to result in critical geographic differences in job opportunities. Thus, job seekers, once they have exhausted their unemployment insurance and/or found a new position, may choose to either commute to work elsewhere (note commuting patterns displayed in Figures 5.9 and 5.10) or to relocate their residence closer to a newfound job.

For the Carolina metropolitan statistical areas, Table 5.8 presents data that shows the size of each of the 22 MSAs. It also provides data on per cent unemployed, for November and December in 2005 and 2006. Considerable variation exits in the unemployment data. Note first that these are years where the Carolinas were two of the leading states in the country in net in-migration, but also, as shown in Figure 5.3, when major losses had recently been experienced in manufacturing industries. Table 5.8 suggests that Carolina Main Street metros are in a much better situation than is true for others, even accounting for South Carolina's near or slightly higher (two percentage points) unemployment rate for the state. It also shows that the recreation-leisure regions along the coast and in the mountains are faring much better within their respective states.

Table 5.8. Civilian Labor Force and Unemployment, by MSA, 2005 and 2006

State and Area	Civilian Labor Force (1,000)				% Labor Force Unemployed			
	2005	2006	2005	2006	2005	2006	2005	2006
North Carolina	**4,378.5**	**4,504.0**	**4,353.4**	**4,488.8**	**5.0**	**4.9**	**4.7**	**4.7**
Asheville	202.4	210.7	201.5	209.7	4.0	3.6	3.7	3.4
Burlington	70.0	72.2	69.5	71.7	5.8	5.3	5.4	5.0
Charlotte-Gastonia-Concord	808.7	828.9	805.6	826.3	5.0	4.8	4.7	4.6
Durham	249.0	257.4	248.8	257.3	4.1	4.0	3.8	3.8
Fayetteville	150.3	153.2	149.3	152.0	5.7	6.1	5.2	5.7
Goldsboro	52.0	51.8	51.6	51.9	5.4	5.3	4.9	4.8
Greensboro-High Point	363.5	367.5	362.0	366.6	5.1	5.1	4.8	4.9
Greenville	83.5	87.2	82.8	87.0	5.4	5.3	4.9	5.0
Hickory-Lenoir-Morganton	176.5	178.2	175.5	177.3	6.3	6.3	5.9	6.1
Jacksonville	57.6	61.0	57.4	60.7	5.3	4.9	4.8	4.5
Raleigh-Cary	514.0	539.8	512.2	537.9	3.8	3.6	3.5	3.4
Rocky Mount	68.7	69.6	68.0	69.4	6.7	6.5	6.2	6.3
Wilmington	166.2	171.6	165.1	170.5	3.9	4.0	3.8	3.9
Winston-Salem	235.1	242.1	234.2	241.7	4.5	4.4	4.2	4.2
South Carolina	**2,094.6**	**2,136.7**	**2,087.1**	**2,137.8**	**6.9**	**6.4**	**6.7**	**6.4**
Anderson	82.7	82.3	82.3	83.5	7.5	7.3	7.3	6.7
Charleston-North Charleston	295.5	303.7	293.9	301.1	5.5	5.2	5.1	4.9
Columbia	358.4	368.9	359.3	371.7	5.8	5.5	5.5	5.3
Florence	93.8	96.6	93.8	95.5	8.9	7.5	8.8	7.7
Greenville	298.4	303.5	297.7	304.8	6.0	5.6	5.8	5.3
Myrtle Beach-Conway-N. Myrtle B	121.6	126.7	120.7	125.8	5.6	4.8	6.1	5.2
Spartanburg	129.8	131.3	129.4	132.0	7.6	6.9	8.0	6.8
Sumter	47.1	48.3	47.2	48.7	8.8	8.1	8.6	7.8

Source: U.S. Department of Labor, Bureau of Labor Statistics, Washington, D.C., 2006.
Web page: http://www.bls.gov/news.release/metro.t01.htm

Information on unemployment at the county level (Figure 5.11, in this case for 2005) further enhances these findings. The difference between the two states appears more striking on the county scale. Thus, 42 percent of North Carolina counties were experiencing an annual average of less than 5.5 percent, as compared to 11 percent of counties in South Carolina. In fact, six counties in South Carolina had over ten percent unemployment. That aside, it is clear that the urban counties are favored with generally much lower unemployment rates, again with the exception of coastal counties in both states and mountain counties in North Carolina. Predominantly rural counties in the urban shadow of Carolina Main Street appear similarly favored. Rural counties in the Coastal Plain and the Sandhills region of the Carolinas are still smarting from an absence of job replacements, as are traditional manufacturing counties in the shadow of the Piedmont metros, if not within their perimeter. Take note, for example, of the case of Greenwood County, a predominantly

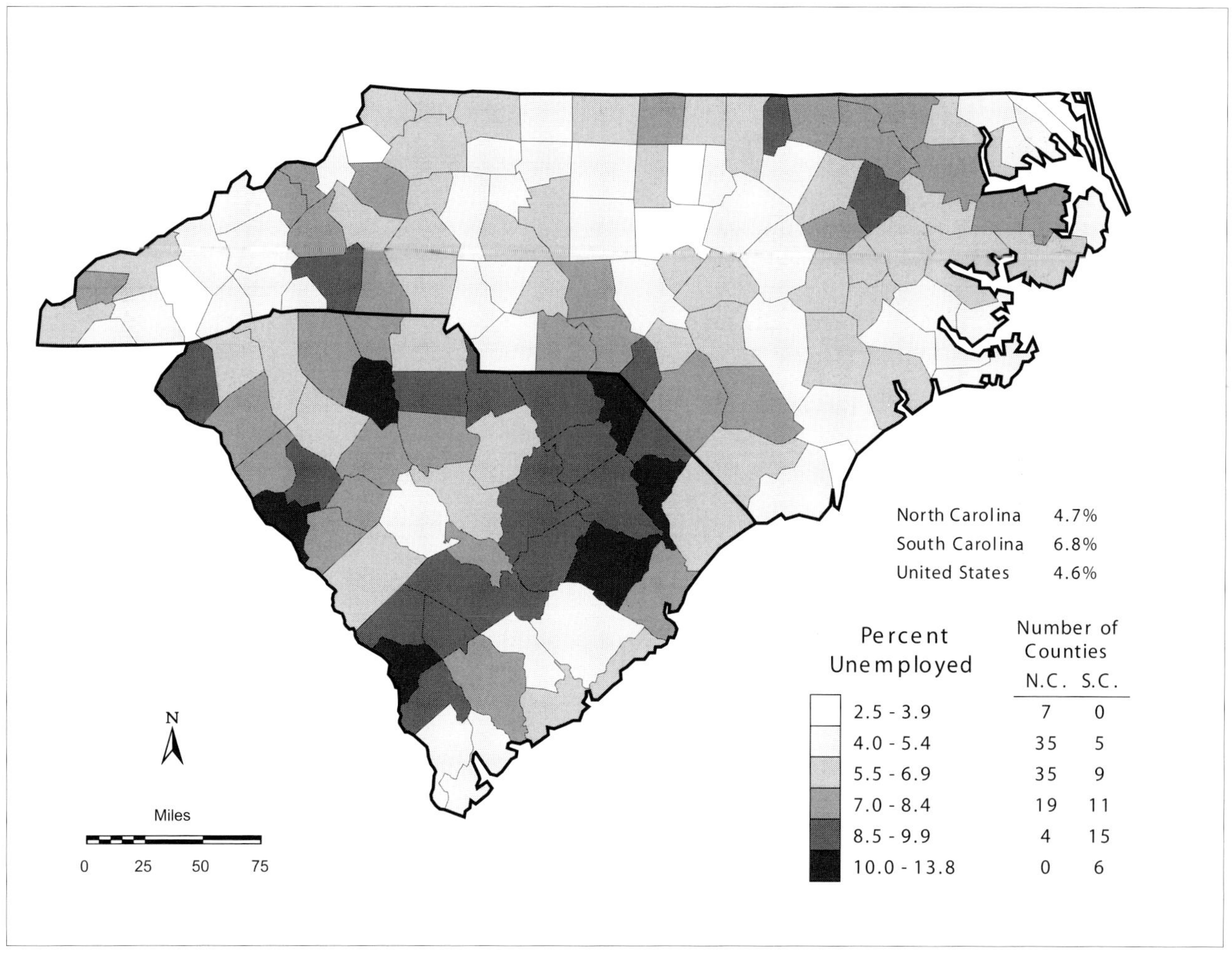

Figure 5.11. Unemployment by County, 2005 Annual Average.
Sources: North Carolina Department of Labor; South Carolina Statistical Abstract.

rural county to the immediate south of the Greenville (S.C.) MSA. Here, as noted by Lord (2001), with the loss of its lead textile industry the county attempted comprehensively to "harness the local to capture the global." With its 2005 unemployment hovering around nine per cent, it appears that this effort has largely failed thus far. For the smaller communities, it is increasingly difficult to avoid the agony of simply being left behind.

So, all in all, it is no surprise that evidence of variable circumstances can be interpreted through the counties' population percentage change. Compare the percentage change by county for the 2000-2005 period (Figure 5.12). Again, there are differences between the two states, with South Carolina having a higher proportion of counties losing people and a much lower proportion of counties gaining over ten percent for the five year period. However, urban areas and their nearby suburbanizing counties throughout the Carolinas experienced the greater population percentage increases, which are compounded by their relative population weight. The most substantial population growth occurred on the coast, from Currituck in the north to Beaufort in the south. The South Carolina Department of Parks, Recreation and Tourism maintains that tourism is South Carolinas leading industry and notes that Horry (Myrtle Beach), Charleston, and Beaufort (Hilton Head) counties, in that order, are by far the leading domestic travel receipt counties in the state. Horry garnered more than 31 percent of the total of $8.5 billion in 2005 (Travel . . ., 2006). The despair of most rural Carolina counties is deepening with their increasing distance in diversity of job opportunities from those available in the metro regions, especially along Carolina Main Street and the coastal zone, though in the latter case employment opportunities are less nuanced with a much lower average wage.

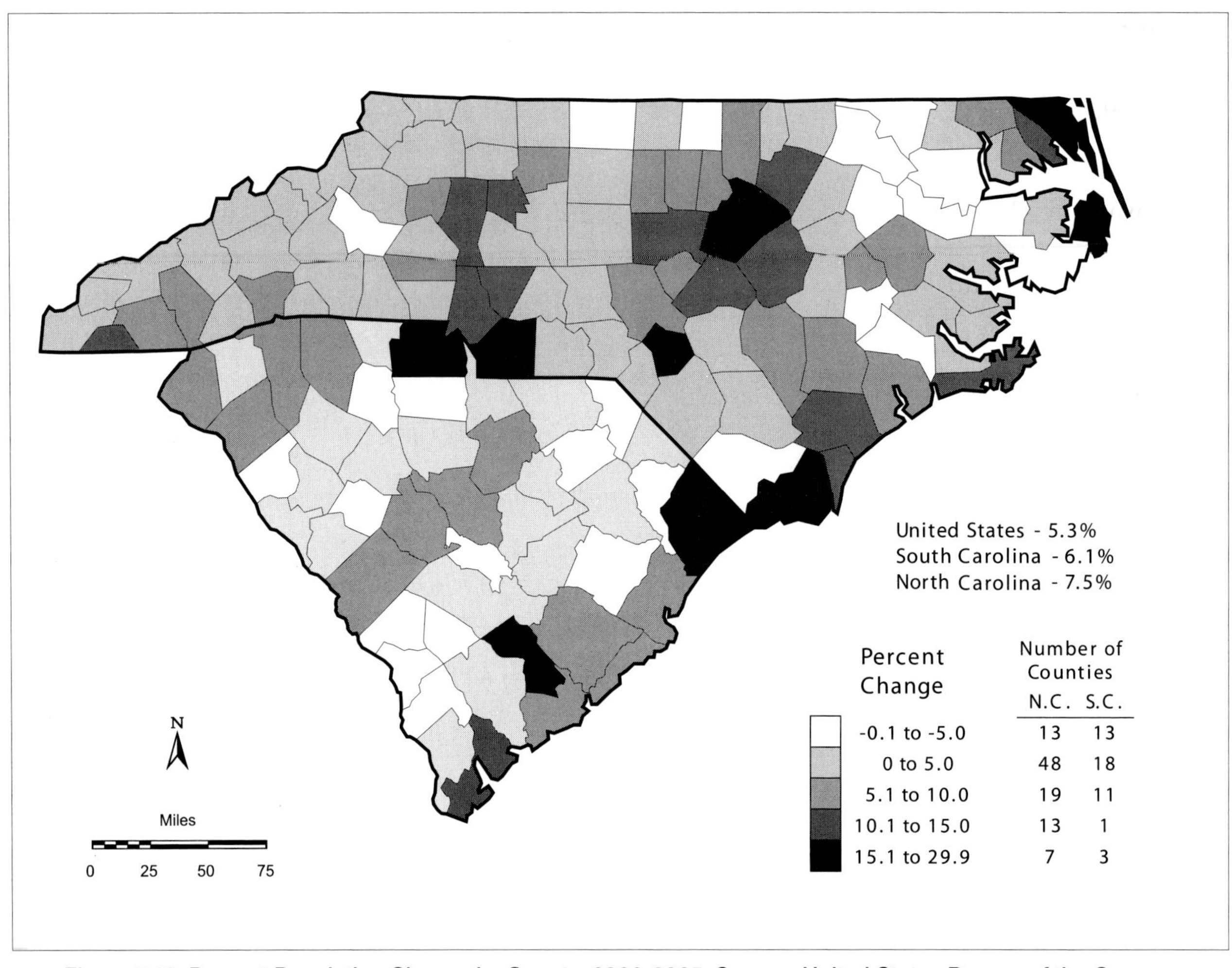

Figure 5.12. Percent Population Change by County, 2000-2005. Source: United States Bureau of the Census.

Summary And Conclusions

Economic and spatial convergence derives from market forces increasingly left to their own devices, if not in fact further so encouraged by state and local governments. Thus, unfettered economies of scale and agglomeration are apt to lead to an increasing regional concentration of capital, labor and output. The gradual evolution of Carolina Main Street is a reasonable example here. It appears currently eminently self-sustaining within a capitalist space economy, which is given relatively free reins in the Carolinas. It is revealing when Google complains about North Carolina state legislators and newspapers not keeping quiet about its yearlong negotiations to locate a server farm in a state where the public's business is legally required to be an open slate. Petty dickering over tax breaks in the deal also was criticized (Cox, 2007). In the end, Google got what most of what it wanted, an anticipated prize of more than $1,250,000 per job. But do consider that by far the largest share of the incentives derive from the relief of taxes that will not be assessed in states also being looked at by Google.

It is equally revealing that while the world is reeling from increasing evidence of human caused global warming, Duke Energy is likely to be given the go ahead to build two new 800-megawatt power plants west of Charlotte, a city already periodically suffering some of the highest levels of air pollution in the country. And it is certainly revealing when very few of the region's city councils decline the option of imposing adequate construction impact fees to pay for additionally needed infrastructure, in the face of irate builders and real estate businesses. In fact, the absence of regional planning and general land use controls is permitting one of the most extraordinary examples of urban sprawl in the country. On a larger scale, this determined capitalistic growth syndrome is revealed in the absence of interstate consultation of present and future water needs, as when the N.C. State Environmental Board approved a controversial water transfer from the Catawba River Basin to the cities of Kannapolis and Concord in support of their explosive growth. Downstream, South Carolina, with communities still smarting from a historic drought from 1998 to 2002, is suing in federal court, since no effort was made to seek their advice and consent (Henderson, 2007). Clearly, broad-based regional and interstate planning authorities are needed in the Carolinas, though most especially in the Carolina Main Street region. In the end, local governments are less and less in control over the destiny of its land and people. Increasing out-of-region ownership of businesses, plants and property, formerly in local ownership, is moving decisions offshore to individuals and corporations having no stake in the welfare of localities.

Is there hope for the evolution of a broader based socioeconomic frame? One within which the Carolinas may equally accommodate all of its residents in the changes that accompany economic globalization? Or will there be the increasing separation of urban-rural opportunities, a widening gap between those who have and those who do not, the persistence of "urban balkanization" in spite of the reemergence of a middle income populated center city, and a hardening of the locational advantages of Carolina Main Street?

Acknowledgments

The Author wishes to express his deepfelt gratitude to Jim Young, Appalachian State University, Boone, North Carolina for the cartographic representation of the figures in this chapter. In addition, Appalachian State University is appreciated for its continuing support of its emeritus faculty.

References

Alderman, D. H., P. W. Mitchell, J. T. Webb, and D. Hanak. 2003. Carolina Thunder Revisited: Toward a Transcultural View of Winston Cup Racing. *The Professional Geographer*, 55 (2), 238-249.

Aronoff, J. 2006. Lenoir Facing Uncertain Future. *Charlotte Observer*, October 4.

Associated Press Release. 2006. State Minority Populations Rise. *Greensboro News and Record*, August 15.

Barron, R. M. 2006a. Passenger Numbers Keep Dropping at Airport. *Greensboro News and Record*, June 20.

_____.2006b. HondaJet Will Enter Crowded Market. *Greensboro News and Record*, August 5.

Batty, M. 2001. Polynucleated Urban Landscapes. *Urban Studies*, 38, 635-655.

Bell, A. 2007. Up to 780 More Job Cuts Seen for Freightliner. *Charlotte Observer*, January 31.

Bell, N. M. 2007. IKEA Moving in, Awaits Neighbors. *Charlotte Observer*, January 18, 8-9A.

Binker, M. 2006. House Budget Cuts Market Aid. *Greensboro News and Record*, June 16.

Bower, E. L. 1996. Mystery and Treasure: The Charleston Complex. In *Snapshots of the Carolinas: Landscapes and Cultures*, ed. D. G. Bennett. Washington, D.C.: Association of American Geographers, 17-22.

Bruegmann, R. 2005. *Sprawl: A Compact History*. Chicago: The University of Chicago Press.

Carlton, D. L. 2003a. The Revolution From Above: The National Market and the Beginnings of Industrialization in North Carolina. In *The South, the Nation, and the World: Perspective on Southern Regional Development*, eds. D. A. Carlton and P. A. Coclanis. Charlottesville, VA: University of Virginia Press, 73-98.

_____. 2003b. Unbalanced Growth and Industrialization: The Case of South Carolina. In *The South, the Nation, and the World: Perspective on Southern Regional Development*, eds. D. A. Carlton and P. A. Coclanis. Charlottesville, VA: University of Virginia Press, 135-150.

Carr, T. 2007. Private Correspondence with Tom Carr, Assistant Town Manager for Development, Hickory, N.C., January 31.

Cherrie, V. 2007. Biotech Subsidy Criticized. *Charlotte Observer*, January 19.

Choe, S. and H. Howard. 2004. Fiber's Wild Ride. *Charlotte Observer*, February 29.

Coleman, T. 2006. What is the Impact of Impact Fees? *The News and Observer*, Raleigh, May 14.

Coleman, T. and D. Bracken. 2006. I-40 Links Cities, Burbs to Region's Economic Engine. *The News and Observer*, Raleigh, July 2.

Cox, J. B. 2007. Google Put Pressure on Legislators to Stay Quiet. *Charlotte Observer*, February 2.

Debbage, K. G. 1996. Air Transportation in the Carolinas. In *Snapshots of the Carolinas: Landscapes and Cultures*, ed. D. G. Bennett. Washington, D.C.: Association of American Geographers, 197-202.

Dodd, S. 2001. The Rise of the Inland Cities: 1790-2000 Census Reports Chronicle Carolinas Changes. Charlotte Observer, April 14.

DuPlessis, J. 2006. Aircraft Parts Plant to Add Jobs. *The State*, Columbia, June 6.

Gade, O. 1996. Furniture in North Carolina. In *Snapshots of the Carolinas: Landscapes and Cultures*, ed. D. G. Bennett. Washington, D.C.: Association of American Geographers, 165-170.

Gade, O., A. Rex, and J. Young (with L. B. Perry). 2002. *North Carolina: People and Environments* (2nd ed). Boone, N.C.: Parkway Publishers.

Glasmeier, A. K. and R. M. Leichenko 2000. From Free-Market Rhetoric to Free-Market Reality: The Future of the U.S. South in an Era of Globalization. In *Poverty or Development*, eds. R. Tardanico and M.B. Rosenberg. New York: Routledge, 19-39.

Graves, W. 2001. Charlotte's Role as a Financial Center: Looking Beyond Bank Assets. *Southeastern Geographer* 41 (2), 230-245.

Haag, J. J. 2006. *North Carolina Motor Vehicle and Heavy Equipment Manufacturing Cluster:* Raleigh, N.C.: Department of Commerce, Division of Policy, Research and Strategic Planning, August 15 *(www.nccommerce.com/categories/statistics.htm).*

Hanham, R. Q. and A. C. Hanham. 2001. The Uneven Development of Manufacturing in the Southeast, 1950-1990. *Southeastern Geographer*, 41 (1), 1-19.

Hart, J. F. and J. T. Morgan. 1995. Spersopolis. *Southeastern Geographer*, 35 (2), 103-117.

Hartshorn, T. A. 1997. The Changed South, 1947-1997. *Southeastern Geographer*, 37 (2), 122-139.

Havlick, D. and S. Kirsch. 2004. A Production Utopia? RTP and the North Carolina Research Triangle. *Southeastern Geographer*, 44 (2), 263-277.

Henderson, B. 2007. State OKs Disputed Water Transfer. *Charlotte Observer*, January 11.

Herring, H. H. 1940. *Southern Industry and Regional Development*. Chapel Hill: The University of North Carolina Press.

Hopkins, S. M. 1998. Why Can't We Move This Stuff? *Charlotte Observer*, August 17.

Ingalls, G. L., and T. G. Moore. 2001. Old, But New: An Inventory of Textile Mill Reuse in the Charlotte Urban Region. *Southeastern Geographer*, 41 (1), 74-88.

Janiskee, R. L., L. S. Mitchell, and J. H. Maguire. 1996. Myrtle Beach: Crowded Mecca By the Sea. In *Snapshots of the Carolinas: Landscapes and Cultures*, ed. D. G. Bennett. Washington, D.C.: Association of American Geographers, 217-220.

Kasarda, J. D. and J. H. Johnson, Jr. 2006. *The Economic Impact of the Hispanic Population on the State of North Carolina*. Chapel Hill: Kenan-Flagler Business School, University of North Carolina at Chapel Hill.

Kennedy, E. A. 1998. Greenville From Back Country to Forefront. *Focus*, 45 (1), 1-6.

Kovacik, C. F. and J. J. Winberry. 1989. *South Carolina: The Making of a Landscape.* Columbia, S.C.: The University of South Carolina Press.

Larsen, L. H. 1990. *The Urban South: A History*. Lexington, KY: The University of Kentucky Press.

Lord, J. D. 2001. Globalization Forces and the Industrial Restructuring of Greenwood County, South Carolina. *Southeastern Geographer*, 41 (2), 184-204.

Mackun, P. J. 2005. Population Change in Metropolitan and Micropolitan Statistical Areas: 1990-2003. *U.S. Bureau of the Census, Current Population Reports*, September.

Manson, G. A. and R. E. Groop. 2000. U.S. Intercounty Migration in the 1990s: People and Income Move Down the Urban Hierarchy. *The Professional Geographer*, 52 (3), 493-504.

Mecia, T. 2003. Pillowtex's Brands May Entice Bidders. *Charlotte Observer*, July 31.

Mitchell, H. and J. B. Cox. 2007. Is It Worth the Price? *Charlotte Observer*, January 29.

Moon, H. 1992. The Interstate Highway System. In *Geographical Snapshots of North America*, ed. D. G. Janelle. New York, N.Y.: The Guilford Press, 425-427.

Morrill, J. and T. Mellnik. 1996. Tobacco PACs Pay to be Heard. *Charlotte Observer*, September.

Morrill, R. 2006. Classic Map Revisited. *Professional Geographer*, 58 (2), 155-160.

Morse, D. 2004. His Aim: Outdo Chinese Imports. *Charlotte Observer*, February 29.

Newman, C. 1995. North Carolina's Piedmont: On a Fast Break. *National Geographic Magazine*, 187 (3), 114-138.

North Carolina: Exports, Jobs, and Foreign Investment. 2006. Washington, D.C.: United States Department of Commerce - Industry, Trade and the Economy, December.

Nwagbara, U., U. Buehlmann, and A. Shuler. 2002. *The Impact of Globalization on North Carolina's Furniture Industries*. Raleigh, N.C.: Department of Commerce, Division of Policy, Research & Strategic Planning *(www.nccommerce.com/categories/statistics.htm)*.

Odum, H. 1936. *Southern Regions of the United States*. Chapel Hill: The University of North Carolina Press.

Opportunities for Automotive Companies in South Carolina. 2006. Columbia, S.C.: Department of Commerce (www.sccommerce.com/wia/RapidResponse.html#report).

Phillips, G. 2006. County's Hispanic Population Exploding. *The Herald Sun*, Durham, August 5.

Pierce, N. and C. Johnson. 1995. *Pierce Report; Recommendations for Our Region's Future*. Charlotte, N.C.: Charlotte City Council.

Porter, M. E. 2000. Locations, Clusters, and Company Strategy. In *The Oxford Handbook of Economic Geography*, eds. G. L. Clark, M. P. Feldman, and M. S. Gertler Oxford, England: Oxford University Press, 253-276.

_____. 2003a. *South Carolina Competitiveness Initiative: Phase I Final Presentation*. Boston, MA: Monitor Company Group L.P.

_____. 2003b. The Economic Performance of Regions. *Regional Studies*, 37.

Price, D. 1998. Along I-85, Industry Begets Industry. *The News and Observer*, Raleigh, July 14.

Rusk, D. 2004. *South Carolina's Cities: Hubs of Progress*. Keynote Speech Delivered at the 64th Annual Meeting of the Municipal Association of South Carolina, Hilton Head Island, July 16.

Smith, H. and W. Graves. 2003. The Corporate (Re)Construction of a New South City: Great Banks Need Great Cities. *Southeastern Geographer*, 43 (2), 213-234.

South Carolina: Exports, Jobs, and Foreign Investment. 2006. Washington, D.C.: United States Department of Commerce - Industry, Trade and the Economy, December.

Stuart, A. W. 1996. The Charlotte Urban Region. In *Snapshots of the Carolinas: Landscapes and Cultures*, ed. D. G. Bennett. Washington, D.C.: Association of American Geographers, 109-114.

Stuart, B. 1979. *Making North Carolina Prosper: A Critique of Balanced Growth and Regional Planning*. Raleigh, N.C.: North Carolina Center for Public Policy Research.

The Drive to Move South: The Growing Role of the Automobile Industry in the SLC Economies. 2004. Atlanta: The Southern Legislative Conference.

Travel Industry Association. 2006. *The Economic Impact of Travel on South Carolina, 2005.* Washington, D.C.: Travel Industry Association of America, August.

Tuttle, S. 1994. Welcome to Charleighboro. *North Carolina,* 52 (7), 4.

Tyer, C. B. 1995. A New Approach to Annexation: Why South Carolina's Cities Are Not Growing. *The South Carolina Public Forum Magazine,* 6, (1), 35-44.

Ulbrich, H. H. and A. L. Steirer. 2004. *Local Governments and Home Rule in South Carolina: A Citizens Guide.* Clemson: Clemson University, Strom Thurmond Institute of Government and Public Affairs.

Vance, R. 1935. *Human Geography of the South.* Chapel Hill: The University of North Carolina Press.

Werner, B. 2006. Small S.C. Banks. *The State,* Columbia, July 30.

Whitacre, D. 1998. Will We Try Another Way? *Charlotte Observer,* March 15.

Whitacre, D. and T. Mellnik. 2003. Charlotte Dollars Pull From Afar. *Charlotte Observer,* September 14.

Wood, P. J. 1986. *Southern Capitalism: The Political Economy of North Carolina, 1880-1990.* Durham, N.C.: Duke University Press.

Ziegler, D. J. 1992. Main Street of the Sunbelt: I-10 and I-20. In *Geographical Snapshots of North America,* ed. D. G. Janelle. New York, N.Y.: The Guilford Press, 26-29.

Zimmer, J. 2006. The Rebirth of Durham's Downtown. *The Sunday Herald,* Durham, September 3.

Web Listings

Hanesbrands, Inc <hanesbrands.com/hbi/en-us/OurCompany/Default.htm>

IGT Corporation <www.burlington.com/companies/>

Philip Morris <www.altria.com/about_altria/01_00_03_philipmorrisusa.asp>

R. J. Reynolds Tobacco <www.rjrt.com/company/profileFactBook.asp>

SaraLee Corporation <www.saralee.com/ourbrands/>

Southern Legislative Conference <www.slcatlanta.org/Publications/EconDev/AutoSouth/DriveTableOfContents.htm>

Chapter 6

AGRICULTURE AND FORESTRY

John Fraser Hart

University of Minnesota

Farmers in the Carolinas must cope with daunting environmental challenges. The tired old Blue Ridge Mountains in the west have soils that are thin and stony, wooded slopes as steep as a cat's face, and precious few areas that are level enough for crop production (Figure 6.1). The Piedmont has been scraped raw by more than a century of cultivation, and it is slowly recuperating under a healing blanket of pasture and second-growth woodland. The soils of the Slate Belt are thin and unproductive, and the porous soils of the Sandhills are extremely droughty. The sandy soils of the Coastal Plain are almost equally thirsty, but paradoxically they may also have to be drained, because the land lies so close to sea level. In the Flatwoods the land is even lower and the water table is high, so extensive areas are poorly drained swamps and marshes, where dark organic matter accumulates.

The harvested cropland of the Carolinas is concentrated on the light-colored sandy soils of the Coastal Plain, but even here only a small fraction of the land is cultivated (Figure 6.2). The soils of the Coastal Plain generally are low in mineral plant nutrients and organic matter, and farmers cannot produce satisfactory crops from them without adding large amounts of fertilizer. The sandy soils have a water-holding capacity of only two inches or so in their upper two feet, and a crop like corn can remove 0.4 inch of water a day during its peak growth period. Small wonder that farmers say, "We need an inch and

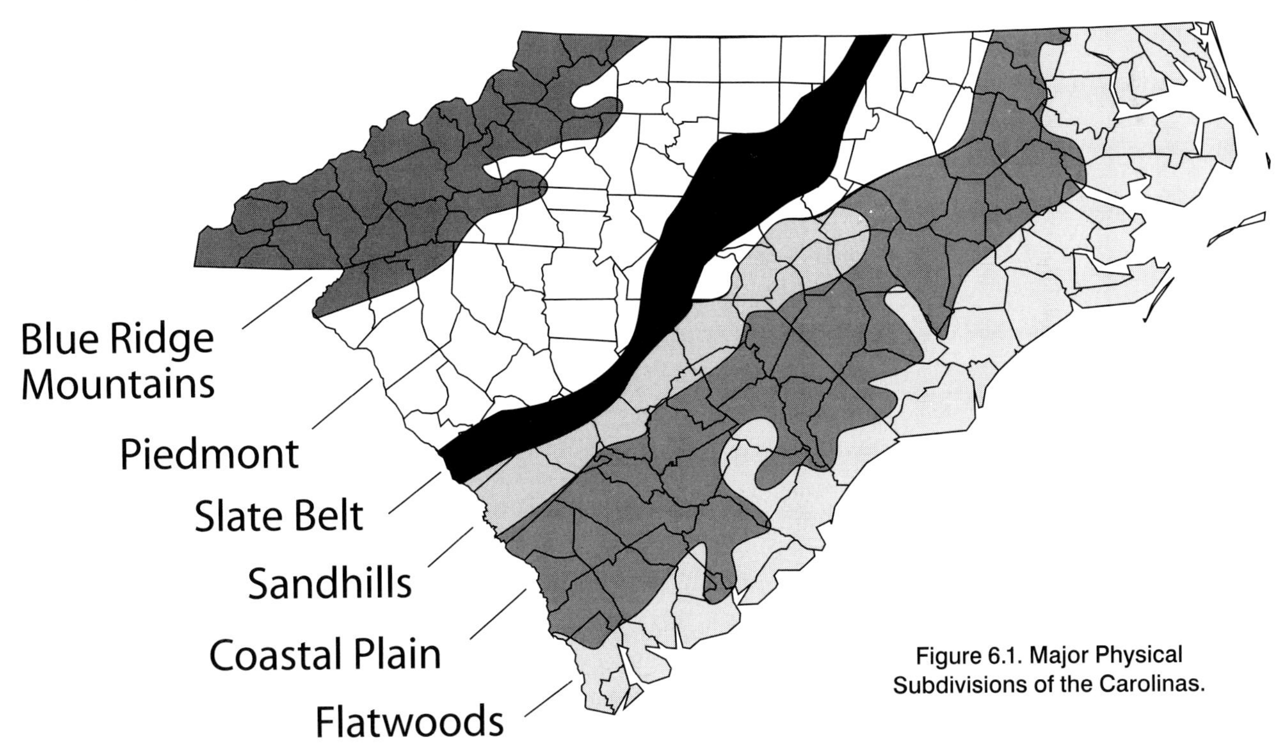

Figure 6.1. Major Physical Subdivisions of the Carolinas.

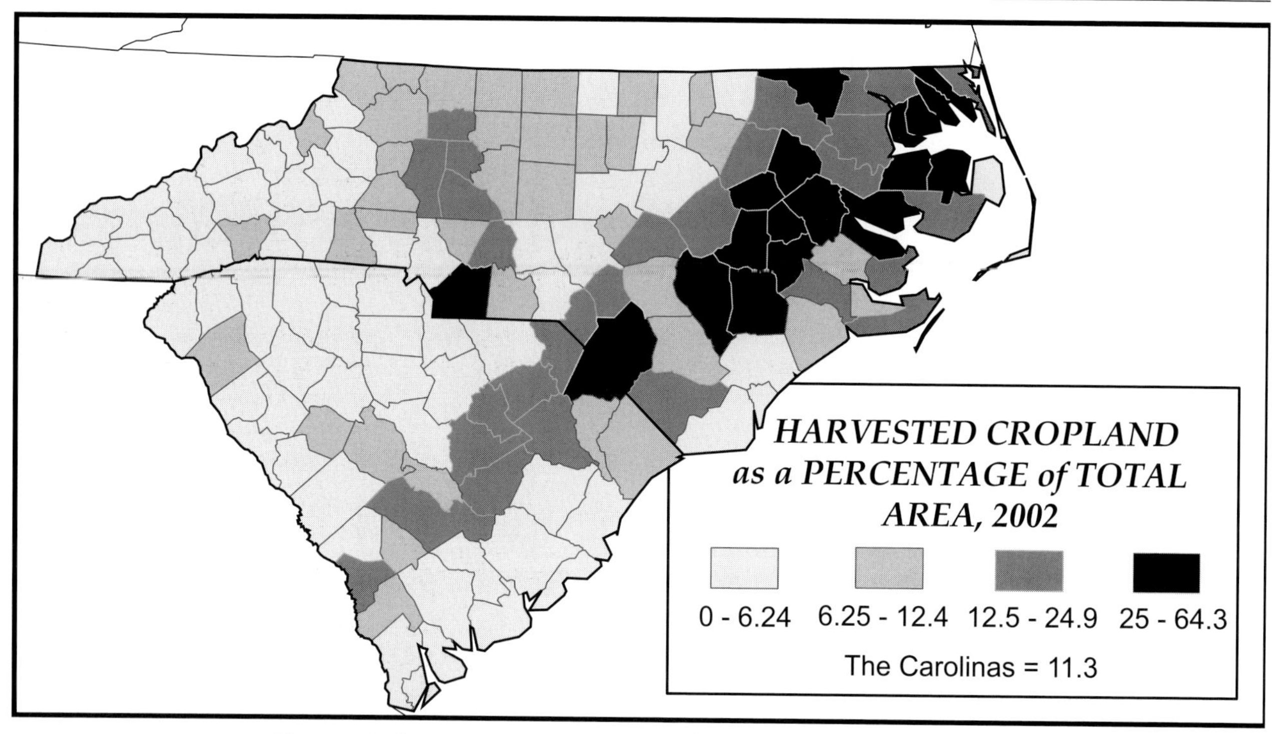

Figure 6.2. Cropland Harvested as a Percentage of Total Area, 2002.

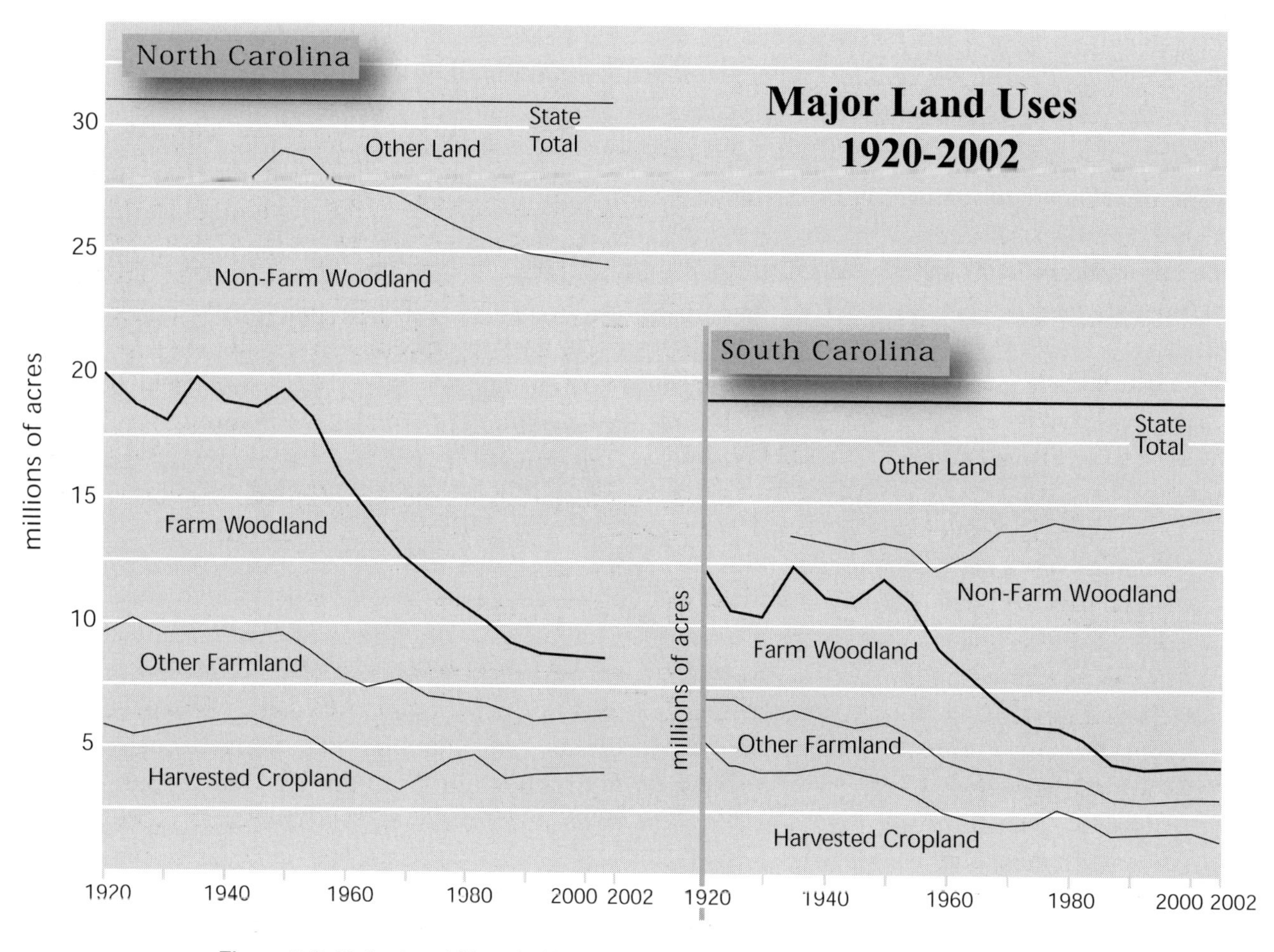

Figure 6.3. Major Land Uses in North Carolina and in South Carolina, 1920-2002.

a half of rain every Saturday during the growing season, but summer droughts of two weeks or longer are common.

The summer rainfall comes mostly from afternoon thunderstorms, which are notoriously unpredictable and spotty. A heavy shower can drench one field with a torrential downpour, while the crops in fields only a few hundred yards away are stressed from lack of water. Pelting gully washers from thundershowers and late summer hurricanes can erode the precious topsoil from fields of row crops even on level land, and high winds from hurricanes can flatten standing crops. Summers are too hot for forage production, and the summer heat is oppressive and enervating. Winters are too cold for tropical crops and year-round production, and springs are too short for cool season crops.

Despite these manifold environmental constraints, in 2002, farmers in the Carolinas (which comprise 2.7 percent of the land area of the conterminous United States) managed to produce 47 percent of the nation's tobacco, 24 percent of its turkeys, 23 percent of its hogs, 11 percent of its broilers, nine percent of its peaches, eight percent of its peanuts, and five percent of its cotton.

The census of agriculture contains information on all, but only, land that is operated and owned by farmers. Uninformed people might be alarmed that the Carolinas have lost farmland steadily since 1950, unless they understand that most of this loss resulted from farmers selling off wooded land that had contributed precious little to their actual farm operations (Figure 6.3). The acreage of harvested cropland, which is a better indicator of effective farmland, has not declined nearly so much, especially since 1960.

Even though the acreage of farmland has been shrinking slightly, the size of individual farms in the Carolinas has been growing quite impressively (Figure 6.4). Before 1950, the average Carolina farm had around 70 acres, of which perhaps 25 were tillable, but by 2002, the average farm had increased to 177 acres, with 114 tillable. Until 1950, corn and cotton were the principal cultivated crops, with tobacco a distant third in North Carolina (Figure 6.5). Soybean acreage boomed in the 1970s and then declined in the 1980s, but in 2002, soybeans were still the leading crop in terms of acreage.

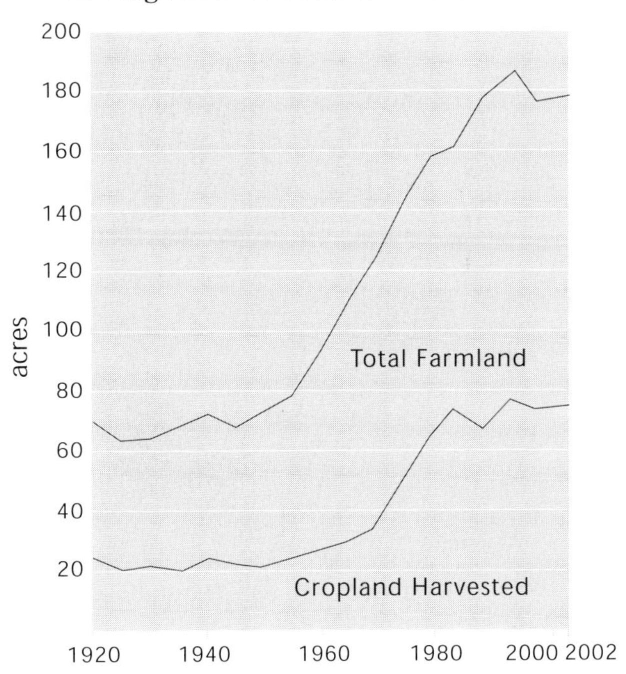

Figure 6.4. Average Size of Farm in the Carolinas, 1920-2002.

The traditional crops of the Carolinas--corn, cotton, tobacco, and peanuts--required intensive human effort. Most farmers grew them on small "patches" of only a few acres carved out of the encircling woods. These patches were intimately intermingled with patches of other crops, pasture, woodland, and swamp, and extensive open fields were rare. James R. Anderson, who directed the land use and land cover mapping program of the United States Geological Survey, once told me that eastern North Carolina has the nation's most complex mosaic of land use polygons.

Tobacco

Tobacco is the traditional money crop of North Carolina, even though it has never been grown on more than one percent of the state's total area, and in South Carolina it has ranked second only to cotton as a money crop (Figure 6.6). In both states, crops have lost some of their former preeminence to poultry, hogs, and other forms of livestock, but in 2002, the Carolinas still produced $736 million worth of the golden leaf,

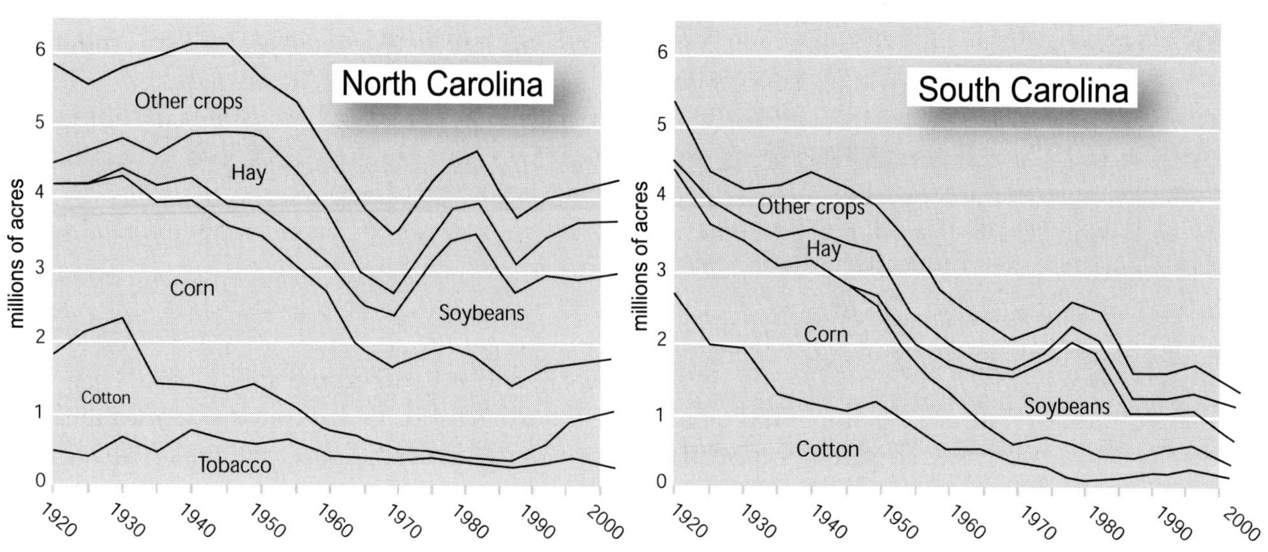

Figure 6.5. Acreage of Major Crops in North Carolina and in South Carolina, 1920-2002.

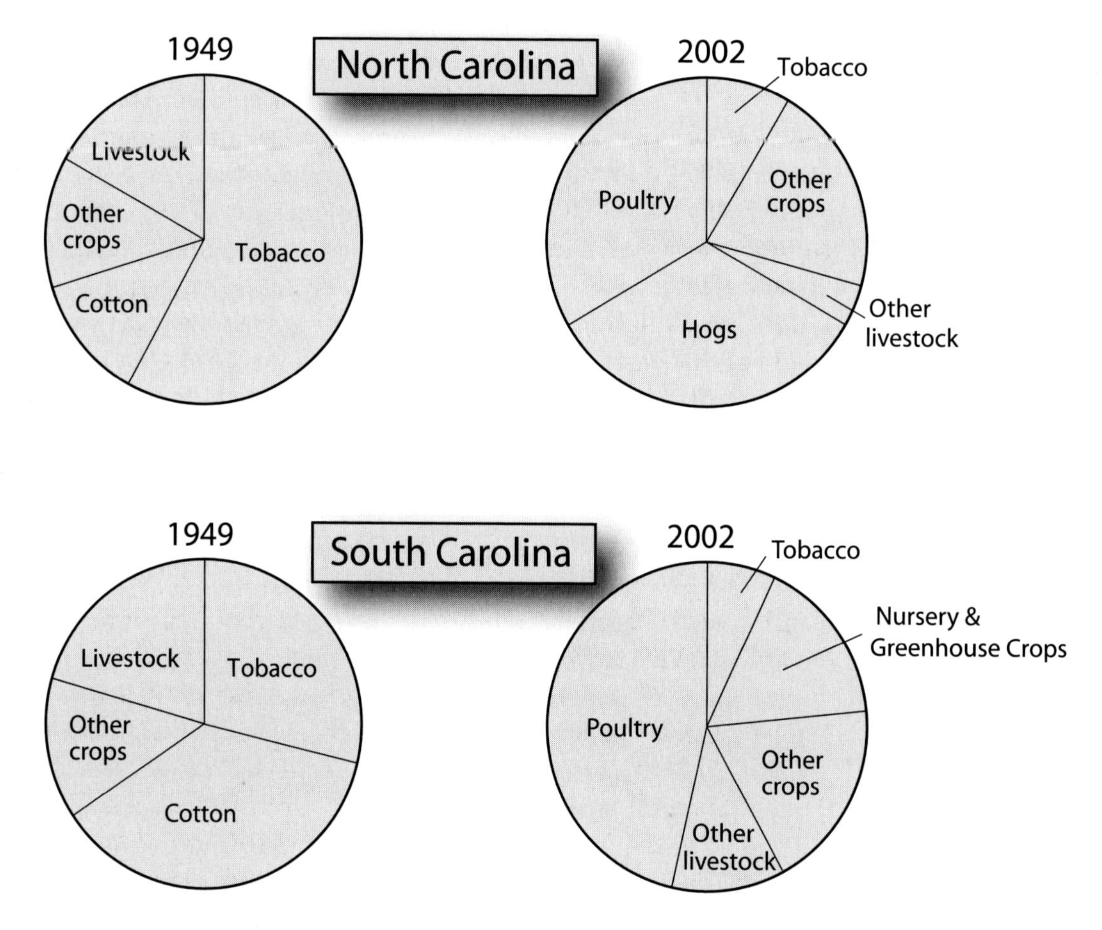

Figure 6.6. Sales of Major Agricultural Commodities, North Carolina and South Carolina, 1949 and 2002.

and it accounted for one-tenth of their total farm income.

Farmers in the Carolinas produce two distinctive types of tobacco. Flue-cured Bright tobacco is the principal type in most of the area, but farmers in the mountains of western North Carolina plant around ten thousand acres of air-cured Burley tobacco, which is the dominant type in Kentucky and Tennessee (Algeo, 1997). Both types are used primarily to manufacture cigarettes, which consist of around one-half flue-cured tobacco and one-third Burley.

The technique of producing flue-cured tobacco developed in southside Virginia around the time of the Civil War when farmers discovered that the seeds of the coarse, heavy, dark-green tobacco plant produced plants with fine, thin, light-green leaves when grown under semi-starvation conditions on sandy soils of low organic content. Farmers on the Piedmont of northern North Carolina avidly seized upon this idea, and Nannie May Tilley (1948, 135) described "the ridiculous spectacle of a controversy between counties as to which could show the greatest amount of poor land."

Farmers produced an even more desirable leaf when they continued semi-starvation with the flue-curing process, which uses high temperatures to dry the leaf rapidly and thus to halt its metabolism. They baked the leaves at temperatures of up to 200° F. for three or four days in small, squat, cube-like barns heated convectionally by flues of sheet metal that crossed their floors. The flue-curing process produced an orange-yellow leaf that was called Bright tobacco because of its color and Virginia tobacco because of the origin of the process. (Farmers in the mountains of western North Carolina air-cure their Burley tobacco by hanging the entire plant in barns and allowing it to dry naturally for several months.)

Farmers on the Coastal Plain of North Carolina began to shift from cotton to flue-cured tobacco production in the late 1880s, when the price of cotton dropped disastrously, and production slowly migrated southward onto the Coastal Plain of South Carolina in the wake of boll weevil depredations in the 1910s. Farmers on the Coastal Plain retained their traditional tenancy system, and they produced tobacco on a large scale with sharecroppers. Their landholdings were much larger than the small family farms of the Piedmont, but the subdivision of their holdings into sharecropper units disguised their true size, because the census of agriculture persisted in treating sharecropper units as separate individual farms.

Flue-cured tobacco production flourished in the Cigarette Revolution that followed World War I, and it had become solidly established in the northern tier of Piedmont counties and on the Coastal Plain of both Carolinas by 1933, when the Agricultural Adjustment Act was passed. This act, and subsequent farm legislation based on it, have guaranteed farmers a minimum price for their crops, but farm policy has fossilized the geography of agricultural production, because in return for guaranteed prices it has placed stringent limitations on the acreage that farmers may plant. The price-support program allots each farmer an acreage "base" on which he may grow a crop, guarantees the price of the crop he produces on this base acreage, and penalizes him severely if he plants a larger acreage.

Tobacco farmers have cultivated their limited acreages so intensively that they have greatly increased their yields, and the government price-support programs have been compelled to supplement their acreage restrictions with poundage quotas. A poundage quota is the number of pounds the farmer is permitted to sell at the guaranteed price. Many farmers plant a larger acreage than they think they will need to make their quota, harvest only the best, and leave the rest standing in the field after harvest. Farmers who want to increase their production may lease acreage base and poundage quotas from owners who do not wish to use them, but it is extremely difficult for a new farmer or a new area to start producing a crop whose acreage and poundage are so strictly controlled.

Tobacco was the nation's last major field crop to be mechanized, and as late as 1960, producing an acre of tobacco required 300 to 500 hours of labor, in contrast to 48 hours for an acre of cotton and only seven for an acre of corn (Hart and

Chestang, 1978). Initially, many flue-cured tobacco farmers were reluctant to retire their mules, because they feared that machines would bruise the valuable leaves, but in the 1960s, they became willing to accept lower leaf quality as the trade-off for their tremendous savings in labor costs when they switched to harvesting machines.

The adoption of harvesting machinery was associated with the adoption of new metal bulk barns in which the leaf is cured by forced air rather than by convection. Bulk barns, which look like semi-trailers, have so completely replaced traditional flue-cured barns that flue-cured tobacco is no longer cured over flues. It is ironical that flue-cured tobacco production has been mechanized and modernized just when the future of tobacco is in such serious jeopardy.

Mechanization has enabled flue-cured tobacco farmers to handle far larger acreages. Traditionally, tobacco farms have been small, because the crop has required such prodigious amounts of hand labor. A large family could cope with no more than five to seven acres, and even then they might have had to hire additional help at the busiest times of the year, but a farm of five tillable acres with a five to seven acre tobacco base could provide a reasonable livelihood for a family.

As late as 1964, the average tobacco farm grew less than five acres of the crop, but by 2002, the average farm grew 23 acres, and many farmers grew one hundred acres or more. They have enlarged their operations by consolidating sharecropper units and by leasing land, tobacco base, and poundage quotas from other owners, but they still require enormous amounts of labor.

Tobacco farmers on the Coastal Plain have shifted increasingly from African American to Hispanic workers. The Hispanic presence is far more obvious on the land than it is in the pages of the Census of Population, partly because it is so recent, and partly because many Hispanic workers do their best to avoid being enumerated by the census.

Five maps are needed to show geographic variations in the importance of tobacco in the Carolinas (Figure 6.7). These maps are standardized for area to compensate for differences in county size. One of the most common blunders made by unskilled map-makers is to shade counties according to the total amount of anything they contain, which gives unfair prominence to larger counties. The area of Horry County, South Carolina, for example, is 1,134 square miles, but the area of Chowan County, North Carolina, is a mere 173, so Chowan County must have more than six times as much per square mile as Horry County to have the same total.

Tobacco has its greatest impact on the landscape in the area within a 50-mile radius of Wilson, North Carolina, with lesser areas extending westward on the Piedmont through the northern tier of counties in North Carolina and southward on the Coastal Plain into South Carolina (Figure 6.7a), but it occupies the greatest share of the harvested cropland on the Piedmont and in the western Mountains, because other crops are more important on the Coastal Plain (Figure 6.7b). Farmers in the Wilson area produce the greatest value of tobacco (Figure 6.7c), but farmers in the peripheral areas rely on the crop for a greater share of their income (Figure 6.7d). Farms on the Coastal Plain have the greatest acreages of tobacco per tobacco farm, and those in the Mountains have the smallest (Figure 6.7e). In short, the Piedmont and the Mountains have the smallest tobacco farms, but they rely more heavily on the crop.

On the Coastal Plain farmers have been ripe for diversification since the early 1960s. They realize that they have relied too heavily on tobacco, which has been under ever-increasing political attack since 1964, when the Surgeon General first asserted that lung cancer might be linked to cigarette smoking. They have been casting about for alternative agricultural enterprises, but they have been handicapped because no other field crop produces even close to the same return as tobacco.

In 1997, for example, the North Carolina Cooperative Extension Service calculated that a farmer could expect an average net return of $1,376 for an acre of tobacco, $158 for an acre of cotton, $121 for an acre of peanuts, $35 for an acre of wheat, $20 for an acre of soybeans, and an actual loss of one dollar for each acre of corn.

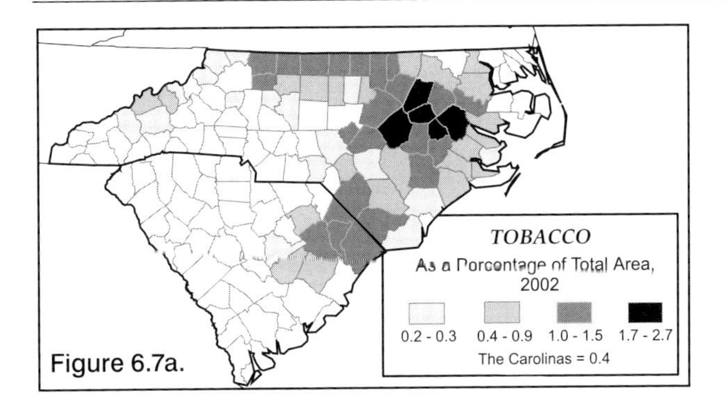

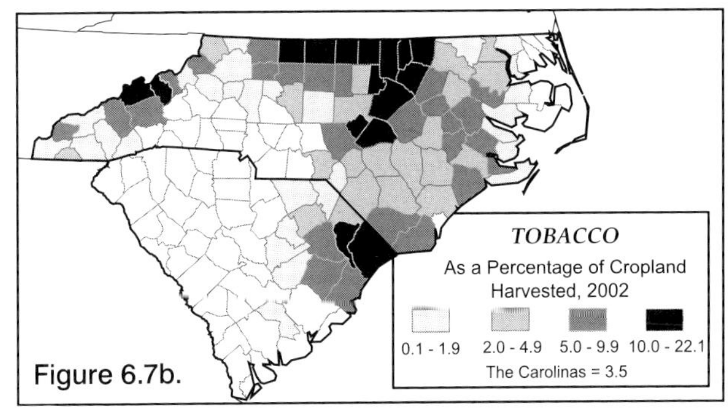

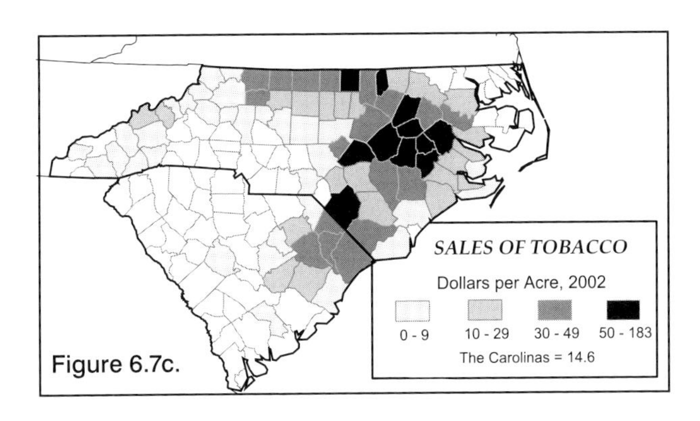

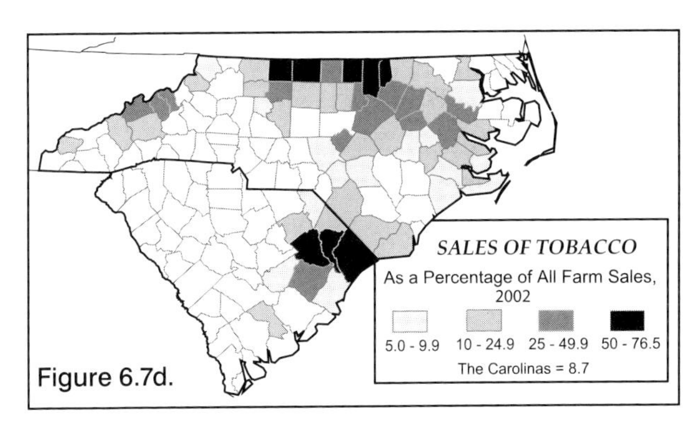

Figure 6.7. Tobacco, 2002.

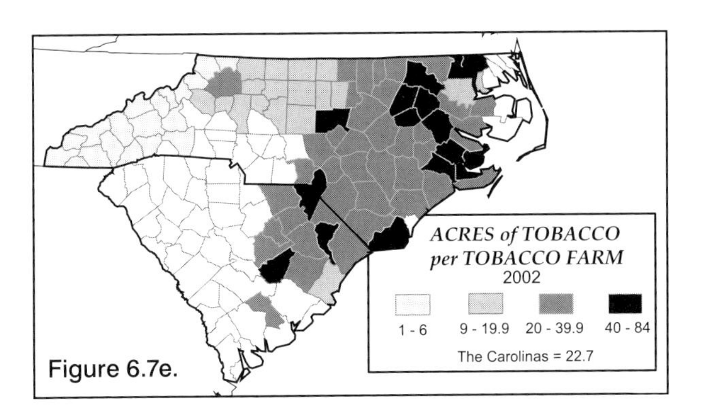

Furthermore, any new agricultural enterprise will require expensive new infrastructural investment in facilities for production, processing, and marketing.

In 1964, peanuts were firmly entrenched as a commercial crop in northeastern North Carolina, but like tobacco, they were tightly controlled by acreage and poundage quotas. A few farmers grew cotton, but it was little more than a curiosity. Everyone grew corn as a subsistence crop. It was a staple in the diet of the people and a feed for their work animals. Other livestock were unimportant. Most farmers had barnyard flocks of chickens for eggs and Sunday dinner, and they butchered a hog or two for home consumption, but the only truly commercial livestock-producing areas were broiler areas near Wilkesboro in northwestern North Carolina and near Siler City in the central part of the state.

Soybeans, Corn and Wheat

Soybeans were the first alternative cash crop on the Carolina Coastal Plain (Figure 6.8A). They required land that was level enough to facilitate the use of large machines and to moderate the risk of soil erosion. They were easy and inexpensive to grow, but they produced such low returns per acre that they needed fairly large acreages. On many farms soybeans were a crop of expediency on land for which the farmer had no better alternative. They could be an acceptable money crop when the price was right, but soybean acreage plummeted in the early 1980s, when a declining export market depressed their price. In 2002, they still were the leading crop of the Carolinas in terms of acreage; they were grown on 1.7 million acres, or nearly one-third of all harvested cropland (Figure 6.8a).

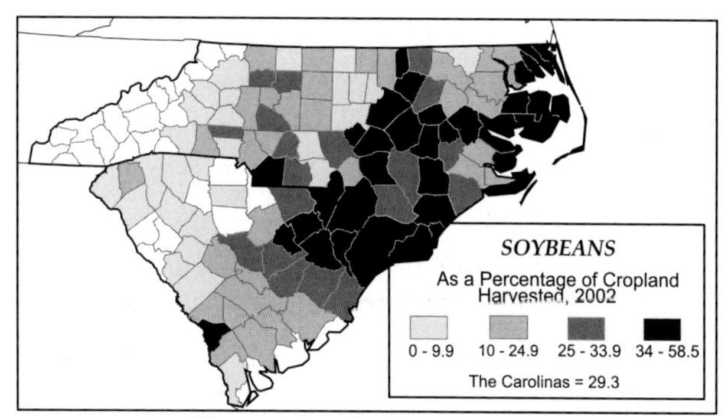

Figure 6.8a. Soybeans, 2002.

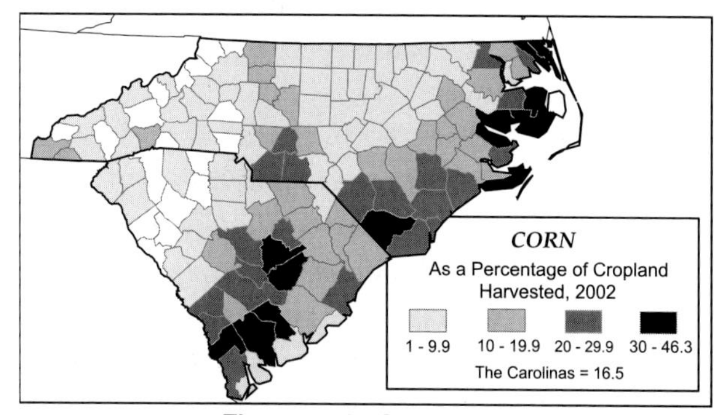

Figure 6.8b. Corn, 2002.

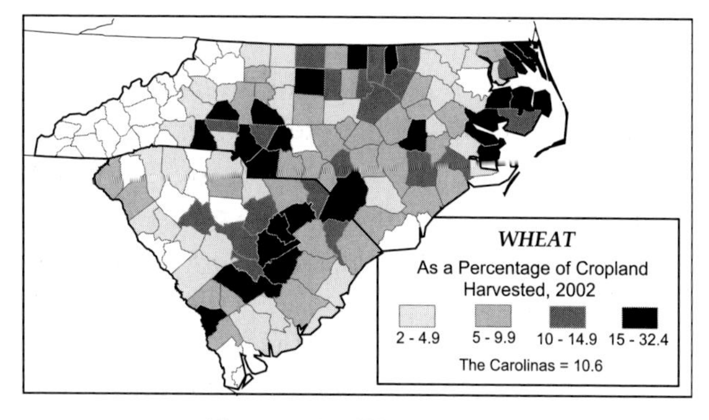

Figure 6.8c. Wheat, 2002.

Farmers could not grow soybeans year after year on the same field because of problems with nematodes, so they developed a two-year cropping system of soybeans with corn and winter wheat. In the old days they had grown corn to feed their mules, and after they had retired their mules they continued to grow it as a cash crop, "just because the land was there," even though corn is not a particularly good crop for the area. In 2002, farmers in the Carolinas grew 0.9 million acres of corn, on one of every six acres of harvested cropland (Figure 6.8b).

Farmers sowed winter wheat in November after they had harvested their corn (Figure 6.8c). They planted short-season soybeans in the stubble immediately after they combined their wheat in the following June, and harvested the soybeans in the fall of the second year. If you keep your eyes open you can often see wheat stubble in fields of young soybeans. This double-cropping system produced three cash crops in two years, but none of them was particularly remunerative. "A lot of farmers were slowly going broke raising corn and soybeans, and they badly needed a better cash crop" (*North Carolina Farmer*, 1995).

Broilers

The production of broilers (chickens less than ten weeks old) has become a significant source of income on many farms in the Carolinas since the 1950s. As early as the 1920s, feed companies tried to increase their sales by encouraging feed-mill operators in small towns to contract with local farmers to feed birds they provided and marketed, but the price fluctuated so much that few farmers were willing to gamble on commercial production, and chicken was a luxury meat.

Vertically integrated contract production has revolutionized the broiler business since World War II. The vertical integrator, usually a feed dealer but sometimes a processor, evens out the risk of price fluctuations by contracting with many farmers to maintain a steady flow of birds. The integrator forges a complete production chain that links hatcheries, feed mills, farmers, and processing plants with marketing and advertising campaigns that have turned names like Perdue and Tyson into household words.

The integrator provides day-old chicks and feed, and the farmer provides the broiler houses, utilities, and labor. The integrator collects, processes, and markets the birds, and exercises complete management control. Farmers do not like being told what to do on their own land, but they like the guaranteed price. The vertically integrated contract producer system has been successful because it enables the small farmer-contractor to make a reasonable income, and thus it helps to preserve the small family farm. Most farmers actually like it, but they also like to grumble about it, just as they like to grumble about everything else under the sun.

Contract production has converted broiler production into an efficient industry that produces high-quality birds at a low cost. It has brought the price down to easily affordable levels, and chicken has passed beef to become the most popular meat in the United States, at least in part because health-conscious consumers are shunning red meat for white.

A broiler house is a long, low, one-story metal structure with canvas upper sides that the farmer can raise or lower to control heat and ventilation. A standard 40-by-400-foot house can hold 22,000 birds that are fed to market weight in six or seven weeks. Corrugated metal grain bins at one end have push-button controls that automatically fill the feed troughs running the length of the floor. In established areas it is easy to borrow money to build a broiler house, which cost around $100,000 in 1995. Much of the work is light enough to be done by women and children, so broiler production can easily supplement other farm enterprises or even a full-time off-farm job.

The first two major broiler-producing areas in the Carolinas were in areas of small farms where the farmers needed to supplement their income (Lord, 1971): in Wilkes County in the mountain foothills of northwestern North Carolina and in Chatham County on the thin soils of the Slate Belt in the central part of the state (Figure 6.9a). "In 1953," said Dennis Ramsey, "they had a lot of little log poultry houses in Wilkes and Chatham Counties."

The third, in Duplin County in eastern North Carolina, owes its origin to Dennis Ramsey, who was invited to Rose Hill to open a movie theater when he left the Army after World War II (Hart, 1976, 65-66). He liked to talk to his customers as they left, and he soon realized that television, which was just coming in, was going to put him out of business. "Farmers used to come in to town on Saturday night to watch the cowboys and Indians," he said, "but now they could stay home and watch them on their television sets, and I knew I had to do something else."

He visited his wife's relatives in north Georgia, and learned from them how to run a broiler business. In 1954, he built a model 40-by-100-foot broiler house that held 4,000 birds, and contracted with nine farmers to raise broilers for him. "I knew the farmers were hurting," he told me, "because the government had cut the tobacco program, and that was the only money crop they had." He bought feed and chicks, contracted with farmers to grow them, and marketed the broilers. In 1954, Duplin County producer 42,000 broilers; in 1959, it produced 10,317,000. In 1956, Dennis Ramsey built his own feed mill, and when he retired in 1973, the Ramsey Feed Company had 250 contract growers.

During the 1960s and 1970s, farmers throughout eastern North Carolina began building broiler house to diversify their income (Hart, 1991, 289-290), and by 1992, poultry (including turkeys and eggs as well as broilers) had passed tobacco to become the leading source of farm income in the Carolinas (Figure 6.6).

Turkeys

Turkey production has followed the broiler model of vertically integrated contract production. Turkey houses are indistinguishable from broiler houses, and only the sign at the farm entrance tells which is which. Farmers start flocks of 12,000 to 14,000 birds in a heated brooder house for six weeks, then divide the flock and move the birds to two grow-out houses. Grow-out takes eight weeks for hens, which are sold at 14 pounds, and 14 weeks for toms, which are sold at 38 to 40 weeks.

In 2002, North Carolina produced 51 million turkeys, one-fifth of the national total. The state led the nation. Ten percent of all U.S. turkeys were produced in a 50-mile radius of the village of Turkey in eastern Sampson County (Figure 6.9b). Marvin Johnson of Rose Hill started the turkey business in Duplin County in 1945. "After the war," he said, "my brother and me didn't had too much to do. We didn't had no money, we didn't own no land, we didn't had no tobacco allotments, we didn't had nothin'."

"My mother raised a few turkeys, maybe 150 or so, and my brother and myself used to sell 'em around the tobacco sales barns. We decided to

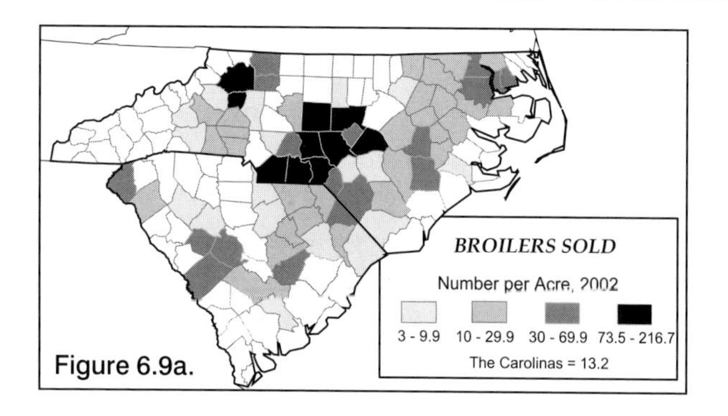

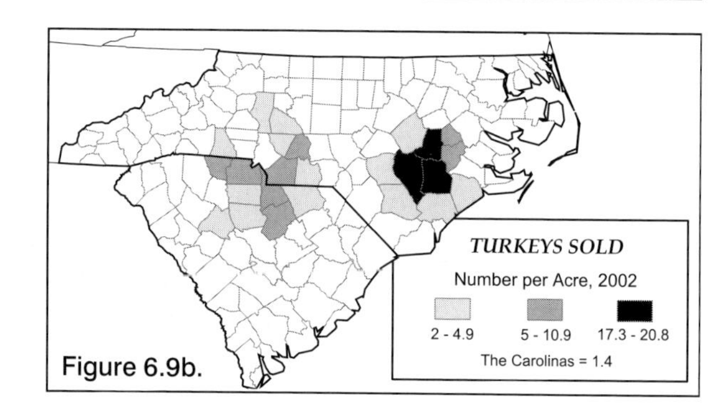

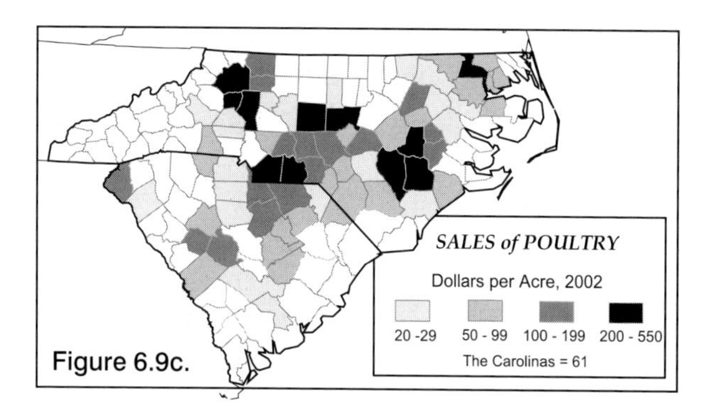

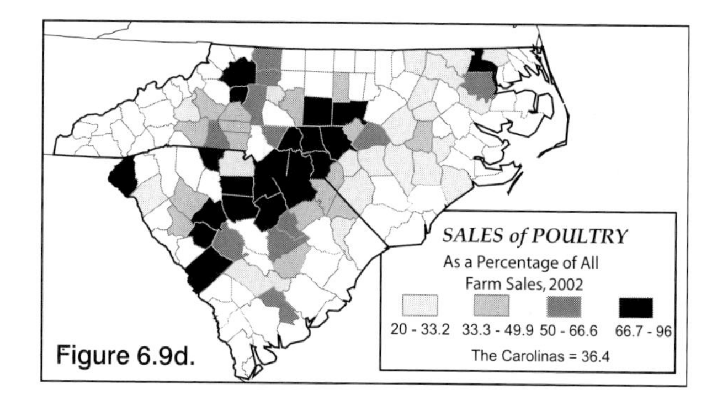

Figure 6.9. Poultry Sales, 2002.

beef it up. We started with a hatchery. We had a little incubator about the size of two refrigerators, and we produced maybe a thousand poults a week. We raised 50-75 thousand turkeys a year on range."

"Our Dad thought we was crazy. We needed a loan of $5,000 to help us get started, but he wouldn't sign a note for us, and we had to go to an old bachelor uncle. Then we wouldn't let Dad in the business with us, even when we was making more money than he was. We started off as Johnson Brothers, but finally, we needed his backing when we built a feed mill for our own use in 1954, and that's when we became Nash Johnson and Sons. In 1956, we started contracting broilers with Dennis Ramsey. We still own all the turkey breeder farms, but we have 200 contractors growing turkeys for us. We had to get into processing because the plant we had been using went broke, and we had to buy it."

Johnson Brothers sold poults to farmers in Union County, southeast of Charlotte, which was a minor turkey-producing area before 1970, when a charismatic Canadian named Bruce Cuddy moved to Marshville in Union County. He bought a turkey hatchery, built a processing plant, contracted with local farmers to grow out turkeys, and developed Cuddy Farms, a vertically integrated turkey-producing company that sparked the meteoric growth of the county's turkey production from 771,000 birds in 1969 to 3,008,000 in 1974 and 11,435,000 in 1992 (Figure 6.9b). Wampler-Longacre has bought Cuddy Farms, but the Cuddy Farms name still adorns many of its trucks.

A third major turkey producer is Carroll's Foods of Warsaw, which is 12 miles north of Rose Hill. In 1939, Ottis S. Carroll opened a grist mill in Warsaw and sold feed to local farmers. In 1953, he constructed 11 broiler houses next to the mill, and expanded by contracting with local growers. In 1967, he realized that he could not compete with Dennis Ramsey, so he sold the broiler business to Tyson and switched to turkeys, with the guidance of Bill Prestage, who later started his own company. In 1997, Carroll's had 240 contract growers who produced 15 million turkeys a year.

Carroll's trucked turkeys to a processing plant in Virginia until 1986, when it joined the

Goldsboro Milling Company, another turkey producer, in developing the processing capacity they needed by forming a jointly owned company, Carolina Turkeys, to build one of the nation's largest turkey-processing plants. This plant is 15 miles east of Mount Olive, out in the middle of nowhere, but it is at the geographic center of their live turkey production, on a piece of property that one of the partners just happened to own.

Carolina Turkeys employs 2,100 workers who process 70,000 turkeys a day, five days a week year-round. Slaughter starts at ten-thirty at night and runs for two eight-hour shifts. The third shift completely disassembles and cleans the plant for sanitation purposes. Slightly more than half the workers are Hispanics who have become permanent residents, and the company has developed mobile home parks to house some of them.

The Census of Agriculture, unfortunately, does not publish separate data on sales of broilers, turkeys, eggs, and other poultry products, but the four leading poultry-producing areas are in North Carolina: broiler areas in Wilkes County and in Chatham County, and turkey and broiler areas in Union and Anson Counties southeast of Charlotte and in Duplin and Wayne Counties on the Coastal Plain (Figure 6.9c). The counties that depend most heavily on poultry sales as a source of farm income, however, are on the Slate Belt and in the Sandhills, where the cultivation of crops is not rewarding (Figure 6.9d).

Hogs

Hog producers in the United States lagged in adopting the broiler/turkey model of contract production, but they have come on like gangbusters once they got started, and North Carolina led the way (Furuseth, 1997). In the 1990s, the state vaulted past Indiana, Minnesota, Illinois and Iowa to become the nation's leading hog producer. It produced 1.4 million hogs in 1964, 5.2 million in 1987, 10.8 million in 1992, and 42.0 million in 2002.

As late as 1980, most of the nation's hogs were produced by hundreds of thousands of small independent farms that marketed a few hundred animals of highly variable quality each year, but the old farm pig pen has been replaced by large new, highly specialized operations that annually churn out thousands of hogs of nearly identical size, shape, and quality. In North Carolina, for example, in 1964, 29,000 producers sold an average of 47 hogs a year, whereas in 2002, the number of producers had dropped to 2,332, but they sold an average of 18,018 hogs a year.

Consumers have forced the change. Those who are health-conscious demand leaner cuts of meat, and most consumers want their meat in convenient, ready-to-use packages, because they are reluctant to spend preparation time in their kitchens. The pork processing companies need a steady supply of uniform high-quality hogs to satisfy these demands. They have developed better breeding stock to produce animals that grow faster and produce leaner meat with less feed, but these superior animals require expensive production facilities and skillful management. Furthermore, it is easier for the processor to deal with a few large producers who can regularly deliver large numbers of uniform animals than with many small producers who deliver heterogeneous mixtures whenever they happen to feel like it.

Modern hog production must be so heavily capitalized that its development has required some truly courageous entrepreneurs who have been willing to take great risks. Wendell H. Murphy, also of Rose Hill (there must be something in the water there!), is the entrepreneurial genius who pioneered the concept of contract hog production. In 1961, he borrowed $10,000 to start a feed mill. To build sales he bought 200 feeder pigs and paid his brother Pete a fee to feed them on a dirt lot behind the mill. He expanded by contracting with local farmers to feed pigs for him. At first they used dirt lots, but in 1974, he began to require them to use buildings built to his own specifications.

In 1997, Murphy Family Farms contracted with 500 farmers to produce $775 million worth of hogs, and was the nation's leading hog producer (Roth, 1997). In second place was Carroll's Foods of Warsaw, which had started contract hog production in 1986 in cooperation with Smith-

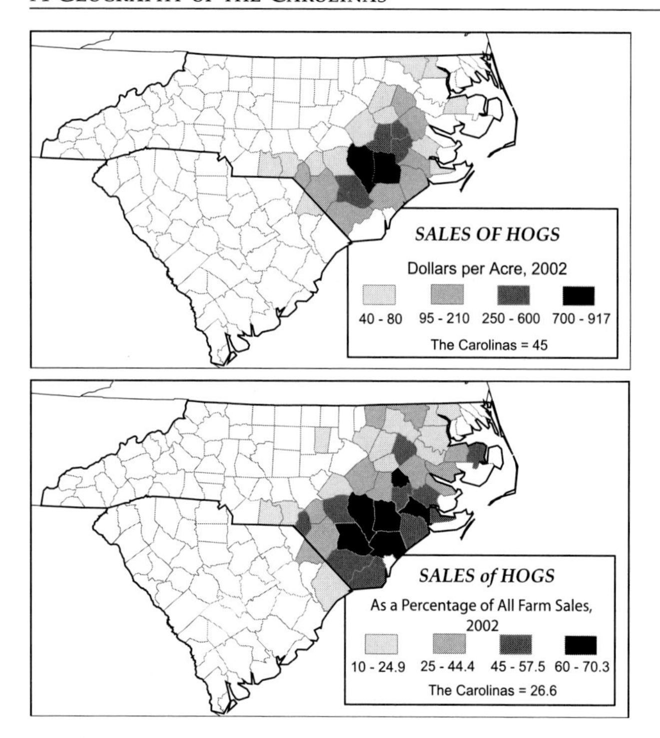

Figure 6.10. Sales of Hogs, 2002.

field Foods of Virginia, a processor, and Prestage Farms of Clinton in Sampson County ranked sixth. They feed corn and soybeans shipped in by unit train from Ohio and Michigan, because North Carolina is a grain deficit area. A shortage of local processing capacity might be a bottleneck to future growth, even though the world's largest hog-processing plant is north of Tar Heel in northwestern Bladen County.

Hog production is widespread in eastern North Carolina, with a truly remarkable concentration in Duplin and Sampson Counties (Figure 6.10). A modern hog farm is a complex of six to fifteen long, one-story buildings in parallel rows, with metal grain bins on stilts next to each building. Each building has large circular openings for ventilator fans or other cooling systems, because hogs generate enormous amounts of body heat, and they are severely stressed by high temperatures.

Three stages of production are in separate structures, and they may even be on separate farms. The piglets stay with their mothers until they are weaned at 21 days when they weigh ten pounds. They go to a nursery for a month or so until they weigh 50 pounds, and then they are moved to a finishing house. Not long ago the standard market weight was "two and a quarter," 225 pounds, but now it is inching up to 300 pounds, or even higher.

Hogs are highly susceptible to a variety of diseases, some of which may even be carried by birds, and most producers enforce a strict policy of "shower in, shower out." Their buildings are enclosed by a chain link fence, and the only entrance is through a locker house where everyone, visitors and employees alike, must strip to the skin, shower, wash their hair, and change their clothes completely before they enter or leave the complex. You shower when you enter because you have to, and you shower when you leave because you want to.

One of the truly distinctive features of a modern hog farm is its manure lagoon. Mature hogs produce two to four times as much solid waste as people do. The easiest way of handling their waste is to flush it into an open-air lagoon lined with tightly compacted clay, where bacterial action breaks it down. Slurry from the lagoon can be sprayed onto crop or pasture land for fertilizer.

Lagoons may leak, they may burst, and they stink to high heaven. Badly constructed lagoons might contaminate groundwater by seepage, and they have generated enormous amounts of negative publicity when they have burst and poured great quantities of waste into nearby streams. Being on the downwind shore of even the best-constructed lagoon is a truly unforgettable experience, and a gentle breeze can waft the fragrance a mile or more.

Lagoons have given critics of the new hog farms an easy and convenient weapon with which to lambaste them. Some critics complain about what they perceive as environmental degradation. Some complain that the new hog farms are "cannibalizing" small farms by squeezing them out of the hog business. Some are animal-rights zealots who abominate all livestock production. And some seem to be mere ideologues who dislike anything that is new, especially if it is also big, and most of all if it involves any massive investment of capital.

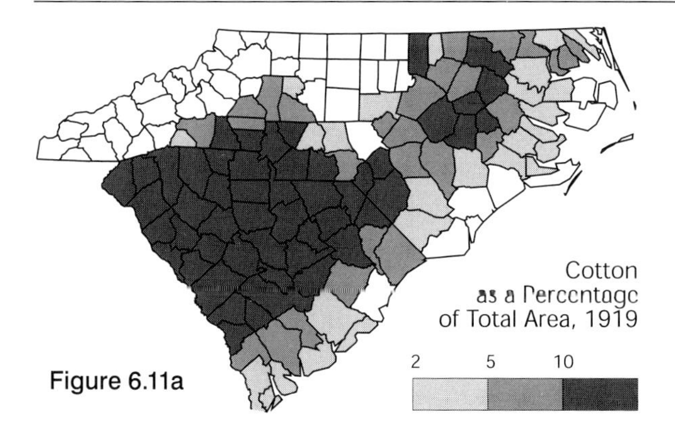

Figure 6.11a

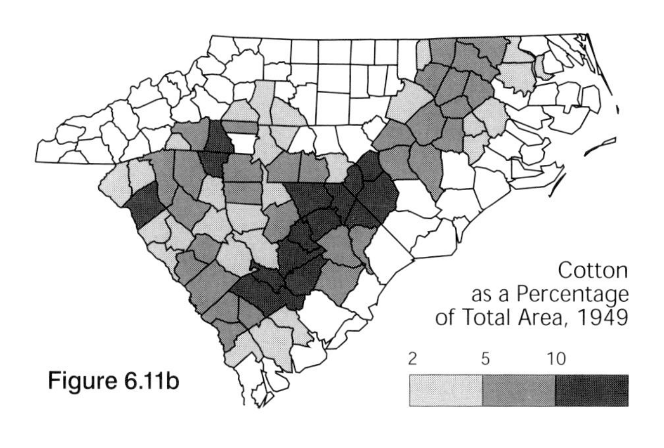

Figure 6.11b

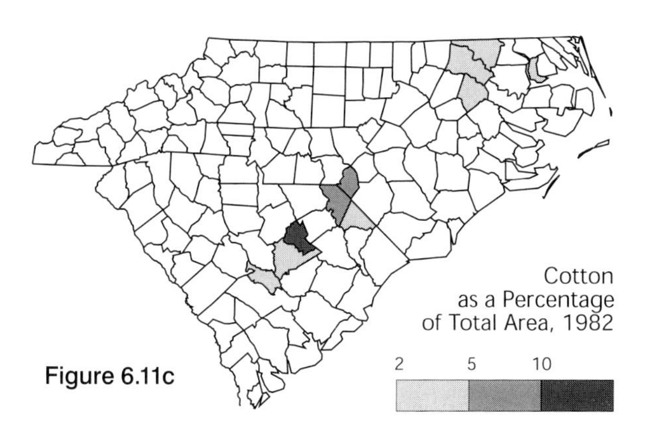

Figure 6.11c

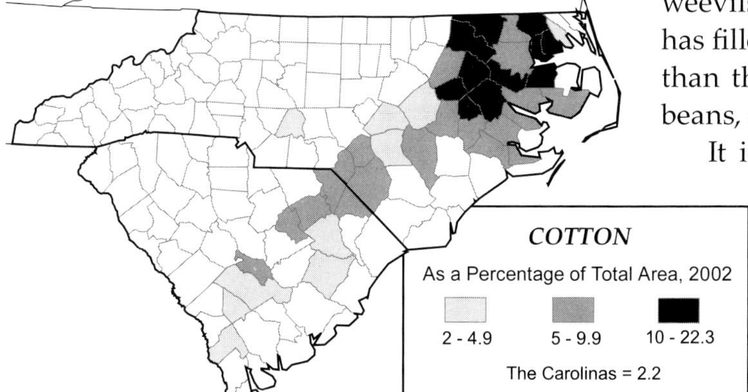

Figure 6.11d. The Regeography of Cotton.

Cotton

One of the most striking developments in Carolina agriculture in the 1990s was the startling resurrection of cotton (Lord, 1996; Nash, 1997; Hart, 2000). The area of cotton in the Carolinas shrank from four million acres in 1919 (Figure 6.11a), when the boll weevil had just begun to devastate the crop, to two million in 1949 (Figure 6.11b), and then to a mere 164,000 acres in 1982 (Figure 6.11c), but by 2002, it had bounced back to more than 1.1 million acres (Figure 6.11d), still only half the 1949 total, and remarkably rearranged geographically.

Two decades ago, I could quite properly write about the demise of King Cotton (Hart, 1977) because cotton had shrunk to the status of a mere curiosity outside a few vestigial island of production in South Carolina and a cluster of three counties at the edge of the peanut country of northeastern North Carolina (Figure 6.11c). In 1978, these three counties (Northampton, Halifax and Edgecombe) had only 13,000 acres of cotton, thirty percent of the cotton acreage of the entire state, but by 2002, they had 171,000 acres of cotton, and their rural landscape was reminiscent of the olden days when cotton had been king. In 2002, however, they had dropped to only 18 percent of the state total, because other areas had increased so dramatically.

Cotton was able to hang on in northeastern North Carolina, if only just barely, because it meshed so well with peanuts (Figure 6.12), the traditional cash crop of the area. The acreage of cotton has increased and the crop has spread so widely because the complete eradication of boll weevils has made it more profitable, because it has filled the need for a more remunerative crop than the combination of corn, wheat, and soybeans, and because the price has been good.

It is tempting to assume that better prices triggered the explosive resurgence of cotton. Certainly an adequate price was necessary, and a good price obviously was an incentive, but the price alone would not have been sufficient without the boll weevil eradication program. In fact, the price of cotton actually had been attractive since 1975 (Lord, 1996, 100), but the acreage of cotton did not begin to surge until boll weevils had been eradicated in northeastern North Carolina in 1982.

The boll weevil eradication program was started in 1978, thanks in no small measure to the initiative and leadership of Marshall Grant, an early chairman of the Southeastern Boll Weevil Eradication Foundation, whose family has farmed near Gaston, just across the river from Roanoke Rapids, since the 1780s (Hart, 2002). An experimental program in Mississippi in 1971 had proved that weevils could be eradicated successfully, but they soon reinfested the test area from all sides. Grant suggested that northeastern North Carolina be picked for a pilot eradication program in 1978, because it was so isolated from other cotton-producing areas. Three years of carefully timed spraying completely eliminated boll weevils from the area, and careful monitoring has detected no signs of subsequent reinfestation.

The boll weevil eradication program was extended to the rest of North Carolina and all of South Carolina in 1983, and by 2004, the boll weevil had been eradicated from the entire South. Marshall Grant told me that the eradication of boll weevils has reduced the cost of growing cotton by at least $50 an acre, because farmers only have to spray twice a year for budworms, rather than ten to 12 times for boll weevils, and their yields have increased by 25 to 75 pounds per acre.

Cotton has replaced the corn-wheat-soybeans combination on the Coastal Plain (Hart 1996). "Corn and soybeans were just not good for us," said Steve Grady, who farms 2,200 acres near Summerlin's Crossroads in the Beautancus community of northern Duplin County. "Cotton has been a better crop." He started with 225 acres of cotton in 1990, and had built up to 1,200 acres by 1997. "I am just a high medium-sized farm," he said, "there are lots bigger than I am." He was also feeding 17,000 hogs on contract for Murphy Family Farms, and was growing 200 acres of tobacco. "Tobacco paid for the hogs and cotton," he said. "If I couldn't grow tobacco I would stop farming."

In 1995, cotton production was concentrated on the Coastal Plain, where the land is level enough for large farm machines, and the farms are large and prosperous enough to be able to afford them (Figure 6.11d). Modern cotton farming requires large operations, and the adoption of cotton required entrepreneurs who were willing to make large investments and to accept large risks. For example, farmers had to grow around 250 acres of cotton to justify the expense of a mechanical cotton picker, but machines have enabled them to handle 500 to 700 acres with only a hired hand or two.

New cotton farmers have had to take the risk of "planting out of program" to "build a base." The program is the government price support program, and the cotton base is the acreage for which a farmer is guaranteed a price. The base is the average acreage of cotton that the farmer has grown in the past three years, so a new cotton farmer had to make a major gamble by growing the crop without a guaranteed price for at least the first three years of production.

The spectacular increase in cotton production has severely taxed existing gin capacity. Steve Grady led a group of his neighbors in an attempt to alleviate their problem by building a cooperative gin near Summerlin's Crossroads at a cost of half a million dollars, but the shortage of gins has transformed the rural landscape by forcing many cotton farmers to shift from wagons to modules.

Traditionally, cotton farmers have unloaded their pickers in the field into cotton wagons with wire-mesh sides

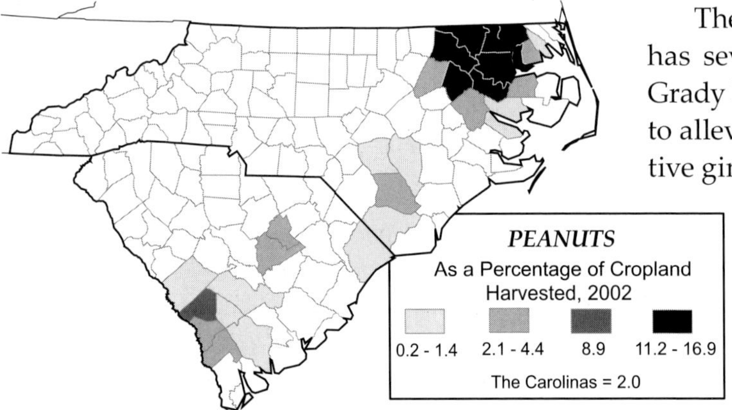

Figure 6.12. Peanuts as a Percentage of Harvested Cropland, 2002.

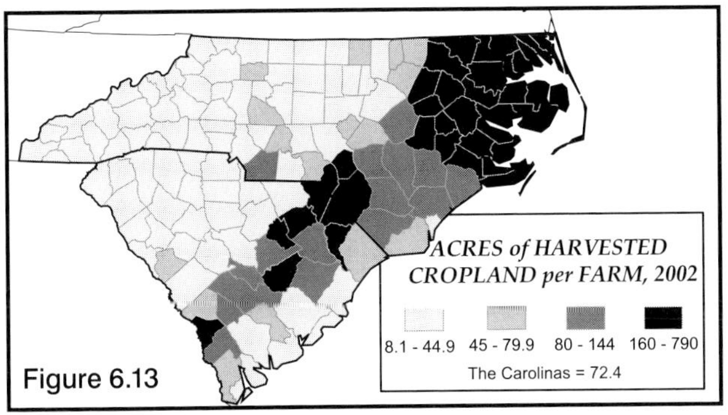

Figure 6.13

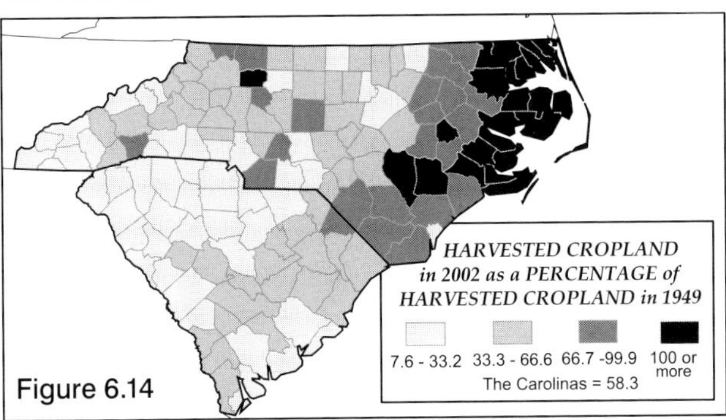

Figure 6.14

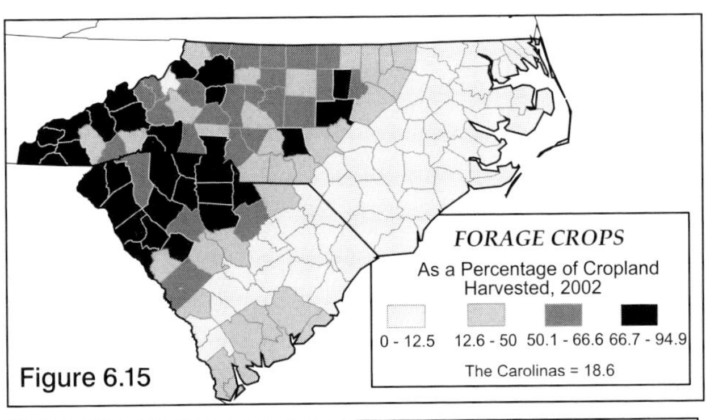

Figure 6.15

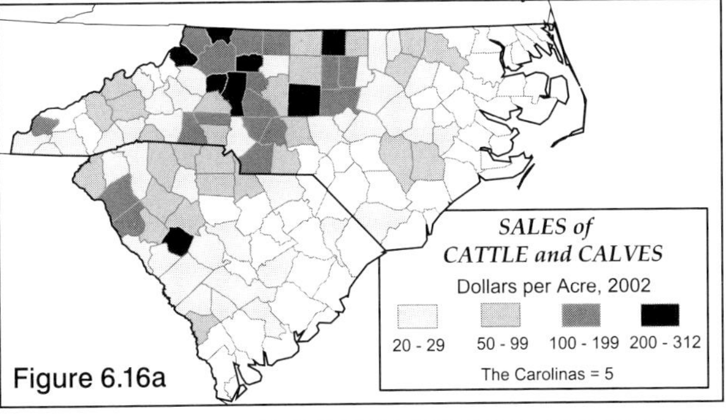

Figure 6.16a

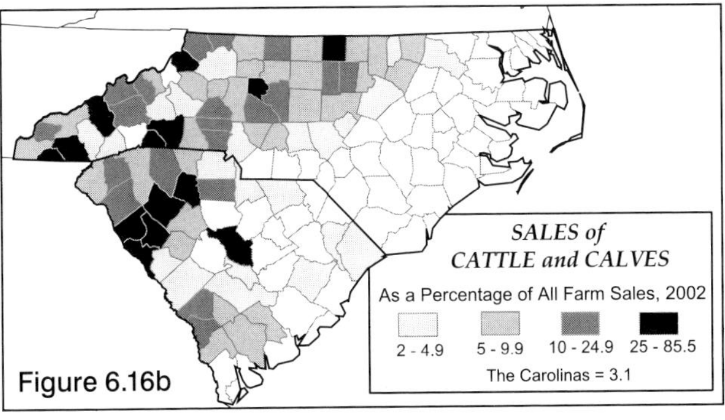

Figure 6.16b

and hauled the wagons to the gin, but the gins could not handle the new volume, and the loaded wagons had to wait in long lines at the gin even though the farmers desperately needed them back in the field. Many farmers have resorted to hydraulic module builders that can compact 6,500 to 12,000 pounds of cotton into rectangular modules the size and shape of semi-trailers. They cover the modules with plastic and leave them in the field until the gin manager says to haul them in.

The Piedmont

Unlike the agricultural economy of the Coastal Plain, which is based on the cultivation of row crops and the confined feeding of hogs and poultry with purchased rather than home farm-grown feeds, the agricultural economy of the Carolinas west of the Fall Line is pastoral; it is a quasi-ranching cattle-rearing economy based on forage crops and pasture.

Small farms and dissected topography severely constrain the options of farmers on the Piedmont and in the Blue Ridge Mountains. In the northern tier of counties along the Virginia line farmers still cling tenaciously to tobacco, which provided more than half of all their farm income in 2002 (Figure 6.7d). They have enlarged their tobacco acreages by leasing (Figure 6.7e), but their farms still are too small to leave them much leeway for experimenting with new ideas (Figure 6.13), and much of the land is too dissected for efficient use of modern farm machinery.

In much of the Piedmont, cotton rather than tobacco was once the dominant crop (Figure 6.11a). The land was worn out growing cotton, and no crop has come forward to replace the deposed king. In 2002, more than half of the land that had been cultivated in 1949 was used only for pasture or woodland (Figure 6.14), and brush was already encroaching on much of the pasture. The rural landscape has a tired, shaggy, overgrown, untrimmed air. Less than one-tenth of the area was harvested cropland in 2002 (Figure 6.2), and more than one-half of this cropland produced forage crops that could only be used to feed cattle (Figure 6.15). The contrast between the Piedmont and Coastal Plain is remarkable. The land on the Piedmont that was once used to grow cotton has been downshifted to forage production and cattle rearing (Figures 6.16a and b).

The development of cattle rearing is one of three major ways in which the agriculture of the Carolina Piedmont has been impacted by the growth of Spersopolis, the vast conurbation that sprawls from Atlanta to Raleigh (Hart and Morgan, 1995). People from the city have acquired rural properties where they can keep hobby herds of beef cattle and enjoy playing cowboy on weekends; their cattle are little more than lawn ornaments. Most of them are pouring money into their properties rather than making any money out of them, but they can bask in the status that cattle confer on their owners.

The second major impact of Spersopolis has been the growth of an urban market. The demand for fluid milk has encouraged the development of large dairy farms (Figure 6.17a and b), and the needs of suburban lawns, gardens, parks, and golf courses has fostered the growth of specialized nursery and greenhouse production in periurban areas (Figures 6.18a and b).

The third major impact of Spersopolis has been the availability of off-farm jobs. Many of the farms of the Piedmont are simply too small to provide an adequate level of living (Figure 6.13). Their operators must supplement their farm income by seeking additional employment, and for many of them farming has become a secondary activity. In 2002, less than half of the farm operators on the Piedmont and in the entire state of South Carolina reported that farming was their principal occupation (Figure 6.19).

Forests

In the Carolinas, timber production and the wood products industry ranks second only to the textile industry in economic importance (Howell and Bischoff, 1996; Johnson and Brown, 1996). The two states annually produce more than two billion cubic feet of primary forest products, mainly pulpwood and sawlogs. Pines and other softwoods comprise roughly 70 percent of the total.

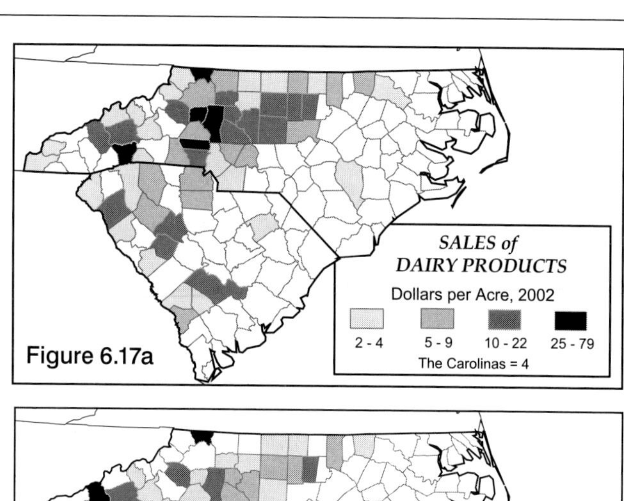

Figure 6.17a

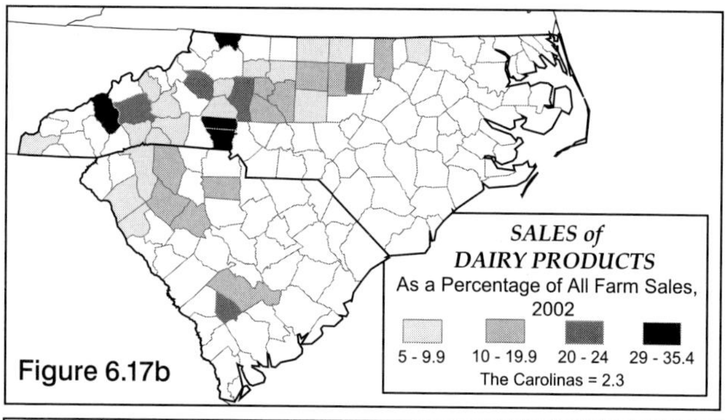

Figure 6.17b

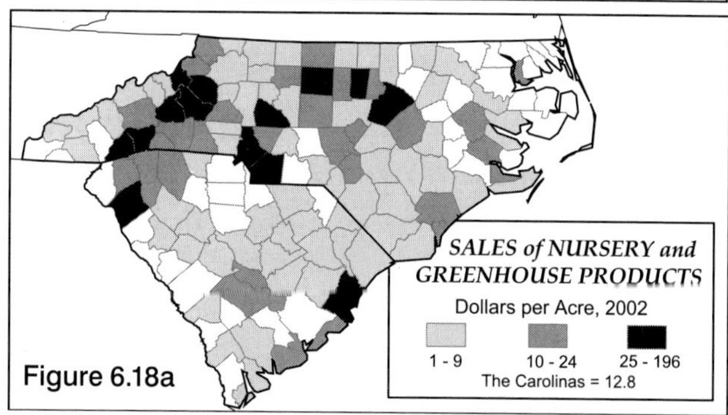

Figure 6.18a

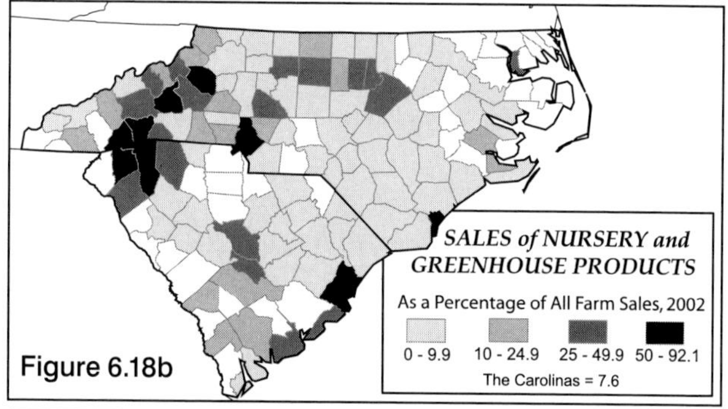

Figure 6.18b

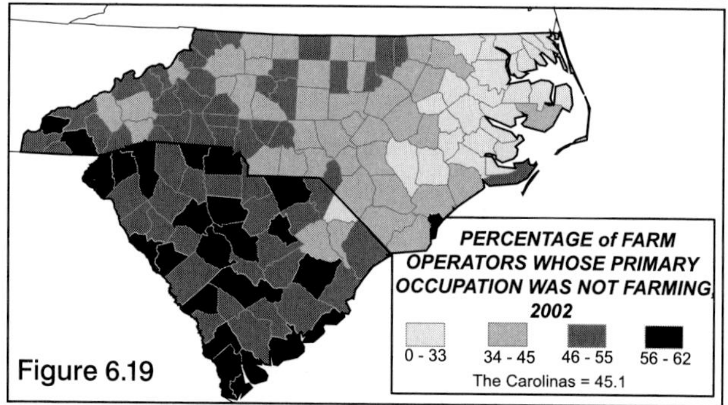

Figure 6.19

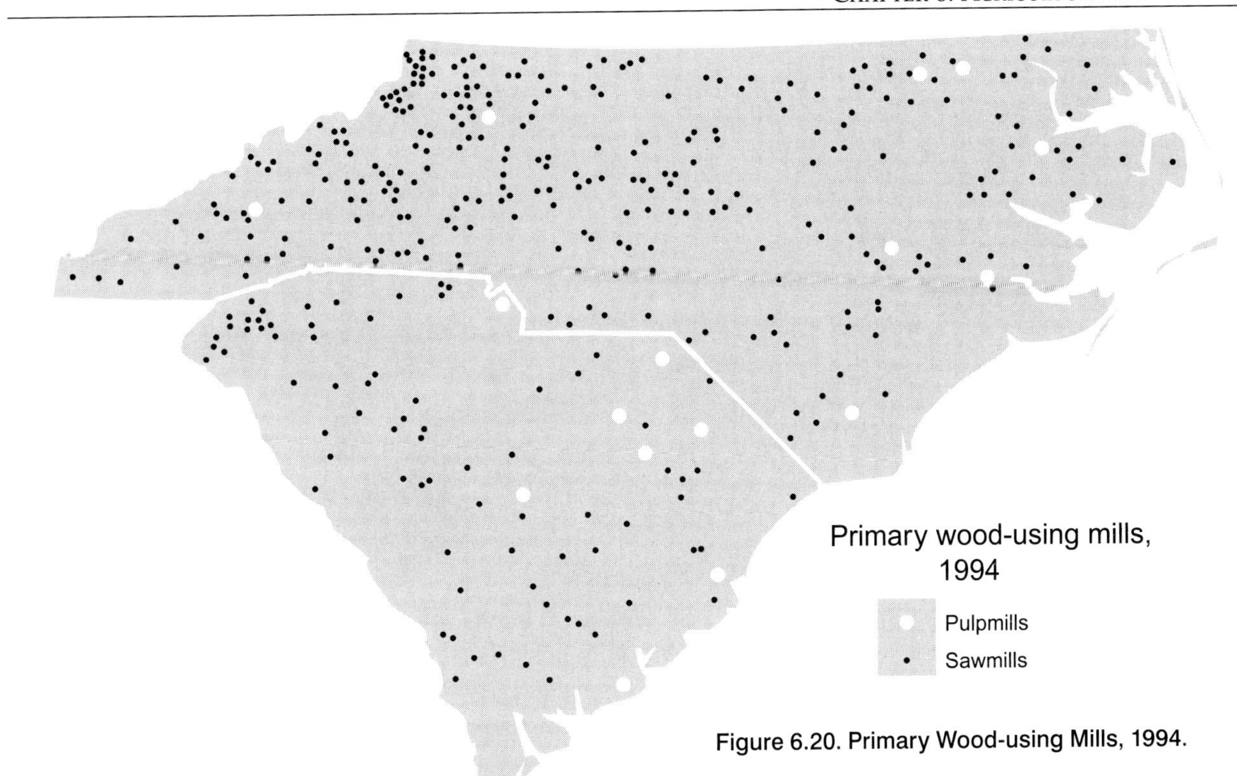

Figure 6.20. Primary Wood-using Mills, 1994.

Five of every eight acres in the Carolinas are wooded (Johnson, 1991; Conner, 1993). Many of the highways in the region are long "pine tunnels" that bore through seemingly interminable stretches of woodland. Naval stores derived from pine trees were the first economic products of the Tar Heel State, and the Palmetto State takes its nickname from the native tree that provided a stout bulwark against British cannon balls during the Revolutionary War.

The traditional cash crops of the Carolinas (tobacco, cotton, peanuts) have been cultivated intensively on small patches of land cleared from the encircling woods. The farmer cleared a new patch when the old patch was worn out, and second-growth woodland quickly colonized the old abandoned patch. Every county had a few small, portable, "peckerwood" sawmills that survived by scavenging this process of rotating land back and forth between cropland and woodland, but in recent years small portable mills have been replaced by a much smaller number of new, larger, and more efficient mills that are permanently sited.

The greatest concentration of sawmills today is in the western mountains (Figure 6.20), whose hardwood forests are the resource base for the wooden furniture manufacturing industry (Gade, 1996). The western mountains are one of the three types of areas in the Carolinas that are extremely heavily wooded because environmental constraints severely restrict any attempt at agriculture (Figure 6.21a): dissected topography and steep slopes in the mountains, thin and infertile soils in the Slate Belt and on the Carolina Piedmont, and sluggish drainage on the low-lying land of the Flatwoods. The only areas where less than half of the land is wooded are the built-up areas of Spersopolis and the heart of the flue-cured tobacco country on the Coastal Plain of eastern North Carolina (Figure 6.21a).

Pine trees colonized the cotton fields that boll weevils devastated in the 1910s. This bonanza forest was the resource base for the rapid growth of the pulp and paper industry throughout the South in the 1950s and 60s (Hart, 1976, 46-56). Most of the pulp mills in the Carolinas are in the east (Figure 6.20), where pulp and paper companies have acquired large acreages of land, whether by outright purchase or by very-long-term lease (Figure 6.21b). In the Carolinas the forest industries own 7.5 million acres, 15 percent of the total area, and two million acres more than the total acreage of cropland harvested in both states.

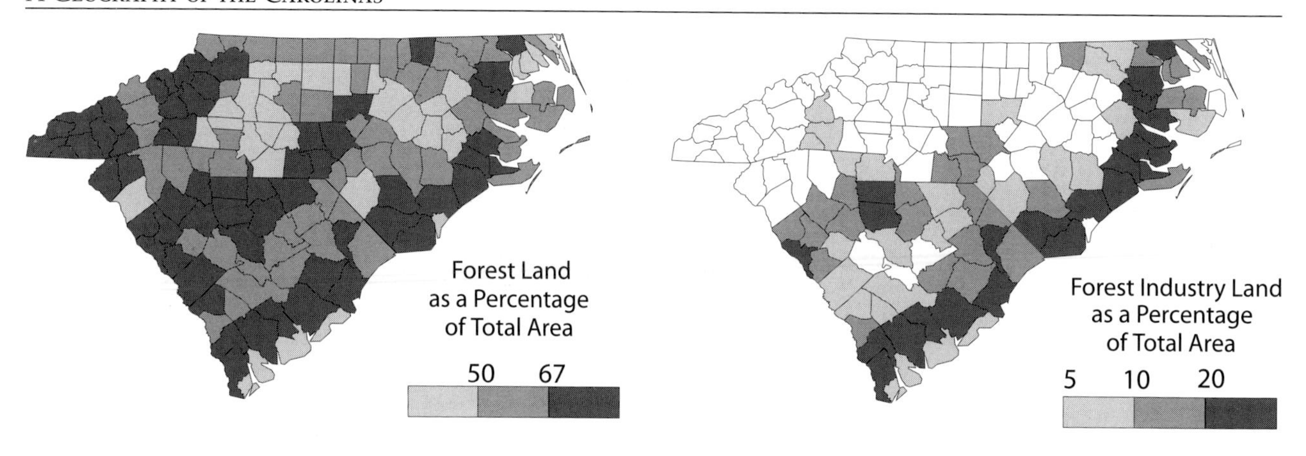

Figure 6.21. Forest Land and Forest Industry Land.

Summary

The production of row crops in the Carolinas is heavily concentrated on the level land of the Coastal Plain, where tobacco has been the traditional crop. Tobacco is still grown on a small scale in the northern tier of counties in North Carolina, and on an even smaller scale in the western mountains, but in the rest of the Carolinas most of the land has been downshifted to cattle-rearing or has been allowed to grow up in second-growth woodland. Many cattle producers have off-farm jobs that enable them to afford to indulge in their hobby.

Flue-cured tobacco production has been mechanized since the 1960s, and sleek new metal bulk barns have replaced the squat old flue-cured barns. Mechanization has enabled farmers on the Coastal Plain to increase their tobacco acreage, but they are under merciless political pressure to shift to other crops. They have grown corn, wheat, and soybeans on the land for which they had no tobacco base, but they have grown these crops by default for lack of anything better, because the corn-wheat-soybeans combination was a sorry cropping system that did not make them much money.

Farms on the Coastal Plain have been large enough to give farmers the freedom to experiment with new ideas, and in the 1990s, they have eagerly latched on to cotton as a new cash crop. A rapid increase in cotton production has been triggered by the boll weevil eradication program that originated in northeastern North Carolina in the late 1980s, and has disseminated therefrom. Cotton production on the Carolina Coastal Plain is hostage to the world price of cotton, which has been exceptionally favorable since the mid-1970s, and it could suffer if that price were to drop.

Local entrepreneurs in southeastern North Carolina developed large-scale contract production of broilers in long, low, purpose-built structures during the 1960s, and turkey and hog producers have emulated the broiler model. Contract feeding of broilers, turkeys, and hogs accounts for the areas that had the greatest net cash farm income (Figure 6.22), but these areas rely almost completely on grain shipped in from the Midwest in unit trains, and they are under heavy pressure from critics who believe that contract feeding causes serious environmental degradation.

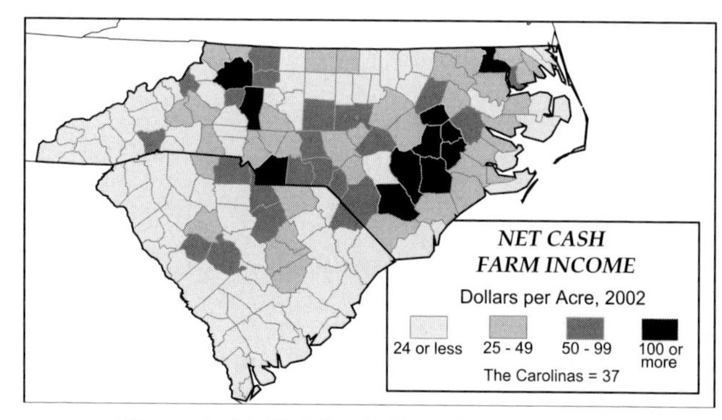

Figure 6.22. Net Cash Farm Income, 2002.

Acknowledgments

I am indebted to all those who have given me so graciously and so generously of their time and knowledge. In addition to those whom I have named in the text, I owe a special debt of gratitude to Dan Blackshear, Ennis Chestang, Curtis Fountain, Marshall Grant, Wendell Murphy, Greg Schmidt, and Jim Stocker. I am grateful to Jodi Larson for having wordprocessed the manuscript and to Mark Lindberg, Hilda Kurtz and Jane Mueller for having converted my rough maps and to Jeff Patton for having designed the final version of the maps.

References

Algeo, K. 1997. The Rise of Tobacco as a Southern Appalachian Staple: Madison County, North Carolina. *Southeastern Geographer* 37 (1), 46-60.

Conner, R. C. 1993. *Forest Statistics for South Carolina, 1993*. Resource Bulletin SE-141. Asheville, NC: Southeastern Forest Experiment Station.

Furuseth, O. 1997. Restructuring of Hog Farming in North Carolina: Explosion and Implosion. *Professional Geographer* 49 (4), 391-403.

Gade, O. 1996. Furniture in North Carolina. In *Snapshots of the Carolinas: Landscapes and Cultures*. ed. D. G. Bennett, 165-170. Washington, D.C.: Association of American Geographers.

Garner, W. W., E. G. Moss, H. S. Yohe, F. B. Wilkinson, and O. C. Stine. 1923. History and Status of Tobacco Culture. In *United States Department of Agriculture Yearbook 1922*, 395-468. Washington, D.C.: Government Printing Office.

Hart, J. F. 1976. *The South*. New York: D. Van Nostrand.

--------. 1977. The Demise of King Cotton. *Annals of the Association of American Geographers* 67(3), 307-322.

--------. 1991. *The Land That Feeds Us*. New York: W. W. Norton.

--------. 2000. The Metempsychosis of King Cotton. *Southeastern Geographer* 40(1), 93-105.

_____. 2002. Boll Weevil Eradication in the United States through 1999. *Southeastern Geographer* 42 (1), 364-369.

Hart, J. F. and E. L. Chestang. 1978. Rural revolution in east Carolina. *Geographical Review* 68(4), 435-458.

--------. 1996. Turmoil in Tobaccoland. *Geographical Review* 86(4), 550-572.

Hart, J. F. and E. C. Mather. 1961. The Character of Tobacco Barns and their Role in the Tobacco Economy of the United States. *Annals of the Association of American Geographers* 51(3), 274-293.

Hart, J. F. and J. T. Morgan. 1995. Spersopolis. *Southeastern Geographer* 35(2), 103-117.

Howell, M. and P. S. Bischoff. 1996. *South Carolina's Timber Industry: An assessment of Timber Product Output and Use, 1994*. Resource Bulletin SRS-7. Asheville, NC: U. S. Forest Service Southern Research Station.

Johnson, T. G. 1991. *Forest Statistics for North Carolina, 1990*. Resource Bulletin SE-120. Asheville, NC: Southeastern Forest Experiment Station.

Johnson, T. G., and D. R. Brown. 1996. *North Carolina's Timber Industry: An Assessment of Timber Product Output and Use, 1994*. Resource Bulletin SRS-4. Asheville, NC: U. S. Forest Service Southern Research Station.

Lord, J. D. 1971. The Growth and Localization of the United States Broiler Chicken Industry. *Southeastern Geographer*. 11(1), 29-42.

--------. 1996. The New Geography of Cotton Production in North Carolina. *Southeastern Geographer* 36(2), 93-112.

Nash, B. J. 1997. Cotton's Comeback. *Region Focus* (Federal Reserve Bank of Richmond). 1(2), 16-19.

Roth, D. 1997. The Ray Kroc of Pigsties. *Forbes* 160(8), 115-120.

Tilley, N. M. 1948. *The bright Tobacco Industry*. Chapel Hill: University of North Carolina Press.

Chapter 7

The Population of the Carolinas

Melinda Meade

University of North Carolina at Chapel Hill

The Carolinas enter the new century with historic population growth and rapid change in ethnicity, age structure, and the population dynamics of birth, death, and migration. With a 2000 population of 12,061,325 in a nation of 281,421,906, Carolinians have grown from 4.1 percent of the nation's population in 1990 to 4.3 percent in 2000 (U.S. Census Bureau, 2000). The population has grown older and more diverse ethnically. It is mostly urban, rural-nonfarm, and metropolitan. All these and other characteristics are quite unevenly distributed across the region. The population of North Carolina, which grew 21.4 percent over the decade compared to South Carolina's 15.1 percent, rose to 66.7 percent of the combined population of the Carolinas. The following overview looks at the current distribution and composition of the population. This is followed by an examination of the developments and dynamics which produced this population, a consideration of its health and well-being, and a discussion of the migration trends which are transforming it. Ten counties are used to illustrate regional patterns within the Carolinas. Finally, of course, the future is prognosticated.

The Current Population

The 12 million people who live in the Carolinas are not staying still. Their historically uneven distribution across the state is becoming more so as they join newcomers in moving to metropolitan areas where the jobs are. The population is growing by natural increase, but it is also growing older. It is also growing by in-migration of all races, enough to have gained an additional Congressional seat for North Carolina in 2000, and so it is growing much more diverse.

Figure 7.1 shows the population size of the counties, and Figure 7.2, the resulting population density. Of North Carolina's 100 counties, 23 have more than 100,000 people and only seven have more than 200,000; of South Carolina's 46 counties, 14 have more than 100,000 people and only five have more than 200,000. Large populations are found in the port cities of Charleston and Wilmington, but population is more concentrated in the central areas of both states. In North Carolina, this takes the form of a contiguous line of counties, known as the "Piedmont Crescent," which is connected along Interstate 85 from the capital, Raleigh (Raleigh-Durham-

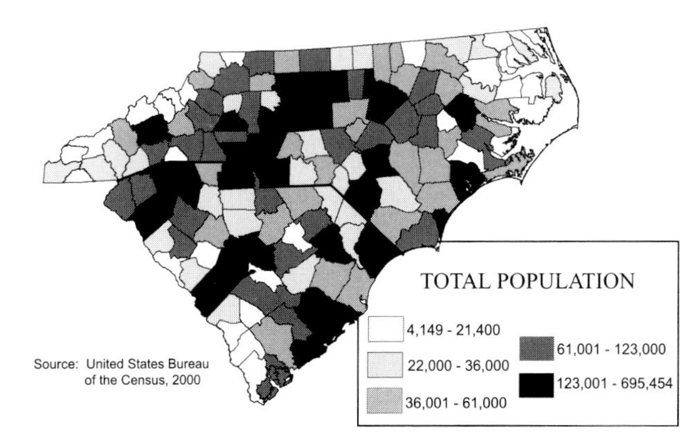

Figure 7.1. Total Population, 2000. Map shows the numbers of people living in each county, according to the 2000 census.

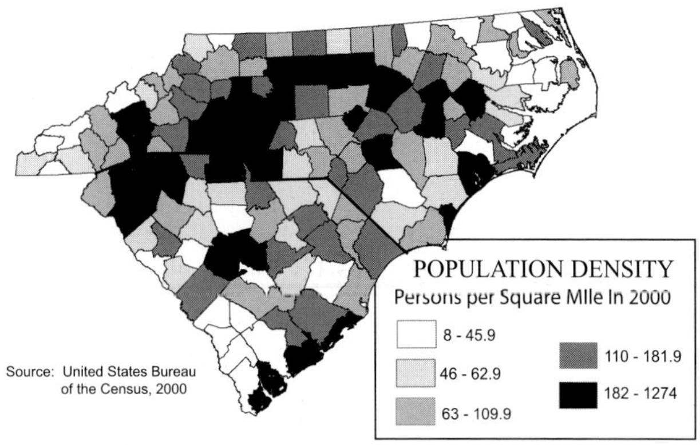

Figure 7.2. Population Density (2000). Map shows the number of people per square mile of land area.

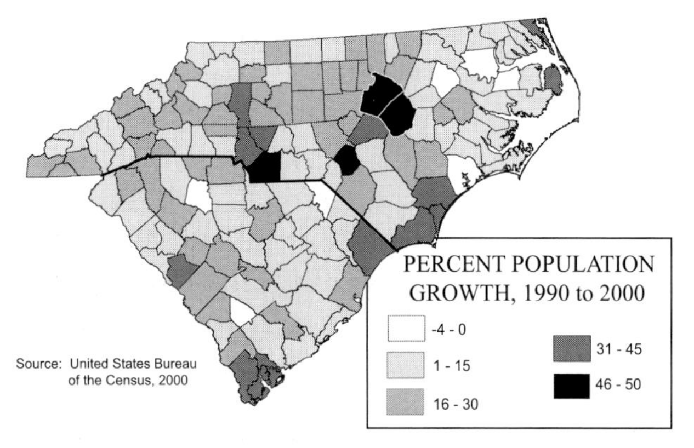

Figure 7.3. Percent Population Growth, 1990-2000. Map shows the percent of population growth over the decade, according to the 2000 census. Only six counties actually lost population, but many grew by only a few percent.

Chapel Hill, the Research Triangle), through the textile/furniture/cigarette industrial centers (the Triad: Greensboro, Winston-Salem, High Point) to the main regional city, Charlotte (Metrolina area), and on into western South Carolina. The densely populated center of that state contains its capital, Colombia. The mountain counties in the west have historically supported small and rather isolated populations, difficult to reach with infrastructure and services. The exception to this, of course, is the city of Asheville rising over the mountain passes. The agricultural Coastal Plain becomes almost empty in northeastern North Carolina, where counties on Albemarle Sound have only a few thousand people. Tyrrell County has 4,149; Hyde County, with 5,826, has hundreds of times more hogs; and Dare County's population lives mostly on the Outer Banks, leaving most of the county's land for wildlife refuge empty enough of people that red wolves have been reintroduced there. These counties have been losing population for several decades. A few still are, but most have had a slight increase this past decade.

The counties that have been growing in population (Figure 7.3) mostly fall into one of three categories. The first group are those with an attractive "amenity," mostly beaches or boating, but also serious golf in the Sandhills (e.g., Pinehurst), and mountain coolness and scenery. The coastal development along the beaches from north of Wilmington south through Myrtle Beach has been phenomenal. Its seed pearl and crown jewel is Hilton Head and the growth of Beaufort County in the southeastern corner. The two counties in South Carolina which grew at over 30 percent the past decade are those which contain Myrtle Beach (Horry) and Hilton Head (Beaufort). Such development and relocation of population poses a greatly increased risk to life and property from hurricane landfall and other storms. The coast is attractive to state residents, as well as to retirement of those from the "Frozen North." Attraction for retirees is not limited to ocean front property, however. The possibility of playing golf and tennis every month while living near the cultural amenities, educational opportunities, and health care provision of major universities and medical centers has similarly attracted retirees to central Piedmont counties, as in the Raleigh-Durham-Chapel Hill-Cary metropolitan area. Smaller numbers of retirees in the more sparsely populated mountain counties, especially near the Great Smokies and Georgia, have made a significant impact.

A second group of counties growing rapidly in population have large institutions, mainly universities or military bases, which have large numbers of participants as well as big payrolls that support other community businesses and jobs. Universities, such as The University of North Carolina at Chapel Hill, North Carolina State, North Carolina Central, and Duke (the Research Triangle) not only attract retirees to a quality of life and health care, but also create whole new industries in this information age and are large employers themselves. Thus, Greenville, N.C. (Pitt County) emerged into a rapidly growing

MSA with the development of East Carolina University, including a second state medical school, and Watauga County in the northwestern mountains has grown apace with Appalachian State University, as well as the winter amenity of skiing. Columbia, S.C. also is that state's capital, but the growth of the University of South Carolina there has played a similar role. Fewer in number and less subject to predictable economic growth are the major military bases such as the U.S. Marines' Camp Lejeune (Jacksonville MSA, Onslow County) or the Army's Fort Brag (Fayetteville MSA, Cumberland and Hoke counties). Military counties offer a variety of demographic anomalies. They tend to have relatively high birth rates, for example; yet many babies may migrate out before they are a year old. In some years during the past decade, Onslow County lost population; but it managed to come out even by the end of the decade.

Most of the growth has occurred in metropolitan areas, which include many of the university, military, and amenity counties. Metropolitan Statistical Areas are located in Figure 7.4. (Metropolitan Statistical Areas (MSAs) are defined by the Census Bureau to contain a city of 50,000 or more population within legal city boundaries, or an Urbanized Area (UA) of densely settled territory containing 50,000 and a total metro population of over 100,000. Thus, a MSA may be one county containing a central city, such as Jacksonville or Fayetteville (military counties) or Greenville (university county); they may contain a city and its suburban county ring; or they may, in the emerging southern form, sprawl and leap over low density areas following major highways.)

One-third of North Carolina's population growth occurred in the urbanized counties of its three largest cities: Mecklenburg (Charlotte), Guilford (Greensboro), and Wake (Raleigh). More significantly, however, the six counties that grew more than 30 percent over the decade belonged to the sprawling economic and residential growth of metropolitan Charlotte (itself in Mecklenburg

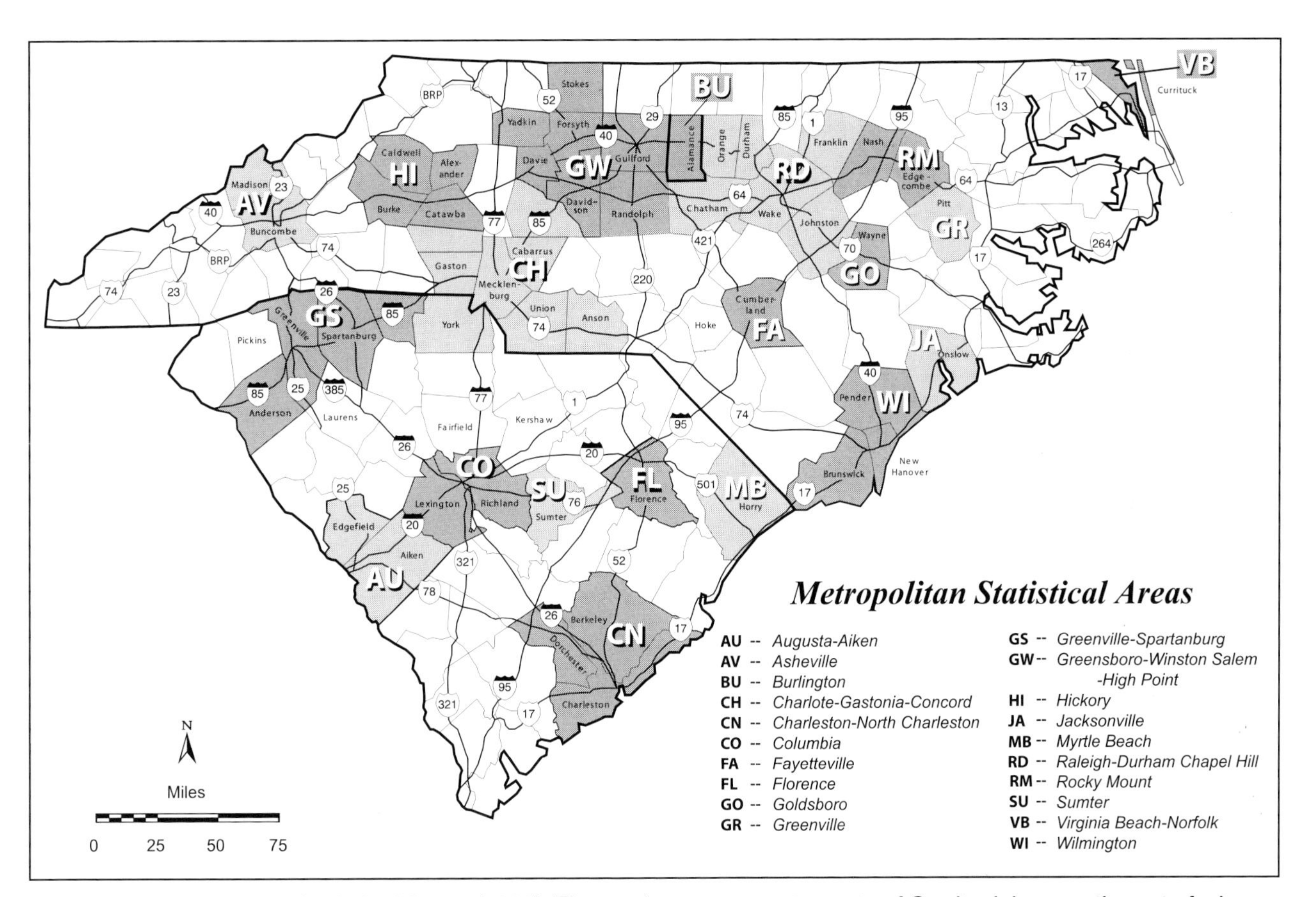

Figure 7.4. Metropolitan Statistical Areas (2000). The northeastern-most county of Currituck is recently part of a large MSA in Virginia. Aiken County on the Georgia border has become part of greater Augusta, GA.

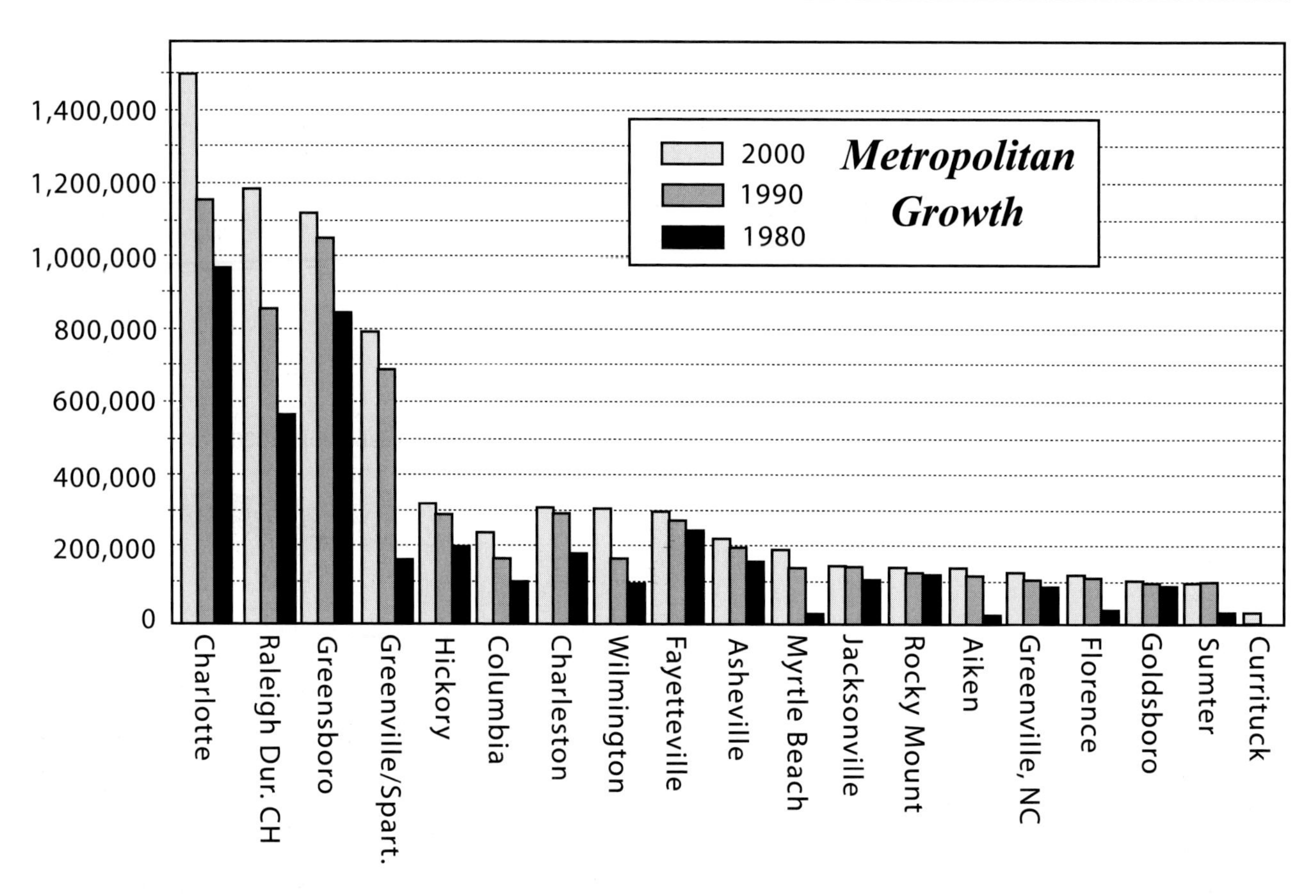

Figure 7.5. Metropolitan Growth (1980, 1990, 2000). Data are from the decennial censuses. The depicted MSA populations for 1980 are not exactly comparable as later population annexations by cities and connections between counties were part of MSA growth. County populations of 1980 were used in several cases to make the MSA comparisons as similar as possible.

County, 36% growth; also Cabarrus, 32.5%; Iredell, 31.6%; Union, 46.9%) and Raleigh (itself in Wake, 47.3% growth, including the mushrooming suburban city of Cary; and Johnston County, 50%). The population of North Carolina is now 66.8 percent metropolitan, that of South Carolina is 53.9 percent, and the combined population of the Carolinas is 62.5 percent.

This impressive metropolitan growth can be seen in Figure 7.5. Metropolitan Charlotte, sprawling across the state border and anchoring the I-85 Piedmont Crescent with its banking, is half again as large as it was twenty years ago. Then, Greensboro was the second largest metropolitan area, but in the past decade Raleigh-Durham-Chapel Hill, also known as the Research Triangle, has exploded beyond it. Other metropolitan areas also changed relative ranking. Wilmington and its coastal amenities have surpassed Asheville and its mountain ones. Myrtle Beach has emerged, seemingly from the sea. Despite some jostling for ranking, however, the urban hierarchy of the Carolinas remains remarkably regular and dispersed throughout both states.

Population Composition

The composition of the population in terms of race, age, and gender shows strong regionalization within the Carolinas. The physiographic regions of the Carolinas are strongly associated with historic economic and cultural differences, as well as topography and climate, and serve to frame the patterns of population distribution. The regions, mapped in Figure 1.4, follow the convention most common in North Carolina: Mountains, Piedmont, Coastal Plain, Tidewater. In South Carolina, "the Mountains" are merely the Blue Ridge. More than half the state is Piedmont, so the Tar Heel regionalization has been less useful. South Carolina convention subdivides the Piedmont and Coastal Plain into several vernacular or agricultural regions (Kovacik and Winberry, 1987, 210-215). The Piedmont is subdivided into the Upper Piedmont manufacturing region which the textile industry used to dominate (Greenville-Spartanburg-Anderson MSA), and the Midlands, which contain an urban-industrial complex focused on the state capital, Colombia. The Coastal Plain is subdivided, vernacularly, by river basin or by landscape and production into the Pee Dee tobacco, agricultural coastal plain, and plantation forest regions. The Tidewater becomes, simply, the coastal zone. These regions of land use and history are today reflected also in population distribution and composition.

The Carolinas remain overwhelmingly a composition in white and black, despite the recent increase in diversity. In 2000, the population of the Carolinas was 70.5 percent white and 24.2 percent black (North Carolina 72.1% white and 21.6% black; South Carolina 67.2% white and 29.5% black). The patterns of concentration and distribution of the white and black populations within the states continue to be almost mirror images, reflecting relatively little African-American movement into the Piedmont industries (Figures 7.6 and 7.7). The impact of white retirees on Currituck County (Norfolk-Virginia Beach-Newport News MSA), New Hanover, Pender and Brunswick counties (Wilmington area) and Horry and Georgetown counties (Myrtle Beach area) and other ocean front amenity locations is evident. Only 1.3 percent of North Carolina's and 1.0 percent of South Carolina's population reported two or more races, compared with 2.4 percent nationally. With one or two percent reported across counties, there is surprisingly little spatial variation to map.

Native Americans of several types can be found in the military counties, but across the Carolinas they constitute 1.2 percent (or 99,551 people) of North Carolina's population and 0.3 percent (13,718) of South Carolina's (Figure 7.8a). The eastern band of the Cherokee Nation, located in the southern mountains of North Carolina, is the only federally-recognized tribe and the best known, but they number fewer than 10,000 in the state. They descend from the Cherokees the federal government failed to evict from this land to the sorrowful long march to Oklahoma. The State of North Carolina recognizes the Lumbee Indians concentrated in its rural southeastern counties. Composing almost two-thirds of the state's American Indian population, the Lumbees are the largest non-federally recognized tribe; but there are many small ones. In South

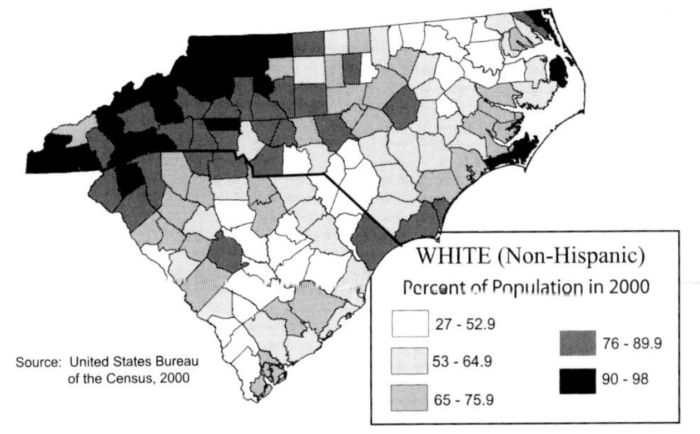

Figure 7.6. Percent of Population White (Non-Hispanic), (2000).

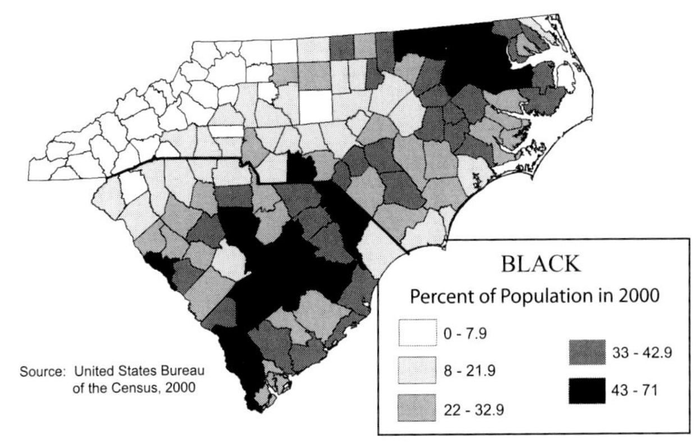

Figure 7.7. Percent of Population Black (African-American), (2000).

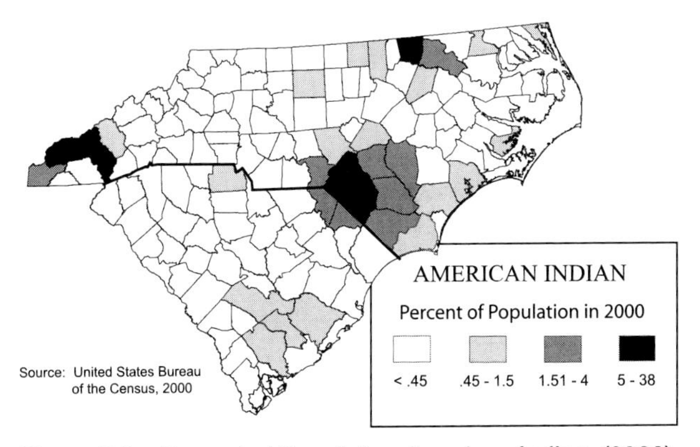

Figure 7.8a. Percent of Population American Indian, (2000).

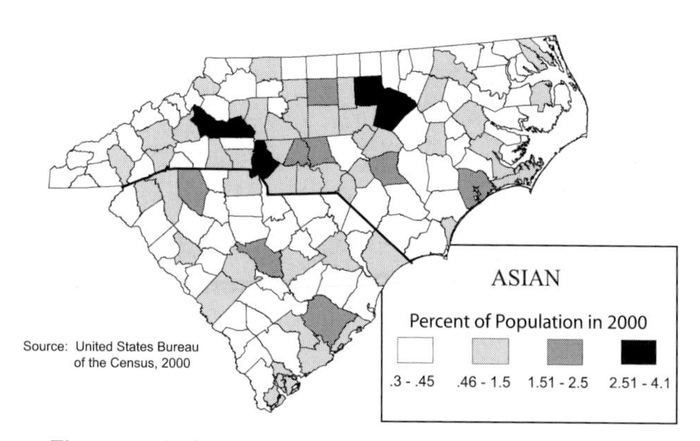

Figure 7.8b. Percent of Population Asian, (2000). This population includes former Southeast Asian refugees and their descendents, Chinese and Indians and Japanese, Filipinos and other nationalities.

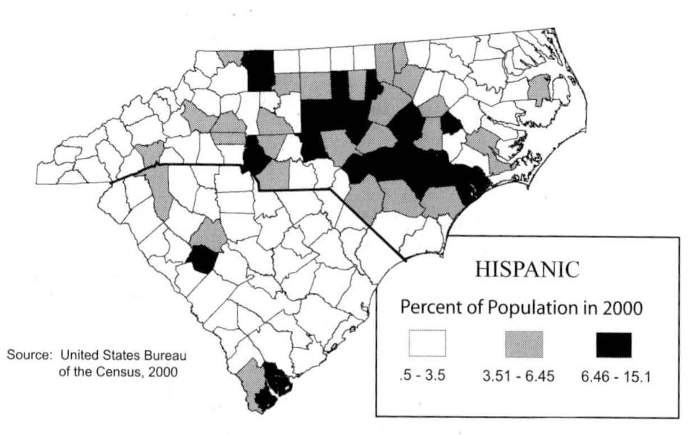

Figure 7.8c. Percent of Population Hispanic, (2000). Hispanics may be of several races.

Carolina, a few hundred to a couple thousand people identify themselves as Catawba, Edisto, Pee Dee, or Santee Indians (Kovacik and Winberry, 1987, 62-63). These non-recognized tribes have no operative treaties, reservations, or official status and over the centuries have become racially mixed. Self-identification is increasing, however, and tribal gatherings and ceremonial centers are seeking to maintain and re-establish cultural characteristics.

The Asian population has grown rapidly at the end of the century, but the 150,000 Asians remain a tiny proportion (S.C. 0.9%; N.C. 1.4%). The Asian population resides almost entirely in the metropolitan counties (Figure 7.8b). There are several different components. Asian Indians constitute 23 percent and Chinese 17 percent of the Asian population in both states. Vietnamese, 12-13 percent of the Asian population, were resettled in the big cities, and Hmong refugees from Laos were resettled in North Carolina's mountains. Both states' Asian populations are 10-11 percent Korean, and only 5-6 percent Japanese. The biggest proportional difference is in North Carolina's 8 percent Filipino to South Carolina's 18 percent, probably because of the historic importance of U.S. naval connections with the Philippines and the location of the naval base near Charleston. Military counties' populations are diverse.

The Hispanic population has been the big surprise of the 2000 census. In North Carolina, it surged to 378,963 from 76,745 in 1990, a 394 percent in a decade, to constitute 4.7 percent of the state's population. Of these, 46,141 were born in the state during the decade (OSP 2000), mostly concentrated in non-metropolitan counties in which chicken broilers are economically important. South Carolina's Hispanic population grew to 95,000, or 2.4 percent of the total. The Hispanic population is only two percent Cuban and eight percent Puerto Rican. It is two-thirds Mexican, and the remainder mostly from Central American countries. In South Carolina, Hispanic people work in Hilton Head, in the chicken broiler factories of Saluda County (near the state center), and in the construction and service industries of Greenville (Figure 7.8c). In North Carolina, Mexican migrants were first attracted to field work in the agricultural counties of the Coastal Plain, which saw significant growth in this group in the 1980s. There were also early concentrations in the military counties. During the past decade, labor has been directly recruited for work in the chicken factories (Johnson and Webb, 2000). The impact on the small, rural towns has been enormous. Some small towns have changed from being 95 percent white to less than 50 percent non-Hispanic white, with the majority of elementary school children not understanding English. By 2000, Hispanic workers seemed indispensable to the construction, landscaping, and hotel/restaurant work of the booming metropolitan counties.

Despite the influx of youthful Hispanic families and laborers, the population of the Carolinas has continued to age along with the rest of the country. From a median age of around 20 before World War II and 23 until the sixties, it has risen continuously to 35.3 in N.C. and 35.4 in S.C. in 2000 (NC-SC: 1970, 26.5-24.7; 1980, 29.9-28; 1990,

33-32.1). A high proportion of elderly people can result from three distinct influences. Population can age in place as fertility (birth) levels fall and mortality levels fall, increasing the number of people living more years while decreasing the number of children born. Large numbers of retirees moving into a county can rather quickly raise the proportion of seniors. Less obviously, large numbers of young adults moving out of a county to look for work or education can increase the proportion of the remaining population that is older. Similarly, a larger proportion of children has been common in rural areas where people have large families; or it can result from young adults moving into an area for work and subsequently starting their families.

All of the migration influences on age structure and fertility can be seen in the opposition of the youngest and the oldest in Figures 7.9a and 7.9b. The population over 65 is proportionately higher in the counties that have lacked economic development and lost young adults for decades. It is also high in the amenity counties that do not have a large population, as in some of the moun-

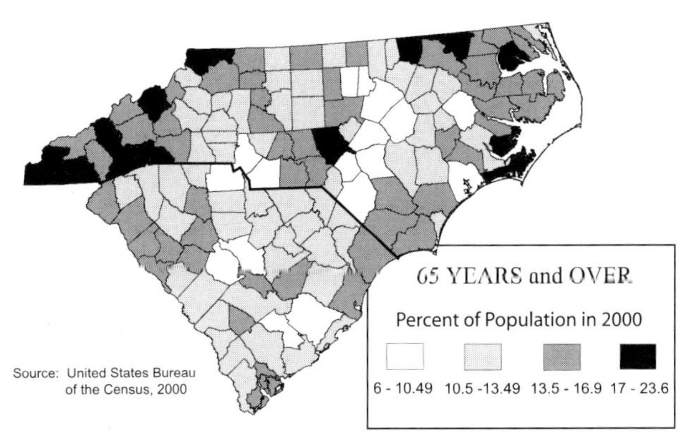

Figure 7.9a. Percent of population 65 and over, (2000).

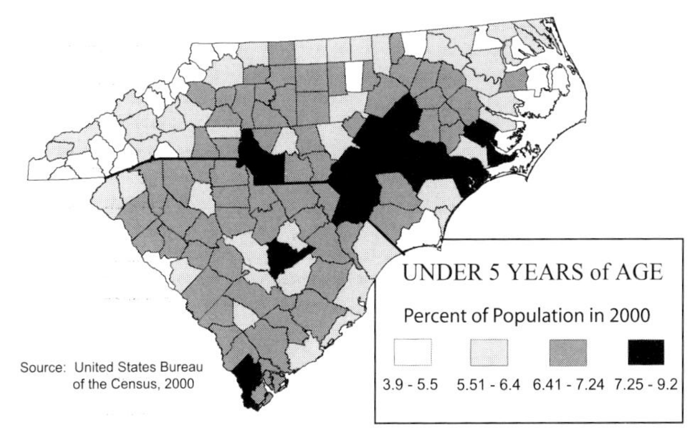

Figure 7.9b. Percent of Population under 5, (2000).

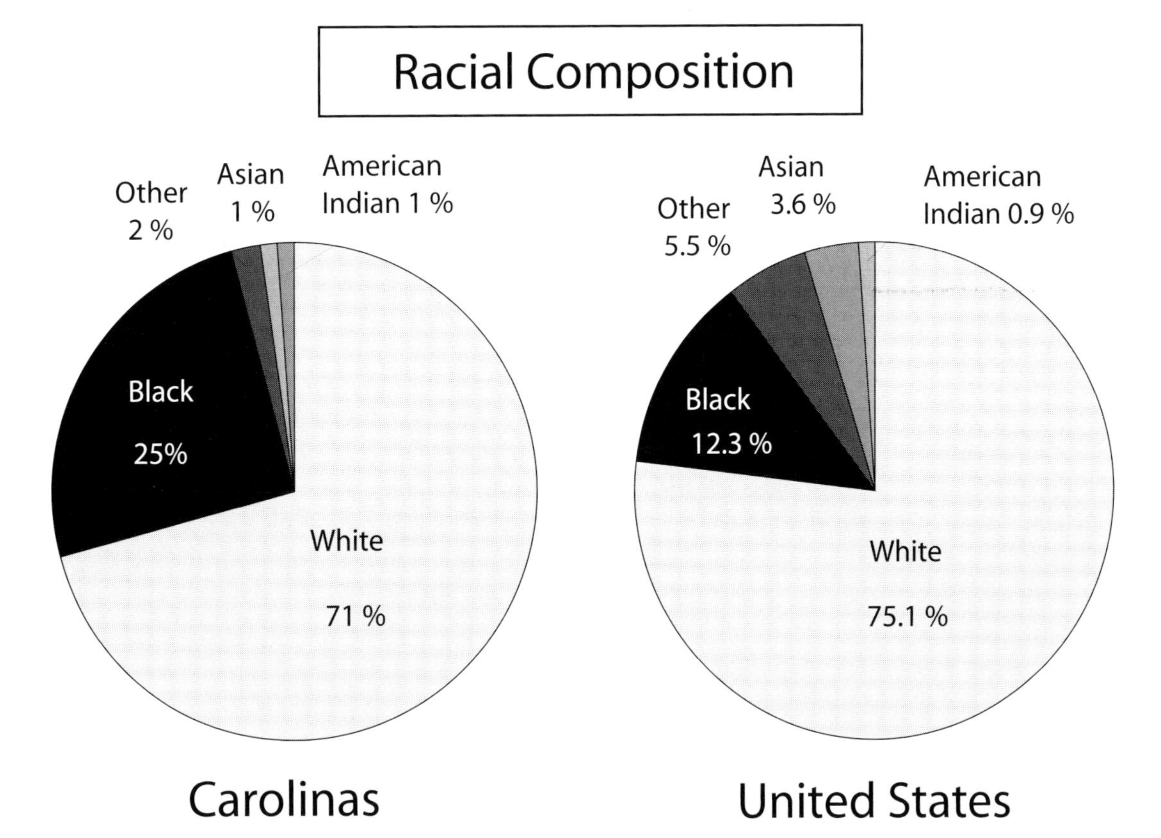

Figure 7.10. Racial Composition of the U.S. and Carolinas, (2000).

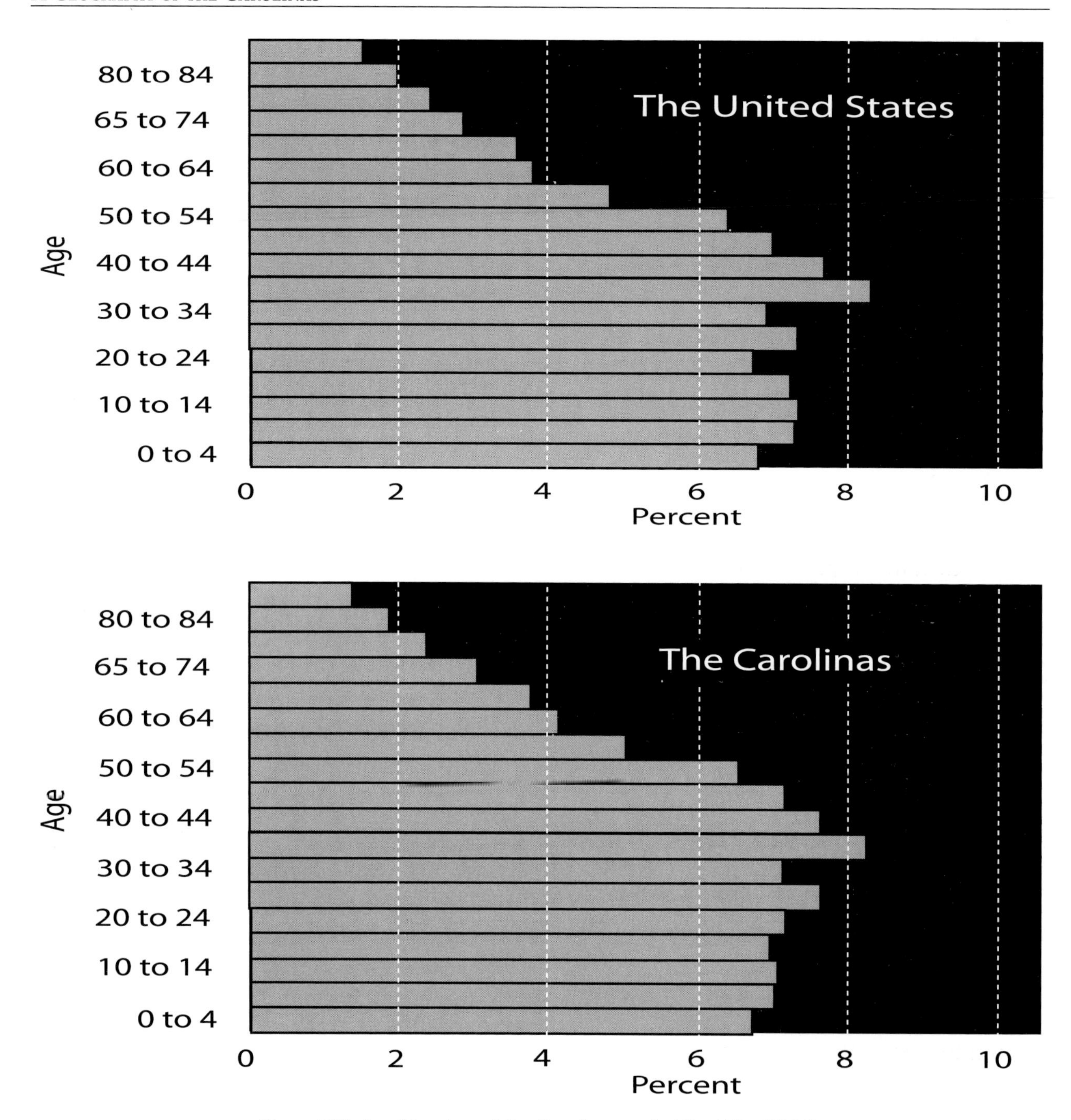

Figure 7.11. Age Structure of the Carolinas and of the U.S., (2000).

tain counties and those attracting golfers and military retirement. The only counties where old and young coincide are in the poor, rural northeastern counties of North Carolina, which have higher than usual fertility rates, heavy out-migration of young adults, and a remaining senior population.

The summation of these patterns of county and regional population composition in the Carolinas is compared with the nation for race in Figure 7.10 and for age structure in Figure 7.11. The African American population is twice as large, proportionately, and this remains the biggest difference. The proportion other than black or white, about 10 percent for the country, is now a little more than four percent in the Carolinas. Hispanics, who may be of any race or combination, have grown to almost five percent of the population of North Carolina, a transforming diversification. This still remains less than half of the 12.5 percent averaged nationally. The Carolinas clearly parallel the national age structure, with the baby boom generation dominant now between ages 35 and 55, its echo sounding especially in the 25-30 year olds, and the grand echo now in school. Rather than the impact of the retirement in-migration that many people in the Carolinas would expect to see, the growing in-migration of young adults starting families and rural fertility rates that continue slightly higher than national levels have instead produced the same median age as the nation, 35. A slightly lower percentage of the population is 65 or older (12.0 for N.C. and 12.1 for S.C., compared with 12.4 for the U.S.). The sex ratio, at 96.0 males per 100 females, is almost the national average, 96.5. How these patterns emerged, why, and where they are going, is discussed in the following section.

Historic Population and Dynamics of Change

The settlement system, types of land use, and economic institutions established in the Carolinas during the first decades of national existence established the mold for the patterns of population that exist today, portrayed in Figures 7.1-12. (This history is explained and developed with great richness and complexity of detail in the sources used for this background description: Goldfield, 2000; Kovacik and Winberry, 1987; Petty, 1943; Powell, 1989.)

From the first days of their founding, the two Carolinas have both shared and diverged. King Charles II, in 1663, issued a grant for an enormous swath of unexplored land, extending from the Atlantic coast all the way to the Pacific, and south to the Spanish holdings. After 66 years, with very little settlement developed, the Carolinas were divided into two provinces in 1729. Georgia was subsequently separated in 1729. There were 40,000-50,000 native Americans occupying the area when Europeans arrived, but their numbers, settlements and occupance had already been drastically reduced by Old World diseases spread indigenously from those introduced early in the sixteenth century by the Spanish in the Caribbean. Both Carolinas were first settled by English speakers along the coast. After the first attempts at settlement by Sir Walter Raleigh and John White had failed, North Carolina was first populated by settlers expanding along the coast from Virginia and bringing their crop, tobacco, with them. In contrast, the key English settlers came to South Carolina from Barbados and settled near their port city, Charles Town. Exports of naval stores and deer skins, and soon the major crop of rice, had made Charleston the fourth largest city by the time of the founding of the U.S. The crops of rice, and later cotton, established the demand for the heavy labor of African slaves. Both Carolinas found the topography, estuarine swamps and tidal marshes made movement to the interior difficult. Separately from the coastal, export-cropping, plantation dominated settlements, both colonies were populated mainly by Scotch-Irish and others who came south along the Great Philadelphia Wagon Road from Pennsylvania, through the Shenandoah Valley of Virginia, and spread out through the Piedmont and Mountains.

At independence, both Carolinas had largely freeholder, Scotch-Irish and German small-farm settlements in their High Country, a population better connected with Pennsylvania than their own coast. Both had a coastal settlement of plantation economy with a white population of French Huguenots, Russians, and Swiss, as well as English, origin, together with a large population of African slaves. It would be 200 years before they were again so diverse.

Population Foundations

In 1790, the Carolinas had a combined population of 644,078, which included 207,877 slaves (32% of its population) and 23,832 slave holders. This constituted 30 percent of the slaves and 50 percent of the slave owners in the nascent United States, and averaged nine slaves per slave holder (7 in N.C., 12 in S.C.). In 1820, the population had grown to 1,141,500, now including 456,801 slaves

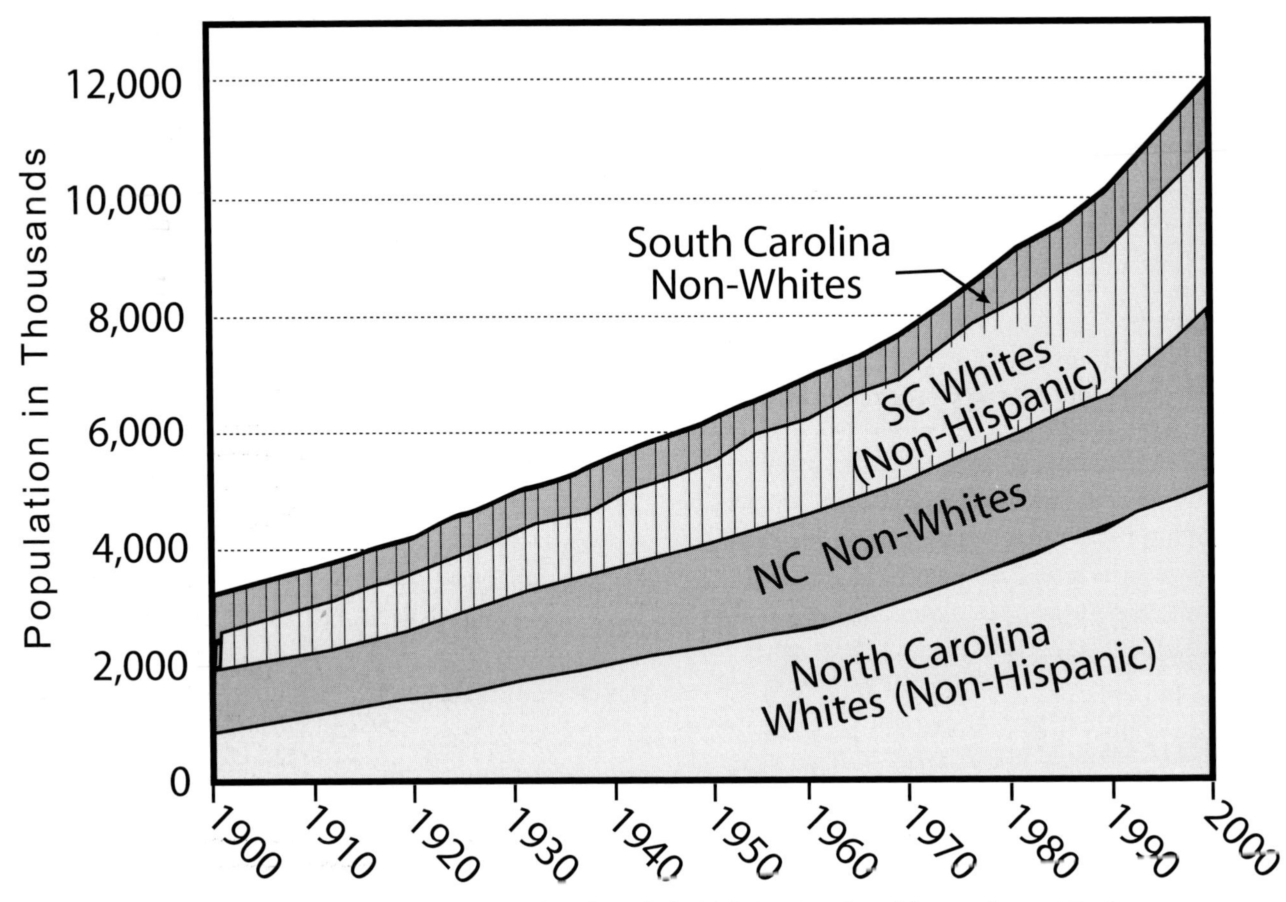

Figure 7.12. Proportionate Population Growth (in 20th Century for whites and nonwhites). Shows the growth of population in the Carolinas during the 20th century, proportionately divided into the white and non-white populations of each state.

ratio of African slave to free white male was 1.3 for the Carolinas (0.7, N.C., 2.0 S.C.). By 1860, with the importance of cotton inland added to that of the coastal rice, the total population of 1,696,330 included 733,465 slaves (43.2% of the population: 57%, S.C.; 33%, N.C.). The slave to free white male ratio had increased to 1.5 (2.6 S.C.; 1.0 N.C.). After the Civil War and passage of the 14th Amendment, the newly enfranchised African American citizens constituted the majority in most coastal plain counties, which were still dominated by the culture of white supremacy.

From 1800 to 1870, there was strong out-migration from the Carolinas, especially South Carolina. As the economy stagnated, soil fertility declined and land eroded, white planter's sons took Negro slaves to settle Georgia, Alabama, Mississippi, and beyond. The white small farmers of the High Country went over the mountains to settle Tennessee and westward. There was almost no European immigration. Even in 1900, most of the natural increase in population left for Georgia (28%), Florida, Arkansas, and Texas, while only 12 percent headed north. This out-migration was more than 60 percent white, although whites were only 40 percent of SC's population; but this was about to change.

The growth and changing composition of the population in the twentieth century is represented in Figure 7.12. The designation "whites" and "nonwhites" is used as the only way to maintain comparability across a century of census categorization. In 1900, only one to three percent of the population was foreign born, despite the enormous wave of European immigration crashing on America's shores. The small cities in the Carolinas had no ethnic European neighborhoods. The white population tended toward homogeneity and opposition to foreigners. In 1880, around four percent of the population was "urban." The two port cities, Charleston and Wilmington, were the largest in their respective states. Beginning in the 1880s and growing until 1930, the textile industry developed along the fast-flowing streams of the Piedmont. The mills used water power, the road and railroad connections of the Piedmont with the north, and the cheap white labor. They built their own small mill towns along the streams and railroads, reinforcing the dispersed settlement patterns rather than forcing urbanization. North Carolina led the nation in the percent of its female workforce in manufacturing (Goldfield, 2000, 65). By 1925, almost 10 percent of South Carolina's population lived in mill towns (Kovacik and Winberry, 1987, 114). As late as 1940, however, only four urban centers in South Carolina and nine in North Carolina had more than 25,000. Charlotte had become the Queen City of this rural region.

As the century turned, African Americans began a heavy, sustained out-migration to Northern cities from the Coastal Plain. This was least true in North Carolina, where labor intensive tobacco farming continued to support farming families on 40-acre farms. Cotton farming actually increased in the 1920s, but it could not be sustained. The black population that had been anchored in place by the entrenched tenant system began to lose their tenancy. After more than a century of row crop cotton cultivation, with little manuring and no conservation tillage or contour plowing, the land was exhausted, eroded, and infertile. The New Deal of the Depression years idled land and reduced the need for tenants, and it also capitalized the mechanization of cotton picking. Blacks had been leaving the Mountains for several decades already, but now they began the Great Out-migration from the Coastal Plain to the industrial cities of the North that transformed the United States. In the 1920s, a quarter of a million people left South Carolina, and only 44,000 came in. Most out-migrants were young adult blacks. North Carolina continued to be the major single destination (28%), and another 25 percent went to other Southern neighbors, but now major streams of black migrants headed up the east coast to Washington, D.C., Baltimore, Pennsylvania, New Jersey, and New York. As can be seen in Figure 7.12a, the proportion of South Carolina's population that was black began a sustained decrease. It was eight percent smaller in 1930 than 1920 alone, despite high rates of natural increase. Half the counties also decreased in white population. North Carolina lost thousands of people, white and black, to the North, but it continued to grow faster than the nation as a whole from natural increase until World War II. Partly because tobacco remained a labor-intensive crop viable on small farms, more blacks stayed home. Between World War II and the 1960s, rural residents continued to leave for economic opportunities in the industrial cities of the North. In South Carolina, an estimated 400,000 blacks left the state between 1950 and 1970 (Kovacik and Winberry, 1987, 139.) Throughout the nineteenth century, approximately 60 percent of the population of South Carolina was non-white. By 2000, the proportion was 33 percent. North Carolina's non-white population declined from 33 percent to 28 percent, but this was due in part to the recent migration reversal.

Between 1970 and 1980, North Carolina saw its first net in-migration of population, including the first net in-migration of blacks. This turnaround has continued. South Carolina (ORSS, 1997) estimated that of the half million in-migrants from 1990 to 1996, 15 percent came from North Carolina, 45.5 percent from states that can be summed as "other southern neighbors," only six percent from foreign countries, and at least 20 percent from the industrial North. This is almost a mirror image of the earlier out-migration, and part of it is composed of those former migrants returning home in retirement. Both Carolinas are now growing faster than the nation as a whole, because of in-migration of both whites and blacks, and now a major influx of Hispanic migrants (Figure 7.3). Most of the growth has concentrated in the rapidly growing metropolitan counties (Figure 7.5).

Demographic Transition and Structure

The population of the Carolinas has made the same transition from high birth and death rates to low birth and death rates as has the nation and an increasing majority of the world. This demographic transition removes the natural increase of a population, the difference between birth and death, as a propulsive force, and it results in an older population structure greatly reduced in the proportion that is children. The relative importance to this process of lowering mortality from infectious disease because of public health sanitation measures, vaccination, improved nutrition, refrigeration, or health care and antibiotics, varies in time and place, though all are involved. The relative importance to the process of lowering fertility (reproduction) that is due to higher age of marriage, marital structures, economic and educational opportunities for women, and means of contraception, varies, though all are needed. The changes of the demographic transition are generally associated with urbanization, movement of labor out of agriculture and into manufacturing and services, improved infrastructures for movement and communication, and increasing levels of education, income, and standards of living. All of these were true in the Carolinas.

At the beginning of the twentieth century, the overwhelmingly rural and agricultural population favored large families and suffered high rates of mortality from infectious disease. It is difficult to estimate comparative rates for this period because the vital statistical registration system of the U.S. was not yet complete. Even after 1930, births and deaths continued to be under-reported and measures needed to be adjusted for this. Table 7.1 uses rates and proportions estimated by Petty (1943) for South Carolina and Hamilton (1974) for North Carolina. Crude rates for death and birth, which are rates based on the total population rather than the age or gender groups at risk of the demographic event, are usually poor for comparison because they may reflect differences in age structure, e.g. the proportion old, rather than differences in conditions relative to health or birth. They have the advantage only of being current rates based on real people (rather than mathematical adjustments), and giving the actual growth. The population age structures of the Carolinas were very similar. The biggest difference between them was the larger proportion of African Americans in South Carolina and the larger numbers and proportion of their out-migration.

Early in the twentieth century, most mortality was from infectious diseases. Although vital statistics, based on systematic reporting, are not yet available, that high toll is indicated by the fact that the median age of the population was 18 and more than 40 percent was fifteen or younger. These are the classic characteristics of a population early in the transition. The Carolinas were highly malarious, both on the Coastal Plain and in the mill towns dependent on impounded water. Tuberculosis was a major cause of death. The state of nutrition for the poor farmers who planted cotton and tobacco to the front door is indicated by the cases of pellagra which filled the state mental hospitals, the insanity resulting simply from deficiency of niacin in the corn-dominated diet.

Births were high enough to produce constant growth, or natural increase, and propel the out-migration. One measure of fertility commonly used before most births occurred in hospitals, where they could be registered, was census based: children under five per thousand women 15-44. This fertility rate was over 800 in 1880, and over 700 at the beginning of the twentieth century, whereas today it is under 300. Racial and rural-urban differences for fertility rates in South Carolina were described by Petty thusly (1943, 118):

> 1920—total, 584/1000 women in their childbearing years; whites, 602/1000; negroes, 568/1000; urban, 355/1000, rural, 645/1000; 1940—total, 445/1000; whites, 397/1000; negroes, 511/1000; urban, 285/1000; rural, 511/1000.

Petty also noted, without surprise or explanation, that

> Death rates in general vary among the counties in relation to the proportion of

Table 7.1. 20th Century Demographic Measures

	Crude Birth Rate		Crude Death Rate		Natural Increase Rate		Infant Mortality Rate					
Year	NC	SC	NC	SC	NC	SC	NC T	NCw	NCnw*	SC T	SCw	SCnw*
1920	31.4	28.2	12.6	14.1	1.9	1.4	14.2	NA	NA	116.0	NA	NA
1930	24.1	23.2	11.3	12.9	1.3	1.0	79.0	67.0	105.0	88.7	69.0	108.0
1940	26.3	30.6	8.9	10.7	1.7	2.0	57.6	49.8	73.8	68.2	51.0	86.0
1950	27.3	30.2	7.7	8.5	2.0	2.2	34.5	30.2	54.2	38.6	29.5	50.2
1960	24.1	25.1	8.4	8.7	1.6	1.6	31.7	22.3	53.3	NA	23.9	48.5
1970	19.3	20.1	8.8	8.8	1.1	1.1	24.3	19.3	36.7	22.9	18.2	31.3
1980	14.4	16.6	8.2	8.1	0.6	0.9	14.5	12.1	20.0	15.6	10.8	22.9
1990	15.8	15.9	8.6	8.5	0.7	0.7	10.6	8.0	16.5	11.7	8.1	17.3
1998	14.8	14.0	9.0	9.1	0.6	0.5	9.3	6.4	16.3	9.6	6.3	15.3

Data from the U.S. Statistical Abstract (various years). Additional South Carolina estimates are from various tables in Petty 1943. Additional North Carolina estimates are from various tables in Hamilton 1974, Vol.2. 1998 data are from National Vital Statistics Report 48(11)2000,83. Crude birth rate is the number of births per thousand total population. Crude death rate is the number of deaths per thousand total population. Natural increase is the difference between birth and death rates, here expressed in the common form per hundred (%). Infant mortality is the number of deaths under one year per thousand births that year. T represents rate of the total population; w the white only; nw the non-white.

*** In most recent data, information was listed as black instead of non-white.**

> the population which is negro, and the proportion which is urban. Since Coastal Plain counties generally have higher proportions of Negro people, these counties have higher crude death rates than piedmont counties. (1943, 122)

These patterns of differentials continue until today. Small though towns and cities were, their economic, educational, and health advantages brought reduction in death and birth rates. Slow and late urbanization of the Carolinas was mirrored by slow and late fertility decline. As can be seen in Table 7.1, infant mortality rates in North Carolina continued through the century to be much higher for nonwhites than whites. In 1998, infant mortality in both Carolinas ranked among the highest in the country, and the over all NC state rate of 9.3 was a weighted average of the 6.4 to 16.3 racial difference. Of course, this is also true for the nation as a whole, as the lower national infant mortality (1997) of 7.1 masked 6.0 for whites and 13.7 for blacks. Birth rates and fertility rates continually have fallen for everyone. The differential between the races for fertility was reduced, but the birth rate for nonwhites remains higher. The "total fertility rate" (TFR) or average number of children a woman would have if she went through her childbearing years experiencing each year today's birth experience, is 1.8 for whites in North Carolina and 2.1 for others, both essentially at replacement reproduction (OSP, 2000). The still small Hispanic population has a TFR of 3.0. South Carolina has a slightly lower birth rate (Table 7.1).

Life expectancy has increased for everyone, but the differentials have actually increased. At mid-century in North Carolina (1950), life expectancy at birth for white males was 66.5 years, white females 72.9, other males 58.5, and other females, 62.8. By century's end, life expectancy for white males had increased to 73.8, white females to 80.5, other males to 65.0, and other females to 75.2 (N.C. OSP, 2001). The increase in other, or minority, or non-white, females to a life expectancy second to white women, living longer than all males, follows the national pattern and world experience when the burden of numerous pregnancies is removed. Even though other males have increased their life expectancy, it remains lower than that of white males fifty years ago. The differential, highest to lowest, has increased from 14.4 to 15.5 years; the differential between males has increased from 8 to 8.8 years. The national experience is very similar, with an overall life expectancy of 76.5 years (1997) composed of white female 79.8, white male 74.3, black female 74.7, and black male 67.3. This national level differential is 12.5 years high to low, and seven years between males by race.

As death rates from infectious disease were reduced, children lived and grew up, and birth rates declined, the population aged and non-communicable, chronic and degenerative, diseases became the major causes of death. That the median age in South Carolina in 1950 was only 23.6 was also due to the high out-migration of young adults that was decades old. By 1970, the turn around year, median age in South Carolina was only 24.7 and in North Carolina 26.5. By 1980, it was, respectively, 28.0 and 29.9, by 1990, 32.1 and 33, and by 2000, 35.4 and 35.3, identical to the national median age of 35.3 (U.S. Censuses). Congruently, the sex ratio of males per 100 females had fallen from over a hundred throughout the nineteenth century to 99 by 1920, 97.0 by 1940, 95.9 by 1970, to 94.3 in 1980, before rebounding with migration turn around to 94.5 in South Carolina, 96.0 in North Carolina, and 96.5 in the United States (calculated from censuses). Since males die at higher rates from conception forward, fewer children and more seniors means a lower sex ratio. The higher out-migration rates of males than females seeking jobs can drastically reduce the sex ratio, and in-migration for work in a prospering, high technology economy tends to increase it slightly (since women are almost equal in that migration).

Rice (1983) concluded that the traditional physiographic/cultural regionalization was not a valid framework for analyzing mortality patterns between 1920 and 1970. Essentially, people were poor and rural all over the state. But using the cross-over year between declining infectious disease deaths from tuberculosis, pneumonia and influenza and increasing degenerative disease rates from heart disease and cancer, she did determine that the Piedmont (and the Tidewater, mostly Wilmington) went through the mortality transition in the early 1920s, a decade before the Coastal Plain or the Mountains, which crossed within a couple years of each other.

In 1920, only 21.5 percent of the population lived in the ten most populous counties. At this time, the Piedmont contained 44.5 percent of the population of North Carolina, as it had a century previously (Steahr, 1973, 31). By 1950, 30 percent lived in the ten most populous counties and 48.7 percent in the Piedmont; by 1970, 38 percent and 53 percent, respectively; and in 2000, 40.1 percent in the ten most populous counties (six of the ten in the Piedmont) and 56.6 percent in the Piedmont. During this time, the proportion of the population living in small towns has changed little. The proportion residing outside municipalities has fallen from 56.9 percent in 1980 to 49.7 percent in 2000, while the proportion residing in cities of over 75,000 has increased from 14.5 percent to 22.2 percent. The growth and concentration of the textile industry and its mill towns in the Piedmont had not changed the dispersed nature of urban settlement in the Carolinas, but it had started the concentration of population on the Piedmont. Although dwarfed by the export of people to the North, the selective nature of migration meant that intra-state movement out of the Tidewater and the exhausted agricultural counties of the Coastal Plain and Mountains to better job opportunities in the Piedmont, left behind in those origin counties higher concentrations of the aged and children, of the less educated, of the less healthy, and of the poorer. These people are harder to reach with health services, good schools, or better jobs.

North Carolina has fared better in this population redistribution than has South Carolina. This is partly because North Carolina's economy, fueled by financial services and high technology research and development, has done better. Much less recognized outside of the region, however, has been the role of the tobacco industry. There is still no large area substitute for tobacco as a crop that can support a family on a small farm today. Perhaps the main reason that the state of North Carolina has defended and supported the tobacco industry so zealously over the decades is that it has kept the population anchored in rural, small-town Carolina. Most of its black population stayed home and did not migrate to the North. Its population did not become a majority urban until the 1980 census. The demise of the cultural and economic dominance of tobacco means relocation of population to the cities of the Piedmont, but at least it has waited for a generational shift out of farming.

The dispersed urban settlement form of the Carolinas, with no dominant metropolis and an exceptionally well-spaced urban central place hierarchy, has supported the characteristic importance of the rural nonfarm population, generally commuting short distances to small factories in local small towns. The broiler industry, among others, has fed on this population distribution. Recent structural changes in the economy have been accompanied by population migration to the better opportunities of the metropolitan counties. This is the process of removing young adults and lower income workers to better futures elsewhere, with effects on the labor force and retail economy of the towns and small cities. This growing gap is being filled today by the unprecedented Hispanic in-migration. Labor is being recruited for the chicken factories, the field harvests of cotton and sweet potato, the heavy labor in the hot sun of road work. The news of the 2000 census in North Carolina is of small towns in rural counties shifting from 90 percent white to majority Hispanic, complete with elementary schools where the majority of children do not speak English. Hispanics are increasing in the cities also, working in construction and hotel, restaurant, and other low-paying jobs, but they are establishing homes and giving birth mostly in rural areas. Their numbers were the biggest surprise of the census, even though there had been several studies of the changes happening. The impact on school systems, health services, and housing needs is locally at critical levels. That the Carolinas are demographically part of the U.S. now is perhaps best illustrated by this, their first mass foreign immigration.

Well-Being

The most fundamental point about the population of the Carolinas today is how closely it resembles the rest of the country. Starting from a foundation of a third to even a majority of African slaves, persisting after losing a war through decades with conditions of economic stagnation and malnourished poverty with high mortality rates, avoiding European immigration, and finally losing over a million of their people to half a century of out-migration to industrial cities elsewhere, the Carolinas today are demographically like everybody else! They do have a higher proportion of African American people than the nation as a whole, and a lower one of Asians, Hispanics, and ethnic Europeans; but this is changing rapidly with in-migration. Probably the most lasting difference is a settlement system based on dispersed urban centers, rural nonfarm driving to jobs, and development of a metropolitan, as distinct from an urban, population. This has become the basis of the characteristic automobile-based, discontinuous green urban sprawl of suburb and manufacturing and research alike. The toll taken in air pollution and traffic congestion has become more obvious than any solution as these metropolitan counties now are among those which lead the nation in growth.

African Americans also remain relatively concentrated on the Coastal Plain, on average poorer and less educated, and disproportionately living in places with poor roads, poor job opportunities, poor schools, poor health care, and more sickness. On average, black babies have lower birth weight and higher neonatal mortality; more black mothers are unmarried heads of household; black teenagers give birth at twice the rate that white teenagers do; and more black children are in poverty. Yet, today, these issues of equity, access, marital behavior and family structure have become national issues. Table 7.2 illustrates how closely the Carolinas have come to resemble the U.S.

About half the women are currently married. A little more than a fifth have never married, although many still will. A quarter of households consist of people living alone, and about eight

Table 7.2. Socioeconomic Characteristics and Population in 2000

	N.C.	S.C.	U.S.
People per household	2.49	2.53	2.59
% householders living alone	25.4	25.0	25.8
% females 15+ now married	53.8	51.6	51.7
% females 15+ never married	21.7	23.0	24.4
% single female hh with children	7.3	8.5	7.2
% births to unmarried mothers	35.7	32.5	29.2
Births per 1000 women 15-19	51	59	41
Births per 1000 women 20-34	111	120	110
Total fertility rate	1.9	2.0	2.1
$ median household income 1997[a]	35,320	33,325	37,005
% children in poverty 1997[a]	18.6	23.0	19.9
% > 24 high school or higher	77.5	78.7	81.6
% > 24 bachelor's degree or higher	22.0	22.9	25.1
% metropolitan	67.5	70.0	c.78

Data are from the 2000 Census Profiles and 2000 Census Supplementary Survey Summary (factfinder.census.gov). Data for a are from Census Bureau QuickFacts,2000, based on Current Population Surveys.

percent of these are seniors. The average number of people per household is now two and a half. The population has reached replacement reproduction.

As Table 7.2 also shows, the Carolinas are still more rural, poorer, and less educated than the nation, but the difference is no longer striking. One of the most significant expressions of the historic lower standards of living is in the higher mortality. In 1998, the age adjusted U.S. death rate per 100,000 was 478.1; that of North Carolina 518.6; and that of South Carolina, 550.8. Another expression of this is a life expectancy a couple years less—but not for white females. The differences among the races and between the genders are so much that probably little difference remains between the Carolinas and the nation when rates are age, race, sex adjusted and standardized. This is a difficult comparison to make at this time. (The U.S. has recently shifted to the Tenth International Classification of Disease, whereas most state data are still published according to the Ninth. More importantly, the National Center for Health Statistics has changed the Standard Million population used for age adjustments from the age distribution of 1940, which has been used for decades, to that of 2000. Mortality data are currently being published by the new standard, and also "comparability modified" to the old standard, for the last years of complete data, i.e. 1998. North Carolina is shifting to this standard, but South Carolina's data is not available in that form.)

Heart disease, of course, is the number one killer, followed by cancer and stroke as a distant third, together accounting for almost 60 percent of deaths. This is the same for all groups of Americans everywhere. For blacks, diabetes then becomes the fourth cause and chronic obstructive lung disease for whites, but the difference is only shuffling among the top seven causes. HIV/AIDS has entered as one of the top ten causes of death for blacks in North Carolina. Motor vehicle injury is a top ten cause of death for everyone and, from the age of one to 24, the leading cause of death. Another measure, potential years of life lost, is constructed from the number of years remaining when a death occurs until the average group life expectancy. Its use serves to target public health and medical interventions where they can have the most economic and family impact. Motor vehicle accidents are very costly, even more than suicide and homicide; lesser causes of death, which are not among the top ten, can collectively cause more years of life lost than even cancer, which accounts for over 21 percent of all years of life lost. AIDS is already annually depriving black males of 50,000 years of life, but only three percent of the total. Although heart disease is the greatest cause of death, it does not have the greatest cost of years of life in fact only 17 percent of the total.

Mortality rates are adjusted for age differences according to the 2000 standard population. Black males have the highest rates for all causes of mortality, and so do males compared to females. The map of distribution of black population, Figure 7.7, could well be mistaken for a map of the highest infant mortality and, especially in South Carolina, some of the highest rates of stroke in the country (on the Coastal Plain), for unknown cause.

HIV/AIDS began to spread in the Carolinas in the mid-to-late 1980s. After an initial attraction to the largest cities, where the largest numbers of cases still occur, it seemed to spread quickly through the Coastal Plain, especially near Interstate 95. The highest rates are now established in the biggest cities and throughout the Coastal Plain. The historic pattern of slavery remains as a map of mortality.

Regional Expressions

The patterns of population distribution, ethnic composition and age structure are presented in Figures 1-11. The history of settlement, economy and migration which produced them has also had the effect of regionalizing the patterns of change into the physiographic/cultural regions. These regions are portrayed in Figure 7.13. On that map also are located ten counties across the Carolinas that will be used to illustrate and explicate the generalities that have been discussed at a localized scale.

The Mountains

Wilkes County is located in northwestern North Carolina along the eastern slope of the Blue Ridge, though most of it is actually in the Piedmont. It was settled mainly by Scotch Irish from Pennsylvania. In 1790, it joined the American union with 8,710 people, including 553 slaves held by 131 householders. By 1820, its ratio of slave to white person was 0.1, and, by 1860, its ratio of slave to white person was still 0.1. The population had less than doubled, as most of the natural increase had migrated westward. The trade connections of its free-holder farmers were oriented toward Pennsylvania, and during the Civil War, most of their sympathy was with the North. At the beginning of the twentieth century, hundreds of the black citizens who did live there left the mountains, further reducing the already small minority population in both absolute and relative terms. In 2000, Wilkes County is only four percent black. It is, more surprisingly, three percent Hispanic (Table 7.3).

Table 7.3. Population Characteristics of Wilkes County, N.C.

2000 Population	Percent Hispanic	Percent 65+	Sex Ratio	Percent children in pov'ty	1860 ratio Slaves/ Whites	Median Income as % of U.S. Median Income
65,632	3.4	14.1	97.2	19.4	0.1	83

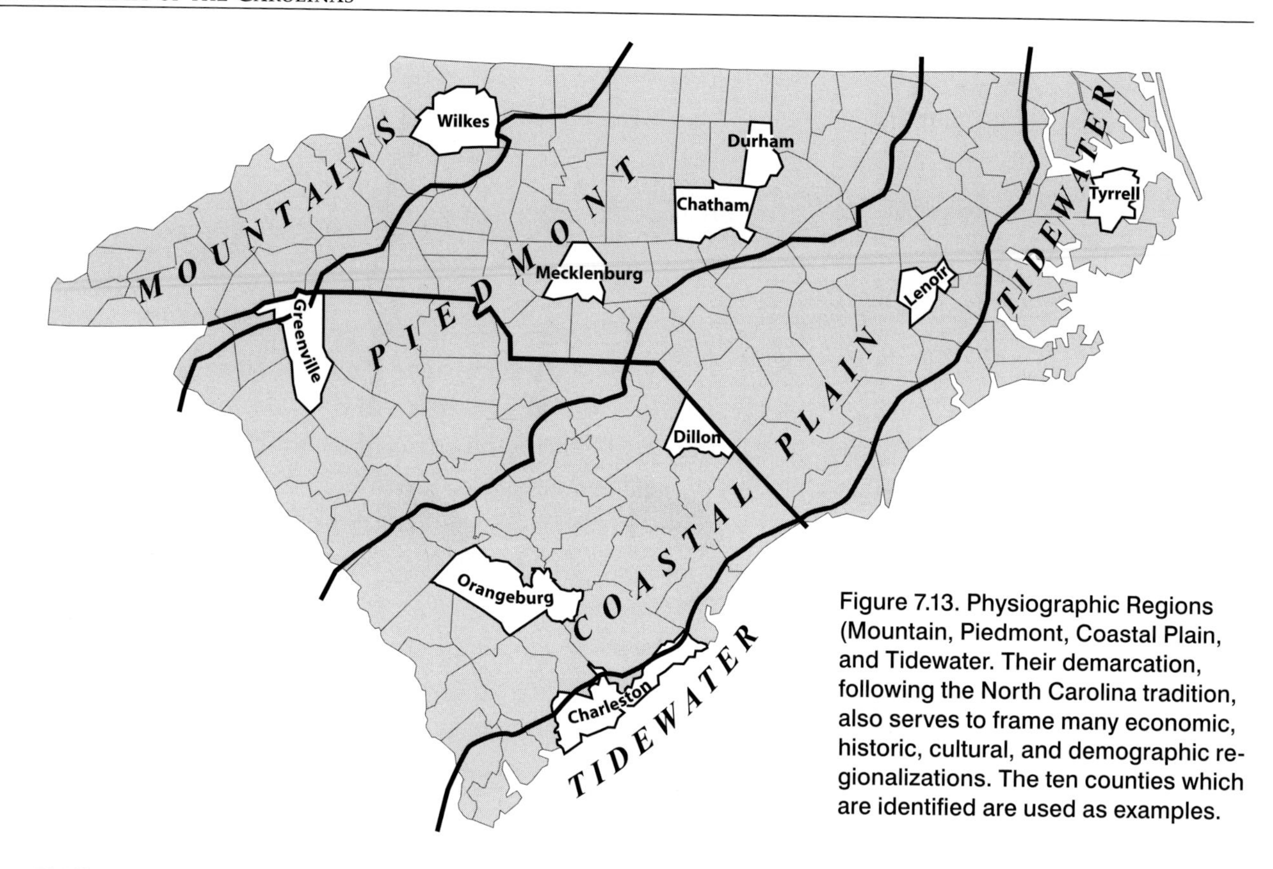

Figure 7.13. Physiographic Regions (Mountain, Piedmont, Coastal Plain, and Tidewater. Their demarcation, following the North Carolina tradition, also serves to frame many economic, historic, cultural, and demographic regionalizations. The ten counties which are identified are used as examples.

Wilkes County continues to be rural and relatively isolated. It grew in the past decade only 10.5 percent (Figure 7.3). It has a median income only 83 percent of the US median, or 87 percent of North Carolina's, and is in the bottom quintile of high school education. It is 93 percent white (Figure 7.14) and older than the state (Figure 7.9a) or nation because of aging in place as young adults left. This can be seen in its age structure (Figure 7.15) of 20-year olds and smaller proportion of children. It has lower age adjusted rates of heart disease, chronic obstructive pulmonary disease (COPD), colon cancer, and AIDS than the state. Wilkes is typical of those western counties in the Carolinas which are not booming from their amenities.

The Tidewater

The Tidewater counties were not hospitable to English settlement from the sea, except where ports could be established to serve inland trade. The amenity value of beach and sound for recreational development and retirement has altered the sea coast in some areas, especially around Myrtle Beach. The two counties chosen to represent the Tidewater are typical of the classic Tidewater patterns. Tyrrell County is in the Tidewater of Albemarle Sound in northeastern North Carolina. Charleston, one of the great cities of the Old South, is still the major port city of the Carolinas (Table 7.4).

Table 7.4. Population Characteristics of Tyrell County, N.C. and Charleston County, S.C.

2000 Population	Percent Hispanic	Percent 65+	Sex Ratio	Percent children in pov'ty	1860 ratio Slaves/ Whites	Median Income as % of U.S. Median Income
Tyrell: 4149	3.6	16.1	114.1	38.7	0.5	58
Charleston: 309,969	2.4	11.9	93.5	26.7	1.3	95

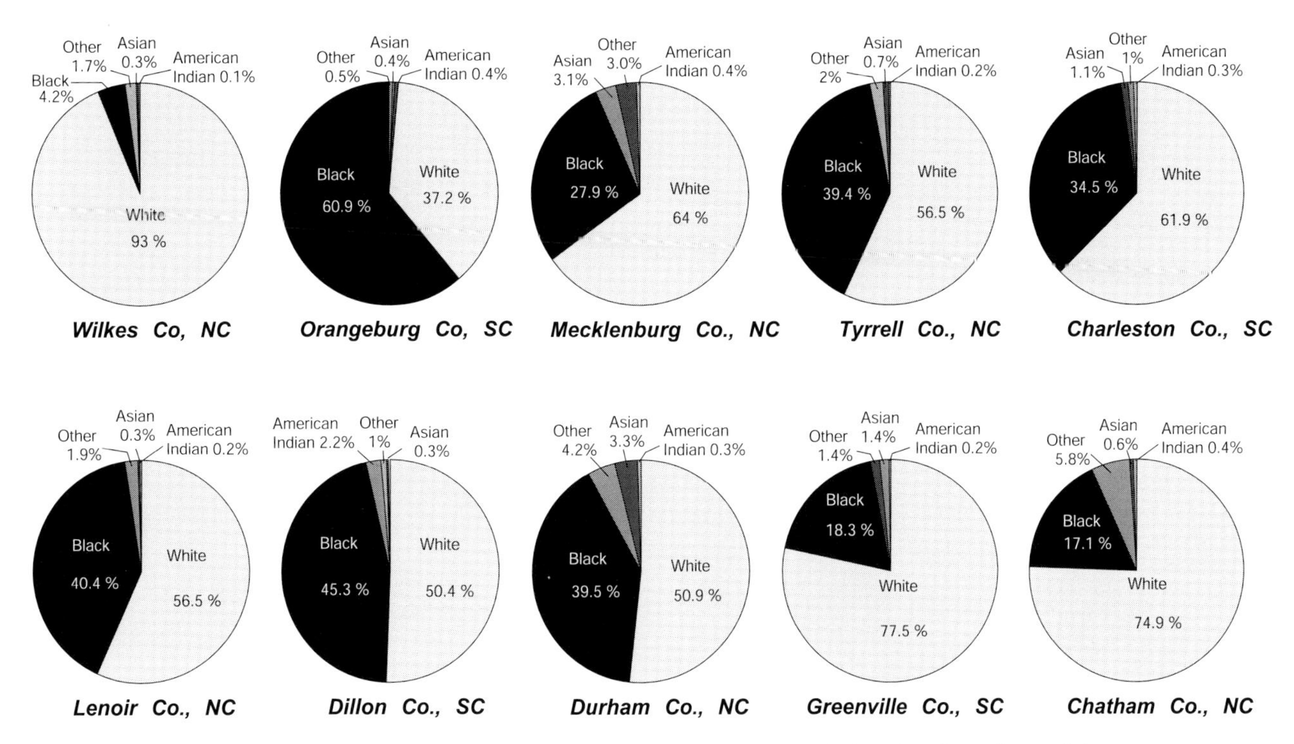

Figure 7.14. (Racial Composition for sample counties, 2000). Graphs show the racial composition of the population in the ten counties discussed as examples.

In 1790, Tyrrell County had a population of 4826, 24 percent African slaves. In 2000, it has a smaller population, 39 percent African American. As the age structure (Figure 7.15) and sex ratio suggest, there is an exceptionally large proportion male 35-45, composed mostly of a small prison (institutionalized census population of 437), which constitutes 10 percent of the county's population. Like other Tidewater counties inland from the ocean front, it is one of the poorest in the state and has lost population for decades. Between 1990 and 2000, its population turned around and grew 7.6 percent. It gained 150 Hispanic people. Its population is in the bottom quintile of proportion with high school education, and it has an unusually small proportion of children to attend elementary school. It has high rates of infant mortality, breast cancer, and liver cirrhosis, but low rates of homicide and AIDS. Tyrrell is typical of the infrastructurally isolated Tidewater counties that have not been transformed by the beach and boat amenity attractions.

Charleston represents the other side of the Tidewater, one of the two (with Wilmington) port cities that used to be the largest in their states. In 1790, with 67,000 people, Charleston was an American giant; but it stagnated with the slave economy and King Cotton. In 1820, there were 2.8 slaves per white person. By 1860, this ratio was down to 1.3, leaving whites the minority. In the late nineteenth century whites migrated in, and blacks began a century of migrating out. Today, it is just over one-third African American. This is somewhat misleading, however. At least since the development of Hilton Head, whites have been moving out to new residential developments in adjacent counties, part of the Southern urban sprawl. Charleston county grew only five percent in the last decade, but its metropolitan area gained more (Figures 7.4 and 7.5). Its economy is not especially prosperous and its median income remains less than the national average. With the Tidewater port cities of Savannah and Wilmington, Charleston exists in a surrounding context of relative isolation and rural poverty until the power of amenity development targets it (as it has, recently, Wilmington).

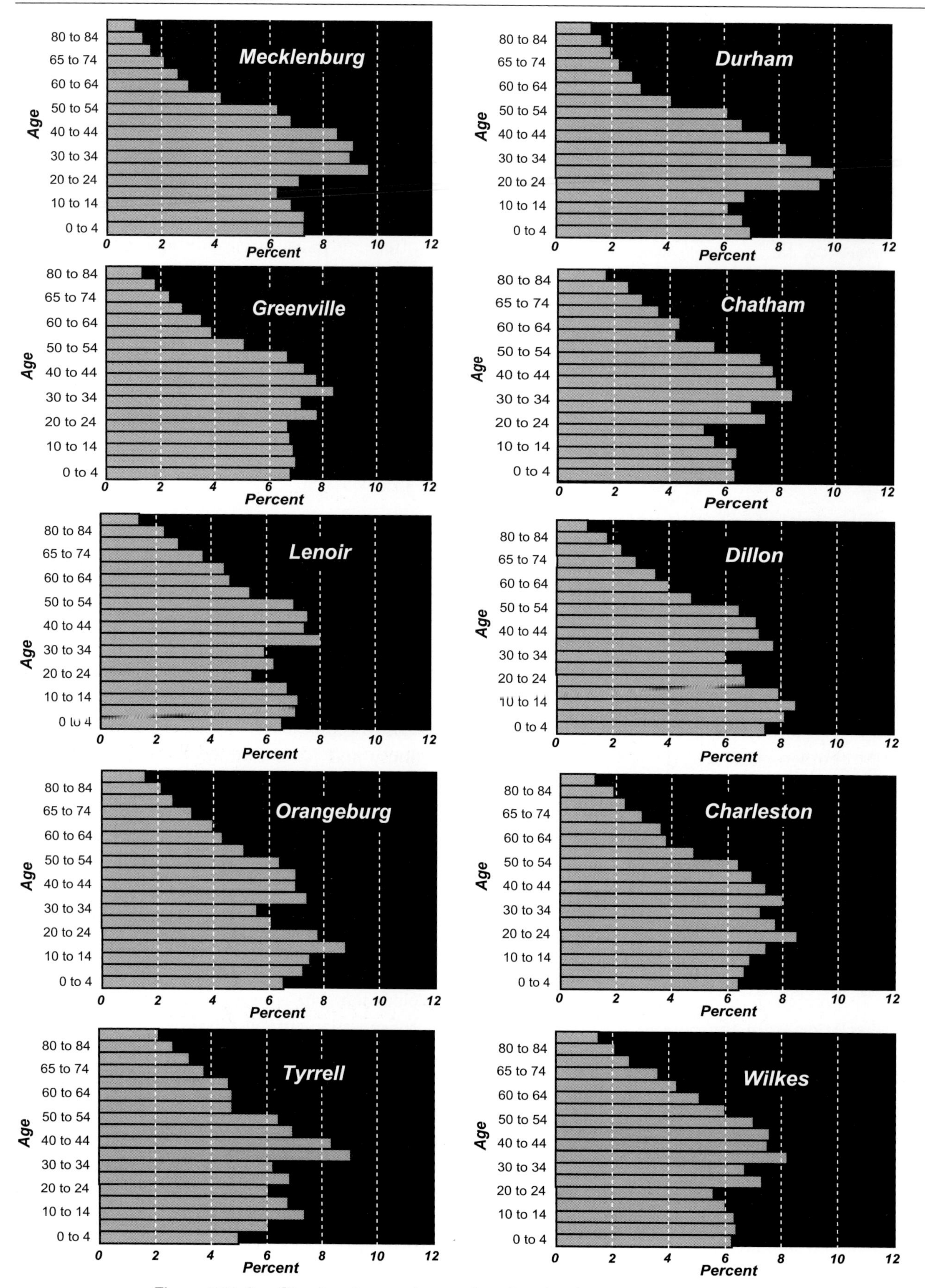

Figure 7.15. Age Structure in sample counties. Graphs show the age structure of the population in ten counties discussed as examples.

The Coastal Plain

The counties of the Coastal Plain, extending over hundreds of miles, are varied in their expression of previous migration and development, but they clearly share the legacy of the plantation economy of the past and of their settlement from the coast. They are represented in this discussion by Lenoir, in North Carolina, Dillon in South Carolina but on the border, and Orangeburg in southern South Carolina (Figure 7.13). The two characteristics they most obviously share are having a small proportion white (Figures 7.6 and 7.14) and being poor. Their age structures closely match each other, marked by the baby boom and its echo. They seem to have recovered somewhat from decades of out-migration, though the drop in 20-year olds and lowered sex ratios suggest it continues. At older ages, the steep step backs still illustrate the missing migrants of decades past (Table 7.5).

The agricultural counties of the Coastal Plain continue to be poor. Their median income varies from two-thirds to three-quarters of the nation's and only a little more of their respective state. More than a fourth, even over a third, of the children is in poverty. Their circumstances are not simply a calculus of race. Orangeburg, which once had two slaves for every white person, remains almost two-thirds black after losing generations to northern cities. Although an agricultural county, it is not as poor as other counties in SC (e.g., Dillon, which was not yet a separate county until the 1920 census). Orangeburg grew eight percent in the past decade, faster than Lenoir, which is only 43 percent black and in a more prosperous North Carolina.

Diabetes, AIDS, cerebrovascular disease, motor vehicle accidents, and homicide are generally higher in the Coastal Plain counties. Except for the counties with institutional cores, i.e. universities or military bases, the Coastal Plain counties continue to stagnate demographically, although the chicken and hog industries have become important. Many counties continue to lose young adults looking for better economic opportunities. Counties in North Carolina, such as Lenoir, have begun to gain from Hispanic migration, but this has not had impact yet in South Carolina.

The Piedmont

The contrast of the Coastal Plain counties with those of the Piedmont is striking. As Piedmont counties provide the jobs for most of those moving out of the other regions, a higher proportion of the population is becoming concentrated there. The most rapidly growing counties, the biggest metropolitan areas, the highest rates of in-migration from other states, and all things prospering are concentrated there. The counties chosen as examples are Mecklenburg, the largest urban county with the biggest city (Charlotte), center of a metropolitan area (Charlotte-Gastonia-Rock Hill, alias Metrolina) of more than a million; Greenville, center of the Greenville-Spartanburg-Anderson metropolitan area

Table 7.5. Population Characteristics of Lenoir County, N.C. and Dillon and Orangeburg Counties, S.C.

2000 Population	Percent Hispanic	Percent 65+	Sex Ratio	Percent children in pov'ty	1860 ratio Slaves/ Whites	Median Income as % of U.S. Median Income
Lenoir: 59,648	3.2	14.6	90.5	26.6	1.1	76
Dillon: 30,722	-	11.5	87.3	36.6	-	65
Orangeburg: 91,582	1.0	13.2	86.9	32.0	2.0	72

in South Carolina; and Durham and Chatham counties, one old urban and one still mostly rural, but both part of the rapidly growing Raleigh-Durham-Chapel Hill metropolitan area of over a million (Table 7.6).

The rise to dominance of the Piedmont metropolises is a fairly recent development. These counties were first settled mostly from Pennsylvania by Scotch-Irish and Germans. There were some household slaves, but no plantations. In 1790, Mecklenburg had only 11,360 people. It gained only 6,000 people by 1860, averaging less than two-thirds of a slave per white. Even by 1900, with a county population of 55,268, Charlotte was not only 60 percent the size of Charleston, but it was smaller than Orangeburg, and the same size as Greenville. It is not until 1940 that Mecklenburg suddenly loomed largest and boomed into the financial heart of the Carolinas (Figure 7.5).

Today, its large population is growing rapidly, 36 percent in the last decade. It has induced 30 and 40 percent growth in all its surrounding counties, too, to create an interstate metropolis of well over a million. That its large population is now more than six percent Hispanic should end any thought that migrants are agricultural workers only. This is seen also in Durham (as well as Raleigh), but again Hispanics have only begun to appear in South Carolina, in Greenville. As Figures 7.7 and 7.15 illustrate, Piedmont counties remain mostly white, but the participation of a black labor force in its industrialization has varied.

In South Carolina, blacks were largely kept out of the textile industry, leaving Greenville County only 18 percent black today. In contrast, blacks were important to the labor force of the cigarette industry from its foundations in Durham, N.C. As African American entrepreneurs developed the insurance industry and financing, Durham became the "Negro Wall Street" and a destination of out-migration from the Coastal Plain.

Chatham County illustrates some of the tensions of the region today. Northern Chatham is growing rapidly as sprawl from Chapel Hill, as bedrooms for work in the Research Triangle and as targeted retirement developments. This is reflected in the high median income. Yet, the county has been very rural until the last 20 years, and the western part of the county remains agricultural. As local labor has been attracted to the jobs of the Research Triangle, the chicken factories of Siler City have recruited from Mexico. The biggest town in the county, Siler City has only 6,966 people in 2000. The news is that its population is now 39 percent Hispanic. All the problems of health care, public education, housing, and community ethnic relations have developed, seemingly overnight.

The in-migration of not only retirees, but working age population, is illustrated in the population structure in Figure 7.15. As earlier maps showed (Figures 7.8b and 7.8c), the diversification of the Carolinas is in full swing here. None of the leading causes of mortality are concentrated here,

Table 7.6. Population Characteristics of Mecklenburg, Durham and Chatham Counties, N.C. and Greenville County, S.C.

2000 Population	Percent Hispanic	Percent 65+	Sex Ratio	Percent children in pov'ty	1860 ratio Slaves/ Whites	Median Income as % of U.S. Median Income
Mecklenburg: 695,454	6.5	8.6	96.5	14.7	0.6	123
Greenville: 379,616	3.8	11.5	94.9	16.2	0.5	105
Durham: 223,314	7.6	16.3	93.1	19.4	-	108
Chatham: 49,329	9.6	15.3	96.9	13.1	0.5	113

but health care and medical research is. Even as Durham closed its last American Tobacco plant, it declared itself the "City of Medicine." The Research Triangle, including Raleigh's sprawl into Johnston County and the mushrooming city of Cary in its suburbs, is growing at a pace to overtake Mecklenburg by the next census.

Trends and Prognostications

The population of the Carolinas today is growing faster than the nation as a whole. That growth is bringing ethnic diversification, but population remains a study in black and white. Demographic trends in longer life expectancy, replacement fertility, older age structure, and more single-person and never-married households, converge on national ones. Heart disease, cancer (especially lung cancer), and stroke (cerebrovascular disease) account for almost 60 percent of deaths. The majority of the population is urban, and most of it is metropolitan. Although the African American population continues to have higher mortality from infant mortality of low birth weight infants, diabetes, homicide, heart disease and stroke, and women continue to have lower rates than males, even these inequities concur with national pattern and process.

The population as a whole continues to be poorer, less educated, and to have higher rates of teenage pregnancies and households with children headed by women without husbands, than the nation as a whole. This is not true, however, in the growing, prospering counties of the Piedmont. The population overall has twice the proportion black, roughly equal proportion native American, and one-third the proportion Asian or Hispanic, as the nation. The most significant change, however, is the rate at which ethnic diversification with Asian and Hispanic in-migration is occurring, especially in North Carolina. Although many migrants are moving to the cities, it is the rural counties of North Carolina that are being ethnically transformed.

More than the slight lag from national mortality and fertility declines, it is migration and its absence that have shaped the population of the Carolinas and made it regionally distinct. Migration from different origins settled the Carolinas with coastal and upland populations more ethnically and economically different than those usually in separate states. From 1790 to 1970, migration selectively removed many of the healthiest, best educated, and most ambitious of citizens. For several decades in the twentieth century, this meant the loss of hundreds of thousands of African Americans from the coastal counties to the cities of the North, changing the black population in all but a few counties from majority to minority. Cheap labor and cultural hostility to ethnic difference kept out European immigrants for two centuries. Then, with the Great Reversal in 1980, people of all races and nationalities began to move into the Carolinas. Whether for high tech work, field work, or retirement, the in-migration is fueling vigorous population increase, as well as changing composition and distribution. The sudden increase in Hispanic population is the first foreign immigration to have an impact in 200 years. South Carolina has been less attractive to Mexican laborers, but it can be only a matter of time.

The other migration that did not occur was the mass movement of rural people to Carolinian cities for industrial jobs. That was partly because the people moved out of the state to Northern industrial jobs instead. It was also because of the distinctive pattern of widespread, evenly distributed, regular urban hierarchy ("central places") and lack of a truly dominant city. New textile mill towns were kept small and dispersed. As good roads developed, the rural nonfarm population could live at home and drive, within a reasonable time, to factory work. Even with the low threshold definition of the census bureau, 2,500 people in an incorporated place or a continually developed area around a city of 50,000, the population of the Carolinas did not officially become majority urban until the census of 1980, decades after the rest of the nation. Today, even as the population becomes more concentrated in the rapidly growing Piedmont counties, the dispersed urban system is such that a large proportion of the population remains rural nonfarm and its centers unincorporated, even as it is counted as metropolitan. The dispersed-center, inter-commuting, sprawling green metropolitan settlement has made the area very attractive in which to live, work or retire. It has also made the population settlement almost impossible to serve with a mass transit system.

The South is growing faster than the nation, gaining political representation and power, and gaining international economic clout. Within the South, only Atlanta is growing faster than the Carolina Piedmont. At current rates, in 20 years, Raleigh-Durham-Chapel Hill will be the dominant metropolitan area of the Carolinas. As African Americans migrate back into the Carolinas, whether returning for retirement or taking advantage of new economic opportunities near family, they too are moving to the Piedmont. Unless some new economic engine can be found for the rural counties of the Coastal Plain and Tidewater, population—even retirement—will become increasingly concentrated in the metropolitan counties of the Piedmont. Management of growth and its impact on air pollution and traffic congestion, school systems and affordable housing, is already growing desperate among planners and governing boards. It can only become an obsession.

References

Bennett, D. G., ed. 1996. *Snapshots of the Carolinas: Landscapes and Cultures*. Washington, D.C.: Association of American Geographers.

Bennett, D. G. and J. F. Florin. 1996. The Population of the Carolinas. In *Snapshots of the Carolinas*, ed. D. G. Bennett, 105-108. Washington, D.C.: Association of American Geographers.

Florin, J. W. and R. J. Kopec. 1973. *The Changing Population of the Southeast*. Studies in Geography, Number 5. Chapel Hill, N.C.: Department of Geography, University of North Carolina.

Goldfield, D. R. 2000. History. In *The North Carolina Atlas*, eds. D. M. Orr and A.W. Stuart, 47-76. Chapel Hill, N.C.: University of North Carolina Press.

Hamilton, C. H. 1974. *North Carolina Population Trends: A Demographic Sourcebook*. 3 vols. Chapel Hill, N.C.: Carolina Population Center.

Ingalls, G. L. 2000. Urbanization. In *The North Carolina Atlas*, eds. D. M. Orr and A. W. Stuart, 103-121. Chapel Hill, N.C.: University of North Carolina Press.

Inter-University Consortium for Political and Social Research. Study 00003: Historical Demographic, Economic, and Social Data: U.S., 1790-1970. Ann Arbor.

Ives, S. M. and A. Stuart. 2000. Population. In *The North Carolina Atlas*, eds. D. M. Orr and A. W. Stuart, 77-102. Chapel Hill, N.C.: University of North Carolina Press.

Johnson-Webb, K. D. 2000. *Formal and Informal Hispanic Labor Recruitment: North Carolina Communities in Transition*. Unpublished Doctoral Dissertation, Department of Geography, University of North Carolina at Chapel Hill.

Kovacik, C. F. and J. J. Winberry. 1987. *South Carolina: The Making of a Landscape*. Columbia, S.C.: University of South Carolina Press.

North Carolina, Office of State Planning, State Demographic Center. 2001: ospl.state.nc.us/demog/pop.

North Carolina, State Center for Health Statistics. 1998. Leading Causes of Death, expanded edition. Vital Statistics, vol. 2. Raleigh: N.C. Department of Health and Human Services.

Orr, D. M. Jr. and Alfred W. Stuart, eds. 2000. *The North Carolina Atlas*. Chapel Hill, N.C.: University of North Carolina Press.

Petty, J. J. 1943. *The Growth and Distribution of Population in South Carolina*. Bulletin no. 11. South Carolina Planning Board. Columbia, S.C.: State Council for Defense Industrial Development Committee.

Powell, W. S. 1989. *North Carolina through the Centuries*. Chapel Hill: University of North Carolina Press.

Rice, G. H. 1983. *Changing Mortality Patterns in North Carolina, 1920-1972: A Regional Analysis*. Unpublished Doctoral Dissertation, Department of Geography, University of North Carolina at Chapel Hill.

South Carolina, Office of the Budget. 2001. <//www.orss.state.sc.us/ //167.7.127.235/population.>

Stear, T. E. 1973. *North Carolina's Changing Population*. Chapel Hill, N.C.: Carolina Population Center.

U.S. Census Bureau. 1909. *A Century of Population Growth: From the First Census of the United States to the Twelfth, 1790-1900*. Washington, D.C.: U.S. Government Printing Office.

_____. 1910-1990. *Decennial Census of Population and Housing*. Washington, D.C.: U.S. Government Printing Office.

_____. 2001 Profiles of General Demographic Characteristics, South Carolina and North Carolina. 2000 Census of Population and Housing. Washington, D.C.: Online <www.census.gov.>

U.S. National Center for Health Statistics. 2000. *National Vital Statistics Report*, vol.48, no.11, table 26.

_____. 2001. *National Vital Statistics Report*, vol.49, no.3, table 2.

Chapter 8

TOURISM AND RECREATION IN THE CAROLINAS

Robert L. Janiskee

University of South Carolina at Columbia

This chapter highlights some of the leisure patterns, trends, and traditions that help to give the Carolinas its distinctive geographic personality.[i] Although leisure is centered on the notion of escaping work and having fun,[ii] it has important consequences. In the case at hand, the fiscal, social, environmental, and other effects of tourism and recreation have heavily influenced the landscape, lifestyles, and economy of the Carolinas. As befits a part of America renowned for its warm and sunny climate, plentiful surface water, fine beaches, beautiful mountains, diverse ecosystems, abundant wildlife, and rich cultural heritage, the region has an abundant supply of recreation facilities and a prosperous tourism industry. The landscape abounds with developed recreational resources, such as golf courses, boat ramps, hunting camps, and parks. Many residents use these assets to enjoy the leisure-centered lifestyles for which people like Carolina "good ol' boys" and places like Hilton Head are justifiably famous. In Myrtle Beach, Charleston, Charlotte, Cherokee, and many other Carolina communities, the visitor industry is a powerful vehicle for creating jobs, generating tax revenues, and improving life quality. In South Carolina, which attracts nearly 33 million visitors a year, tourism ranks as the second-leading industry, adding over $6.6 billion in direct spending to the economy each year and supporting about 121,000 jobs in the hospitality industry.[iii]

Tourists come to the Carolinas from every state and from dozens of different countries, but the region's tourism industry is heavily dependent on the leisure travel of Carolinians and the residents of nearby states. South Carolina strikes a good bargain with its neighbor to the north by sending about three million tourists a year to North Carolina and receiving about four million in return.[iv] Not surprisingly, the largest non-Carolina tourist contingents come from the neighboring states of Virginia and Georgia, which together send more than eight million visitors a year to the Carolinas. Pennsylvania and Florida also send more than a million tourists each year.

Carolina Tourist Destinations

Hundreds of tourist attractions are scattered through the Carolinas, but most tourist activity takes place in the Grand Strand, Hilton Head Island, the Outer Banks, the North Carolina mountains, the historic Charleston area, Charlotte, Pinehurst-Southern Pines, and tourist oases on the interstate highways[v] (Figure 8.1). The distinctive character of each tourist destination attests to the great diversity of Carolina landscapes, the variegated nature of leisure tastes, and the many strategies available for accommodating mass tourism. Myrtle Beach, one of the most popular and affordable tourist destinations in America, is a family-oriented beach resort with a

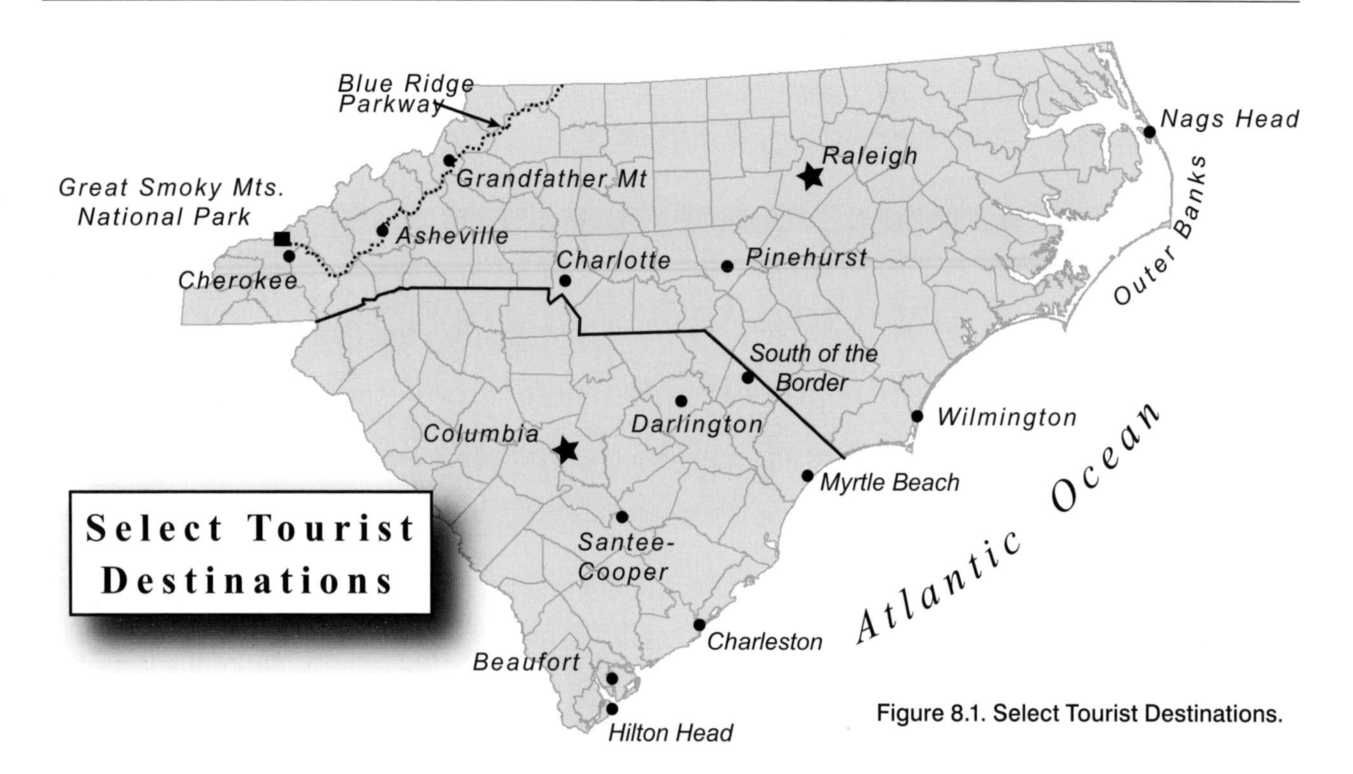

Figure 8.1. Select Tourist Destinations.

"Redneck Riviera" past and Orlando aspirations. Hilton Head Island is a posh retreat for wealthy retirees, conventioneers on generous expense accounts, and affluent tourists seeking splendid golf, tennis and boating. The Outer Banks, 130 miles of unspoiled, uncrowded coastline, offers a wide range of fun things for visitors to see and do on a chain of storm-battered barrier islands. Charleston is one of America's most historic and pedestrian-friendly cities. Mountainous western North Carolina has a thriving tourist industry primarily because it is cool and pretty. Charlotte, the only urban center in the Carolinas with a big city look, offers big city leisure amenities. The Southern Pines-Pinehurst area of North Carolina attracts many thousands of golfers to the world's largest vacation golf facility and an array of challenging courses. Millions of I-95 motorists can tell you about South of the Border, one of the most unusual and successful roadside attractions in America.

Myrtle Beach

The number one tourist attraction in the Carolinas is the 85-mile stretch of coastline between Georgetown, South Carolina and Holden Beach, North Carolina. Although tourism promoters have dubbed it the Grand Strand, most people call it Myrtle Beach. The latter name is actually that of Strand's hub city, but no one really cares. What matters is that this long and narrow coastal strip is one of the busiest and fastest growing beach resorts in America. The Grand Strand entertains at least 14 million visitors a year. No tourist destination outside of Florida claims a bigger share of the American beach resort trade, and few places in the country have a faster rate of population growth, a lower unemployment rate, or a brighter economic outlook (Waccamaw Regional Planning and Development Council, 1997, 1998).

Many site and situation advantages favored the emergence of a great resort at this particular place. Tourists relish the Grand Strand climate, which offers abundant sunshine and warmth, pleasant sea breezes, and a short, mild winter. It is the beaches, however, that have put Myrtle Beach on the map (Photo 8.1). They are wide, clean, sandy, and among the lengthiest in America. Few tidal inlets or rivers interrupt the shoreline, and no large urban-industrial developments or harbor complexes foul the water or spoil the view. All these assets are conveniently

Photo 8.1. The beach at Myrtle Beach State Park.

located about halfway between New York City and Miami, and within a day's drive of most urban centers in the eastern United States. The Grand Strand also welcomes over 300,000 Canadians a year.

Myrtle Beach can be efficiently described as a young, family-friendly, automobile-oriented, heavily commercialized, inexpensive beach resort that is burgeoning in size and complexity, evolving away from the beach, and beginning to choke on its own success (Janiskee, Mitchell and Maguire, 1996). The massive scale of today's Grand Strand tourism infrastructure, incorporating features such as the 350-acre Broadway at the Beach entertainment complex and the 2,200-acre Grande Dunes Resort community, is all the more remarkable in light of the region's brief tenure as a major resort. Tourist facilities remained comparatively small and simple in the Myrtle Beach area through the 1950s and 1960s as the resort catered to a limited regional market with beach recreation and related fun, such as shag dancing to beach music[vi] in Ocean Drive nightspots in North Myrtle Beach. Rising demand finally triggered a burst of tourism development in the late 1960s and early 1970s. Despite a national economic recession and skyrocketing gasoline prices, tourist spending in the Myrtle Beach area increased a remarkable 225 percent during 1970-1975. By the mid-1970s, a rapidly maturing tourism infrastructure was in place and tourism had become a powerful engine for economic development throughout the Grand Strand region (Lewis, 1998). Twenty years later, in the mid-1990s, Myrtle Beach was America's second-fastest growing city and some analysts were projecting that the resort's peak tourist population would soon surge to over one million per day.

Reflecting the transportation technology of its heyday, as well as the pull of the sun-sea-sand vacation theme, Myrtle Beach evolved as an automobile-oriented, beach-and-seafood resort (Brent, 1997). The resort's original Recreational Business District (RBD) has a classic T-shape, the result of tourism-related development spreading along the shore and the main road leading to it (SC 501). The renowned Pavilion Amusement Park, which closed in 2006, occupied an 11-acre tract at the core of the RBD where the arterial highway delivers traffic to the downtown beachfront area. Myrtle Beach has no central promenade like Atlantic City's boardwalk or Virginia Beach's seawall. Instead, Ocean Boulevard serves as an automobile promenade for the younger set.

Photo 8.2. Myrtle Beach's Broadway at the Beach. (Courtesy Myrtle Beach Area Chamber of Commerce)

Tourists drive from place to place to stroll on the beach and piers, the RBD sidewalks, and the walkways of malls and other retail meccas like award-winning Barefoot Landing and the flashy Broadway at the Beach entertainment complex (Duncan, 2005). (Photo 8.2)

The Gay Dolphin curio mega-shop and other attractions located on Ocean Boulevard between 7th and 10th Avenues North serve as the hub for "Motel Row," a miles-long, nearly uninterrupted strip of medium-rise motels occupying narrow lots on both sides of beach-hugging Ocean Boulevard. The great majority of Myrtle Beach tourists find lodging in Motel Row or in the many motels, rental condos, and beach houses that line the shoreline and main highways of the region. Tens of thousands of RV enthusiasts stay in the huge, amenity-rich "campground towns" that punctuate the beachfront in several places. Not surprisingly, few visitors stay in upscale, gated facilities like Grande Dunes and Kingston Plantation. It is Myrtle Beach's renowned affordability that has underpinned its success as a tourist magnet. Modestly priced motels dominate the lodging sector, and all but a tiny fraction of the food establishments are inexpensive family restaurants or fast food stores. Admission fees and other entertainment costs are similarly affordable, and the net effect is to make a Myrtle Beach visit one of the best beach resort bargains in America. In return for cheap prices, visitors are assaulted with commercialism, visual shouting, and tackiness on a truly grand scale. Roadways are lined with billboard forests and "big sign, little box" shops, the beach is equipped with an endless parade of planes dragging advertising banners, and many stores are loaded with cheap products. Some critics have called Myrtle Beach a "Redneck Riviera" or "Gatlinburg by the Sea."

Regardless of lodging mode or location, Myrtle Beach tourists enjoy ready access to the beach (all of which is public) and good opportunities for boating, fishing, shrimping, and crabbing. The array of other leisure attractions is truly impressive. There are over 1,850 eating establishments, several shopping malls, numerous shopping centers and factory outlets, eight piers, two beachfront state parks, a NASCAR SpeedPark, about a half-dozen amusement, water, or go-kart race/ride parks, many water slides, the world's greatest concentration of miniature golf courses, and a wealth of other attractions, such as minor league baseball, the Myrtle Beach Speedway, and cruises on the Intracoastal Waterway. There are more than 100 golf courses in the Grand Strand region, too, and the golfers that help to fill the motels in the spring and fall enjoy some of the cheapest golf vacations available anywhere. Another big draw is the large complement of theaters devoted to live country music performances, a distinction that places it in the company of such country music meccas as Branson (Missouri), Nashville, and Pigeon Forge (Tennessee).

Although any resort can run afoul of crippling problems as it grows larger and more complex (Butler, 1980), tourism planners believe that Myrtle Beach has tremendous growth potential and may even be poised to become the next Orlando. Taking the resort to the next level will require increasing the quality and diversity of visitor attractions, redeveloping the original RBD, siting major new attractions away from the beach, making Myrtle Beach a year round resort, improving highway and airline access, and reducing traffic congestion. Myrtle Beach will finally have a major theme park when the music-themed Hard Rock park opens in 2008, but the resort is unlikely to ever have gambling casinos. There are many live entertainment venues and a wide assortment of top-grade attractions, such as the Palace Theater, an 87,000 square foot Ripley's Aquarium, plus a Hard Rock Cafe, Planet Hollywood, IMAX Discovery theater, NASCAR Cafe, House of Blues, Medieval Times, and Legends in Concert. The development trend away from the crowded beachfront and toward less weather-dependent attractions is well established. All four of the leisure business districts located outside the original RBD—Barefoot Landing, Broadway at the Beach, Restaurant Row, and the new Hard Rock Park—occupy highway-oriented inland sites, and all but Restaurant Row have strong retailing and live entertainment components. The new Carolina Bays Parkway, which parallels the Intracoastal Waterway and U.S. Highway 17, allows through traffic to by-pass congested areas. To further improve accessibility and reduce traffic congestion, projects underway or soon to begin will provide interstate highway access (via I-73), a shuttle system, and greatly improved airport facilities. A chronic labor shortage threatens further growth, so employers are bussing in service workers, importing foreign workers, recruiting part-timers from the area's burgeoning retiree population, and increasing the supply of affordable housing for low-wage workers.

Hilton Head

Myrtle Beach is the biggest and most important coastal resort in the region, but it is not the only one. More than a dozen small resorts in the Carolinas—places such as Wrightsville Beach, Carolina Beach, and Topsail Island—function as beach house communities and provide sun-sand-surf recreation for regional tourists. Many weekenders and vacationers prefer to head north to the Outer Banks or south to the Sea Islands. Either of these popular options offers an attractive, more nearly natural setting, as well as escape from the congestion and commercialism of Myrtle Beach.

The Sea Islands region embraces a lengthy series of coastal barriers and estuaries stretching southward from Charleston to the Georgia border and beyond. While the region has two beachfront state parks, several beach house communities, historic Beaufort, and the nearly pristine ACE basin estuary (Janiskee and Chirico, 1997), it is renowned for upscale island retreats, such as Kiawah, Fripp, Daufuskie, and Hilton Head Islands.

Hilton Head is the largest (42 square miles) and best known of the Sea Island resorts (Kovacik and Winberry, 1989). Hosting about 2.5 million tourists a year, the island resort offers its affluent visitors more than 4,000 hotel rooms, condo rentals and time-share units, 12 miles of excellent beaches, great natural beauty, a manicured cultural landscape, high-quality leisure facilities, uncrowded recreational venues, excellent shopping, and the ego-bolstering effects of a prestigious vacation venue. Hilton Head's hallmark feature is its 13 gated resort complexes and residential club communities. There are ten major gated communities, dubbed "planta-

Photo 8.3. Harbour Town Marina at Sea Pines Plantation, Hilton Head, SC.
(Courtesy Hilton Head Island Convention and Visitor Bureau)

tions." The 4,500 acre Sea Pines Plantation, the first to be built, is still the best known. Among its prestigious attractions are the yacht-filled Harbour Town Marina (Photo 8.3) and the Harbour Town golf course, home of the Verizon Heritage Golf Tournament. The vacationers, conventioneers, and wealthy retirees who pass through the gates at Sea Pines, Port Royal, Shipyard, and the other plantations enter a secure world brimming with higher order amenities, including expensive landscaping, fine hotels, championship golf courses, pro-style tennis facilities, uncrowded beaches, and bicycle paths winding among moss-draped oaks (Mitchell, 1996). Some have joked, "This is what God would have done if He had the money."

Outside the plantation gates (off-plantation), land-use controls that include strict sign ordinances and setbacks have yielded a well-regulated tourism complex in which about 200 retail stores and more than 250 restaurants occupy a prominent position. Among the island's more popular recreational shopping venues are 30 village-like shopping areas, three outlet malls, and the full-service Mall at Shelter Cove. There are seafood and family restaurants in every price range as well as a selection of fast food stores—all of which must have small, tasteful signs and discrete setbacks.

While the recreational opportunities available to Hilton Head visitors include many simple, inexpensive activities, such as walking, bicycling, and swimming, recreation providers lean to golf, tennis, and other pricey or more esoteric options, such as horseback riding, historical tours, sailing, sea kayaking, billfishing, dolphin watching excursions, and sunset dinner cruises. In addition to nearly two dozen high-quality golf courses and more than 300 tennis courts, the island has nine marinas, several riding stables, and two RV campgrounds. Few visitors outside the gated resorts use the island's excellent beaches, since access is problematic and public parking is exceedingly scarce.

It will not be easy to grow Hilton Head's $1.5 billion a year tourism industry, which supports more than 60 percent of the local jobs. Having created an idyllic environment for its pampered residents and guests, Hilton Head must now learn how to sustain economic growth while protecting its environmental resources and leisure lifestyles. Since developable space is nearly gone on the island, crowding problems have worsened and growth has spilled to the adjacent mainland. Traffic congestion and the quickening pace of off-island development (including Del Webb's $1.3 billion Sun City Hilton Head retirement community) has prompted the building of an intra-island bypass toll road (the Cross-Island Parkway), stern measures to protect the island's 2,000 acres of wetlands and the Upper Floridian Aquifer (which is threatened by drawdown and saltwater intrusion), and other initiatives, such as efforts to thin the burgeoning white-tailed deer population, preserve the area's endangered Gullah culture (Winberry, 1996; Leib, 1996), and bolster the supply of low-wage service workers. These and related developments indicate that Hilton Head is a fully mature resort. In recent years, nearly all growth in the island's tourism sector has been attributable to older travelers in the winter season.

The Outer Banks

The Outer Banks, a series of 20 windswept barrier islands stretching along North Carolina's 300-mile coast, do not attract people seeking Myrtle Beach commercialism or Hilton Head pampering. What brings millions of visitors and nearly $500 million a year to this tourism-dependent locale is its uncrowded beaches, seaside villages, interesting historic sites, 900 square miles of bays and marshes, excellent fishing and birding, and the East Coast's biggest dunes and best conditions for surfing, windsurfing, hang gliding, and wreck diving (Morris, 1993).

The bridges and ferries that interlink the barrier islands and the mainland deliver motorists to Highway NC 12, which runs along the shoreline and provides an automobile-convenient, see-it-all-in-one-day means of touring the region and accessing the beaches and marshes, oceanfront settlements, and many scenic attractions and recreation sites. Summer visits predominate, since this is when there are fewer hurricanes and gales to topple buildings into the surf, founder ships in the "Graveyard of the Atlantic," and send tourists fleeing to the mainland. Many vacationers head for rental condos in small beachfront communities at Corolla, Duck, and Kitty Hawk. Many more end up in Nags Head or Kill Devil Hills, busier towns that are rapidly adding to their already large supply of rental cottages, hotels, seafood houses, kite shops, fishing piers, and related tourist businesses.

Although many people visit heritage attractions, especially the Wright Brothers National Memorial at Kitty Hawk and the old lifesaving stations and lighthouses, the primary appeal of the Outer Banks is the beaches. Outer Banks beaches are wide, clean, and great for swimming, saltwater fishing, surfing, windsurfing, kayaking, and kite flying. The best ones are in the two national seashores whose extensive protected areas are the pride of the Outer Banks. Anchoring the southern end of the Outer Banks is the 55-mile long Cape Lookout National Seashore, which is isolated (no bridges), wild, and used mostly for fishing, hunting, bird watching, shelling, primitive camping, and scientific research. Much nearer the main settlements is the 70-mile long Cape Hatteras National Seashore, the oldest (established 1935) of the ten seashores in the National Park system, and the lengthiest stretch of undeveloped shoreline on the Atlantic Coast (Photo 8.4). Among the delights of the Seashore's three main islands—Hatteras, Ocracoke, and part of Bodie—are the Pea Island National Wildlife Refuge, wild pony herds, the historic Chicamacomico Lifesaving Station, the picturesque village of Ocracoke, and three lighthouses, including America's tallest, the 208-foot Cape Hatteras Lighthouse, which was moved a quarter-mile inland to shelter it from beach erosion (Gares, 1996).

Tourists accessing the Outer Banks via U.S. Highway 64 and Roanoke Island can stop at the Fort Raleigh National Historic Site, a reconstruction of the original colonial fort on the Carolina coast. Nearby Manteo stages the Lost Colony, a production reenacting the story of the colonists

Photo 8.4. Cape Hatteras National Seashore and the Hatteras Lighthouse, the nation's tallest. (shown before being moved inland)

who came to Roanoke Island in 1587, tried to establish the first permanent English settlement in America, and mysteriously disappeared three years later. This pageant, which gave TV's Andy Griffith his start in show business, is America's longest running outdoor drama (established 1937) and has been viewed by more than four million people.

Heritage Treasures

Since more than 15 million people a year take a trip that includes a visit to a museum or historic site in the South Atlantic region, the abundant supply of heritage resources in the Carolinas is a major asset. The colonial era is well represented in Charleston's showpiece historic district, in museum villages like Brattonsville, and in other places such as the Tryon Palace at New Bern, which was considered America's most elegant government building in the 1770s. Many Revolutionary War battles were fought in the Carolinas, so there are battlefields, reenactments, and museums at sites like Fort Moultrie, Kings Mountain, Cowpens, Ninety Six, Historic Camden, Moores Creek, and Guilford Courthouse. Old Salem, a museum village in Winston-Salem, shows what early 1800s life was like in a Moravian congregational town (Enscore, 1996). North Carolina's Reed Gold Mine State Historic Site commemorates America's first gold rush in 1799 (Hines and Smith, 1996). Since the Plantation South culture originated in the Carolinas, the region abounds in cotton and rice plantations, house museums, and antebellum buildings, such as the architectural treasures of Charleston and Beaufort, South Carolina. The Civil War left a strong impress on the Carolinas, yielding visitor attractions such as Ft. Sumter, Ft. Fisher, and the Bentonville Battlefield. Among the many other history- and nostalgia-themed attractions that draw visitors are such diverse examples as the Moses H. Cone Memorial Park (Appalachian culture), Durham's Duke Homestead (early tobacco processing), Asheville's Biltmore House (Gilded Era opulence), Wilmington's U.S.S. North Carolina (a World War II battlewagon), the Carl Sandburg Home National

Historic Site in Flat Rock, New Bern's Birthplace of Pepsi store, and Andy Griffith's home town of Mount Airy, North Carolina, which is the prototypical "Mayberry" of TV fame (Janiskee and Drews, 1998). Several government programs and grassroots initiatives promote history-themed tourism on a regional basis in the Carolinas. The South Carolina National Heritage Corridor program, for example, promotes tourism by interlinking natural, cultural, and historic attractions in a 240-mile long corridor extending from the mountains to Charleston.

Of all the places in the Carolinas where people gather to enjoy history, none attracts more visitors and sends them away happier than Charleston. Founded by the British in 1670, Charleston played an important role in American history, especially in the colonial era, the Revolutionary War, and the Civil War. Today, thanks to scrupulous historic preservation and a refusal to surrender the city's livability to mass tourism, it is one of America's most beautifully preserved architectural treasures and a "living museum" that is perennially rated among the country's most interesting and friendly places to visit (Bower, 1996). Nearly four million visitors a year converge on the Holy City (181 churches, 25 denominations) to tour its historic district, sample its trendy restaurant scene and other delights, and make forays to area plantations, gardens, and barrier islands. Being confined to a narrow peninsula "where the Cooper and Ashley Rivers meet to form the Atlantic Ocean," Charleston is a compact walking city. Visitors can park their cars and walk or take a carriage ride through an historic district that has the densest concentration of colonial and antebellum buildings in America—including 73 pre-Revolutionary ones, 136 from the late 1700s, and over 600 built before the 1840s. The city has many museums, including America's first municipal museum (the Charleston Museum), house museums like the Heyward-Washington House (1772) and the neoclassical Joseph Manigault Mansion (1803), and a proposed museum for displaying the H. L. Hunley Confederate submarine, the first submarine to sink a warship. A popular gathering place is the historic Market, a dining and recreational shopping area adjacent to the plush Charleston Place hotel and convention center. Among the many other showpieces that visitors enjoy are historic churches like St. Michael's (where locals met to debate joining the revolution), cemeteries with the graves of founding fathers, the old Exchange and Provost Dungeon, the Dock Street Theater, Rainbow Row, and the waterfront Battery.

Special events like the Southeastern Wildlife Exposition and the lavish 17-day Spoleto Festival USA, one of the world's most comprehensive arts festivals, attract tens of thousands of visitors to Charleston each year. Regardless of their reasons for coming, many tourists use the city's highly regarded small hotels and bed & breakfasts, and a high proportion remain in the area for several days. Many take a water tour of the harbor, and about 330,000 of them each year disembark to visit Ft. Sumter, which absorbed the opening shots of the Civil War, fell into Confederate hands, and was subsequently battered into a heap of rubble by Union gunships. Many more visit the South Carolina Aquarium or tour area attractions like Charles Towne Landing (initial site of the city) on the Ashley River, Patriots Point (the world's largest naval and maritime museum), Fort Moultrie, Kiawah Island Resort, and several historic plantations, including Middleton Place, Magnolia Plantation & Gardens, Drayton Hall, and Boone Hall Plantation. Visitors are usually eager to see Boone Hall's magnificent live oak avenue, the inspiration for Tara in "Gone with the Wind."

The Highlands

The rugged western part of the Carolinas has eastern America's highest mountain (6,684-foot Mt. Mitchell) and highest waterfall (411-foot Whitewater Falls), a four-season climate (Soule, 1996), beautiful scenery, vast forests, many whitewater streams, interesting towns, two hugely popular national park units, Appalachian culture, heritage treasures like America's largest historic mansion, and other appeals that have made tourism important throughout the highlands (Summerlin and Summerlin, 1997). The climate is a major attraction. During the summer, oppressive heat sends thousands of weekenders and vacationers fleeing to the cooler highlands and sea breeze coastlines. Spring and fall also have many comfortable days, and winter brings just enough cold and snow to support the southernmost ski resorts in the eastern U.S. Since they operate within the confines of the country's shortest skiing season, the eight ski areas in this area (Richards, 1998) are heavily dependent on snow-making technology. Following the national trend, some also offer warm-weather attractions, such as golf, tennis, alpine slides, and scenic chair lift rides.

Sightseeing, especially the driving-for-pleasure variety, is very popular in this part of the Carolinas. North Carolina has the highest and most scenic stretches of the Blue Ridge Parkway, including the renowned Linn Cove Viaduct. At the southern end of the immensely popular parkway lies Great Smoky Mountains National Park, the most heavily visited National Park.[vii] Visitation peaks in the summer throughout the highlands, and again in October when "leaf peeper" tourists enjoy the brilliant reds and golds of the fall foliage, clog scenic roads, including Newfound Gap Road, the Cherohala Skyway, and the Foothills Highway (SC 11), and fill every motel, inn, and bed & breakfast (Boyle, 1998). The thousands who want to ride the Great Smoky Mountains Railway excursion train at this time of year must book their reservations far in advance.

The wooded slopes, rocky crags, whitewater streams, waterfalls, and wildlife of the highlands are well suited for sightseeing, camping, hiking, backpacking, rock climbing, canoeing, mountain biking, off-highway vehicle riding, gem and mineral collecting, hunting, fly fishing, and many other recreational activities. Most outdoor recreation takes place on the more than one million acres of multiple-use public recreation lands in the Pisgah and Nantahala National Forests and in the Andrew Pickens Ranger District of the Sumter National Forest in South Carolina (including the Chatooga Wild and Scenic River corridor). Nevertheless, the private sector has played a very important role in developing nature-based recreation and ecotourism resources. Some of the most popular scenic attractions in the region are privately owned, including Chimney Rock Park (Photo 8.5) and Grandfather Mountain. Private outfitters also make it convenient, affordable, and safe for several hundred thousand people a year to raft the highland rivers, especially the "frisky, not risky" Nantahala, the potentially dangerous Chatooga (of "Deliverance" fame), the Nolichucky, and the French Broad (Mayfield and DeHart, 1996).

Photo 8.5. Chimney Rock. Courtesy Chimney Rock Park)

There are many tourist destinations in the highlands, but the Cherokee, Asheville, and High Country areas of North Carolina merit special mention. The Cherokee Reservation (Jones, 1996), which is formally named the Qualla Boundary, is a 44,000-acre tract adjacent to Great Smoky Mountains National Park and not far from the popular tourist town of Maggie Valley. While functioning as the capital of the Eastern Band of the Cherokee Indians (Moore and Ross, 1996), who are about 11,500 strong, Cherokee is also the eastern gateway to the park and hosts

Photo 8.6. Linn Cove Viaduct carrying the Blue Ridge Parkway across the southeastern flank of Grandfather Mountain. (Courtesy Daniel Stillwell)

over three million visitors a year. Heavily commercialized, and often derided for its "rubber tomahawks and roadside chiefs" approach to tourism development, Cherokee provides over 1,700 rooms, 28 campgrounds, and essentially the same gas-food-souvenir services as its eastern gateway counterpart, Gatlinburg. There are some significant differences, however. Cherokee has trout fishing, the Oconaluftee Indian Village, the Museum of the Cherokee Indian, an outdoor drama (Unto These Hills) that has entertained over six million people since 1950, a Tribal Bingo with large cash jackpots, and 24-hour gambling and big-name entertainment in the $82 million Harrah's Cherokee Casino. Harrah's, with its nearly 2,000 video gambling machines, 1,200 employees, and a footprint the size of three football fields, was North Carolina's number one tourist attraction in 1998.

Asheville is the largest and most cosmopolitan city in the mountains. Rated one of the most livable cities in America, it has much to offer visitors. Among the Asheville area's appeals are a revitalized, pedestrian-friendly downtown, many fine bed & breakfasts, a thriving arts community, professional baseball and hockey, an arboretum, the Southern Highland Craft Guild's Folk Art Center, convenient access to the Blue Ridge Parkway and several national forests, and a good selection of museums, upscale shops, restaurants, and nightspots. The city also has the Southeast's second-finest collection of Art Deco buildings (after Miami) and many other historically significant structures, including the Biltmore Estate (1895), the Thomas Wolfe Memorial, the Zebulon Vance birthplace and 1830s working farm, and the Grove Park Inn (1913). The 510-room, four-diamond Grove Park Inn is one of the oldest and most famous grand resorts in the South, but Asheville's premier tourist attraction is the Biltmore Estate. Just before the turn of the century, George Vanderbilt bought 125,000 acres of timberland (including Mt. Pisgah) and constructed a 250-room French Renaissance mansion that he used to entertain a long list of world leaders and captains of industry. Today, Biltmore is still the largest private residence in America and one of the country's most visited historic residences. Many thousands travel the three-mile access road to tour the building, gardens, winery, and other attractions of the 8,000-acre estate, which was landscaped by the renowned Frederick Law Olmsted.

The northern mountains, also known as the High Country, comprise the third major tourist destination. The High Country's key assets are gorgeous scenery, abundant outdoor recreation, and the coolest temperatures in the entire South. Among this area's many tourist attractions are four ski areas, the Blue Ridge Parkway, the Moses H. Cone Memorial Park, Grandfather Mountain, Linville Falls, numerous craft shops, interesting Appalachian cultural festivals, several luxury resorts, the only public caverns in the Carolinas, family attractions, and the pretty-but-touristy towns of Boone, Blowing Rock, and Banner Elk. Many feel that Grandfather Mountain represents the essence of the High Country's appeal. Located just off the Blue Ridge Parkway near Linville, Grandfather Mountain is a 4,000-acre privately owned park that offers spectacular views from the 5,964-foot crest, a Mile-High Swinging Bridge, 13 miles of alpine hiking trails, the second-largest Highland Games and Scottish Gathering in the country, a nature museum, and abundant wildlife. In recognition of its rare species and natural diversity, the United Nations has designated Grandfather Mountain an International Biosphere Reserve—the only one of the more than 300 such reserves that is privately owned. On the southeastern flank of the mountain is the Blue Ridge Parkway's 1,243-foot long Linn Cove Viaduct, one of the world's most beautiful and environmentally harmonious bridges (Photo 8.6).

Southern Pines—Pinehurst

North Carolina's Southern Pines-Pinehurst resort region is oddly situated. Although not far from Interstate 95, and therefore not hard to get to, it is 100 miles from the sea, nowhere near the mountains, and imbedded in the decidedly rural Sandhills just outside the busy Fort Bragg military base. It is not scenery or excitement that beckons tourists to this place, but extraordinary golf and pottery. English potters of the colonial era used the rich clay deposits of the region to establish a pottery industry whose beautiful wares, produced by over 90 potteries in a four-county area centered on Seagrove, attract hordes of visitors to more than 40 local shops. It is the high quality standard of the area's many golf courses, however, that have put Southern Pines-Pinehurst on the national tourist map and prompted boosters to proclaim Pinehurst the "Golf Capital of the World."

Golf history runs deep at Pinehurst, and golfers everywhere associate the name with the best course designers and with golf's highest standards and finest traditions. In 1898, only ten years after America's first permanent golf course was founded (at Ardsley, New York), Pinehurst No.1 became North Carolina's first golf course and Pinehurst soon became the country's first real golf vacation destination. Today, the Southern Pines-Pinehurst area has several first class golf resorts, many golf-oriented hotels, inns, and restaurants, and more than three dozen championship golf courses that attract golfers from all over America and the world. Pinehurst's crown jewel, the elegantly traditional Pinehurst Resort and Country Club, is the world's largest golf resort complex. This venerable operation boasts eight signature courses, including the highly ranked No. 2 and No. 7 layouts. Pinehurst No. 2, which was designed by the legendary Donald Ross in 1903 and has been rated America's second-best resort course, has hosted many prestigious tournaments, including the U.S. Open, U.S. Senior Open and the PGA Tour Championship. Among the many other notable layouts in the vicinity are the Pine Needles Resort, site of the 1996 U.S. Women's Open, and Pit Golf Links, one of America's top public courses.

Many who visit Southern Pines-Pinehurst are attracted by the area's fine tennis facilities and clinics and by the ambience of the area's historic inns, fine restaurants, quiet country lanes, and friendly little towns. The New England-style village of Pinehurst has a "wagon wheel" street plan designed by Frederick Law Olmsted in the late 1800s. Aberdeen has a classic turn-of-the-century train station and is loaded with antique and collectibles shops. The entire town of Cameron, which looks much as it did over a century ago, has been declared an historic district.

Tourists can easily drive to the nearby Raleigh-Durham-Chapel Hill locale (the Triangle) to enjoy attractions such as Raleigh's nationally-acclaimed North Carolina Museum of Art and the postcard-pretty campus of UNC Chapel Hill. The 500-acre North Carolina Zoological Park, the world's largest natural habitat walk-through zoo, is also only a short drive away at Asheboro.

The Queen City

Charlotte is America's second-largest financial center and the most populous city in the Carolinas (Stuart, 1996). It is also the only city in the Carolinas with big-city features that include an airline hub, major league sports, and a sky-scraper silhouette. Charlotte's reputation is that of a work-oriented, nose-to-the-grindstone town. Nevertheless, as befits such a large and rapidly growing place, the Queen City's metro area has a wide range of diversions for day-trippers, business travelers, conventioneers, and others. A heavily promoted attraction is Discovery Place, a science museum for kids that has hundreds of hands-on exhibits, an OMNIMAX theater with a 79-foot domed screen, and the world's most so-

Photo 8.7. Charlotte's Mint Museum of Art in former U.S. Mint Building. (Courtesy Mint Museum of Art)

phisticated planetarium. Charlotte is also very proud of its Mint Museum of Art (Photo 8.7), which has a world-class collection of ceramics and pre-Columbian art, and the Blumenthal Performing Arts Center, which is home to an opera company, a dance troupe, and the Charlotte Philharmonic and Symphony Orchestras. Other popular attractions include the Museum of the New South, the Afro-American Cultural Center, and various recreational, shopping, dining, or performance venues, such as Pineville's Carolina Place Mall, SouthPark Mall, Eastland Mall, and North Davidson Street (NoDa). Funding has been sought for an aquarium, which might not be economically viable. Since Uptown Charlotte is smallish and has few substantial tourist attractions besides convention and sports facilities, much of the metro area's visitor industry is accounted for by outlying attractions, such as the Lowe's Motor Speedway 20 miles to the northeast at Concord and Paramount's Carowinds theme park on the South Carolina state line.

Since a city's professional sports teams play a big role in establishing where it fits in the urban pecking order, Charlotte's lofty aspirations have made it hungry for a full complement of franchises, especially major league ones. In 2006, the city was home to the NBA Charlotte Bobcats, the NFL Carolina Panthers, the WNBA Charlotte Sting, three major NASCAR Nextel Cup races, and several minor league sports franchises. The city has long wanted a major league baseball franchise, but is not likely to get one. Charlotte's major sports events are hugely important to the local economy. Bank of America Stadium seats over 73,000 football fans for each home game, and the Lowe's Motor Speedway draws 167,000 NASCAR fans to each of its three Nextel Cup races and additional hundreds of thousands to the 300 or so miscellaneous races, auto shows, and related events held at the track each year.

Pedro's Interstate Oasis

No tourism inventory of the Carolinas could be considered complete without mentioning South of the Border (SOB), one of America's most successful roadside attractions. SOB is a 135-acre, tourist-oriented business complex located on Interstate 95 at Dillon, South Carolina just south of the North Carolina border. Since SOB is about halfway between New York City and Miami, millions of motorists each year run the lengthy gauntlet of SOB billboards and see SOB's landmark 200-foot Sombrero Tower. About seven million tourists a year, most of them repeat visitors, drive through the 97-foot mustachioed Pedro that straddles the entrance ("the largest free-standing sign east of the Mississippi") to experience at first hand a place that has been described variously as a tourist trap, a wayside wonderland, and a neon Tijuana. The tongue-in-cheek Mexican theme is carried to wretched excess in purveying food ("you're always a wiener at Pedro's"), gasoline, over 300 motel rooms (including 20 "heir-conditioned" honeymoon suites), RV campsites, Pedroland Park amusement rides, arcade games, carpet golf ("the Golf of Mexico"), other family activities ("something for every Juan"), America's largest selection of fireworks, 14 gift shops with souvenirs, such as "Pedro-phernalia" T-shirts, a dozen weddings each summer weekend, the Club Mex Convention Center, a night spot called Club Cancun, and glass elevator rides to the Sombrero Tower platform (from which one can view only SOB, the interstate, and countless pine trees). Alan Shafer, the entrepreneurial genius who launched SOB in 1950, died in 2001. The current operators embrace the intentionally campy SOB business model and have followed through with Shafer's decision to eliminate the culturally insensitive aspects of SOB "Mexican speak." In the 1990s, Shafer gradually toned down the cartoonish Mexican-speak of the more than 250 Pedro billboards that have enticed motorists to SOB for over 50 years.

Recreation and Sports, Carolina Style

This section explores leisure activities that give the Carolina recreation and sport scene its interesting, regionally distinctive character. Most recreational likes and dislikes of Carolinians are generally similar to those of Americans at large,

Photo 8.8. Congaree River, SC.

but there are some salient differences and interesting nuances. Among the most important of the activities defining the region's "recreational personality" are fishing and hunting, resort camping, community festivals, stock car racing, college football and basketball, and ice hockey.

Gone Fishin'

Fishing is a signature outdoor recreational pursuit in the Carolinas (Photo 8.8). In addition to hundreds of miles of tidewater shoreline and vast reaches of estuarine marshes and tidal inlets (2.2 million acres in North Carolina alone), the region's fishable waters include tens of thousands of reservoirs and farm ponds, plus thousands of miles of rivers, streams, and blackwater swamps.

Fresh water fishing predominates. Carolina angling traditionally has meant panfishing, a time-honored pursuit that involves repairing to a farm pond, river, blackwater swamp, or other fishy place to relax, commune with nature, and dangle crickets or worms in front of bream, crappies, redbreast, catfish, and other small fish that are easy to catch and good to eat. These days, however, the archetypal outing involves trailering a specially equipped bass boat to one of the region's numerous impoundments to use artificial baits and a remarkable array of fishing-related gizmos in pursuit of largemouth bass and techno-fun. Many decades of dam building have left the Carolinas with a huge supply of reservoirs and ponds providing excellent bass habitat and insuring that Carolinians will account for a healthy share of the nearly $10 billion that America's nearly 11 million bass anglers spend each year on their sport (U.S. Fish and Wildlife Service, 2001). This is more than is spent on golf and tennis combined. Carolina anglers wet their lines in tens of thousands of impoundments (48,000 in North Carolina alone), including many large reservoirs in the Piedmont that collectively have thousands of miles of shoreline, many lake-oriented residences or vacation homes, and a large complement of marinas, picnic areas, campgrounds, state parks, and related leisure facilities. Many of these big lakes provide venues for bass fishing tournaments, including events that attract the top touring pros, entertain huge crowds, and pump over a half-million dollars each into the local economy. Both major bass fishing tours, the BASSmaster Tournament Trail and the Wal-Mart FLW Tour, have held nationally televised tournaments in the Carolinas.

The Carolinas offer some other interesting choices for fresh water angling. After South Carolina's Santee Cooper project inadvertently created one of America's best largemouth bass and crappie fishing resources, fisheries biologists also turned it into a perennial producer of giant catfish and the world's first fishery for landlocked striped bass (rockfish). Today, 40-pound rockfish can be caught from the Congaree River within sight of the state Capitol in Columbia. A number of Carolina lakes have been stocked with rockfish, hybrid bass, or other game fish, and at least one—Lake Jocassee in the Appalachian foothills of South Carolina—has trophy-size trout in the deep, cold waters of its triple-story fishery. Trout fishing of the more conventional variety is also available in many mountain streams as well as in the cold-water tailraces of several big dams.

Many Carolina anglers prefer salt water fishing. At the apex of the salt water pursuits is billfishing, the quest for elusive blue marlin and sailfish in the deep blue water about 60 miles out (Janiskee, 1994). There, where the Gulf Stream overlaps the continental shelf, billfish migrate and feed above the craggy bottom like hawks soaring along a ridge line. Many billfishing boats (including charters) operate out of marinas at Oregon Inlet, Georgetown, Charleston, Hilton Head, and several other Carolina ports. These expensive boats, many of which follow the billfishing tournament circuit, have deep-vee hulls, turbocharged diesel engines, sophisticated electronics for navigating and fish-finding, and other special features, such as flying bridges, outriggers, and fighting chairs. Concerned about dwindling billfish stocks, most fishermen now use the tag-and-release approach to billfishing.

Most salt water angling in the Carolinas is simpler, cheaper, and more productive than billfishing. Fishermen throng the Grand Strand piers to catch spot, croakers, and other common food fish. Surfcasters dot the shoreline in many places when the tide is right. Head boats operating from many marinas haul fee-paying customers to nearshore honey holes and artificial reefs. Elsewhere, boaters cast for sea trout in the inlets, troll for Spanish mackerel, or choose from a long list of other salt water options.

One popular choice is shellfishing, a traditional activity that now has a greater element of risk and an interesting new twist. Like the generations before them, coastal residents and visitors continue to amuse and feed themselves by gathering oysters and catching shrimp or crabs in the tidal inlets and estuaries. These days, however, gatherers must take care to avoid oyster beds contaminated with health-threatening pollutants. Many of South Carolina's recreational shellfishers have taken up shrimp-baiting, a technique so productive that it has been highly regulated in South Carolina and outlawed in some states.

Gone Huntin'

Hunting is not as popular as it used to be, but hunting traditions have been preserved throughout rural Carolina (Marks, 1991). (See Photo 8.9) Today, hunting not only thins wildlife populations, protects crops, puts meat on the table, and helps urbanites stay in touch with nature, but also preserves habitat, boosts local economies, and provides indicators for life-style choices, ethnicity, economic status, and social rank. From the recreational, economic, and cultural perspectives, the most important game species in the region are deer, quail, doves, turkey, raccoons, and ducks.

White-tailed deer nearly disappeared from the Carolinas in the 1920s, but restocking and habitat improvement (especially cropland to forest conversion) enabled them to flourish again. Today, there are nearly 800,000 white-tails in South Carolina alone, and deer hunting is popular in the Coastal Plain and Piedmont counties of both states (Mason, 1996). In South Carolina, 141,000 deer hunters harvested 244,000 white tails

Photo 8.9. Hunters after dove shoot in South Carolina Low Country.

in 2005 while enjoying the nation's longest deer season and most liberal bag limits. There are extensive areas of public hunting lands in the Carolinas, mostly in national forests, wildlife refuges, and wildlife management areas. Private landowners commonly lease deer hunting rights to hunt clubs, some of which use dogs (where legal) to roust the deer out of thick bottomlands. Back roads motorists in deer-rich Carolina locales can easily spot numerous hunt club signs at woods road entrances, as well as deer stands on the edges of fields, forest openings, and power lines.

Quail hunting dominated the Carolina shooting sports for many generations. By the 1960s, however, most of the region's small, inefficient farms had disappeared, and with them went the seemingly endless supply of weedy fields and fence lines that had provided the best quail habitat on earth. Plenty of quail are still around, but hunting them now is mostly a sport for the fortunate minority with access to high-quality habitat. It is very helpful to have friends, relatives, or business associates who own or lease prime quail cover. The very best shooting, however, is enjoyed by those paying steep fees for package hunts on shooting preserves or plantations with many coveys of pen-raised birds dispersed in carefully managed, easily accessible cover.

Another great Carolina hunting tradition, the dove shoot, has remained more egalitarian. Every fall, in the early morning and late afternoon, flocks of mourning doves descend on grain fields, game food plots, and waterholes in the Carolinas. Hunters positioned at intervals along field edges, power line clearings, and other flyways can sometimes bag their daily limits of about a dozen birds each. Although this is not a simple task—the doves often fail to show up and are very hard to hit—it is comparatively simple to gain access to hunting land. Many hunters take part in shoots on state game management lands. More and more hunters these days are leasing dove rights on private land (as by joining a dove club) or paying hunting fees to enterprising farmers and other commercial providers. Carolina dove hunters especially covet invitations to the "society shoots" hosted by wealthy landowners, since these are attended by the community elite and offer niceties such as barbecue, shuttle service to the shooting stands, and post-shoot libations.

Carolinians have hunted raccoons at night with dogs for many generations. In the old days, it was the dirt-poor sharecroppers and farm laborers who pursued raccoons most avidly, and they did it as much for the meat and fur as for the fun of it, just as they hunted squirrels, rabbits, and other "low-status" game. With the passage of time, coon hunting left its table fare origins behind, moved up a notch on the socioeconomic scale and became a sport rooted in the joy of pursuit and the quest for bragging rights. Today, as always, the only way to gain status in coon hunting circles is to own talented hunting dogs, the better examples of which now cost $1,000 or more. Competition coon hunting, which tests striking, treeing, and barking skills, offers a vehicle for bringing dog owners together to socialize and objectively determine which dogs are best. It is a source of considerable Carolina pride that the American Coon Hunters Association has been holding the ACHA Grand American Coon Hunt, one of the largest and most important of the competition hunting events, at Orangeburg, South Carolina since 1965.

Like white-tails, turkeys once nearly disappeared from the Carolinas. And like the white-tail population, the turkey population of the region's river bottom forests and piney uplands is now so large that it offers some of America's best hunting and most liberal bag limits. (It is no mere coincidence that the National Wild Turkey Federation is headquartered in Edgefield, South Carolina.) Many outdoorsmen feel that turkey hunting offers about as much challenge and excitement as hunting trophy white-tails. As an added bonus, most turkey hunting takes place during the spring mating season, a time when the weather is pleasant and the deer hunting season is closed.

Waterfowling offers yet another example of hunting conditions that are very different from what they once were. Generations ago, the wetlands and rice field impoundments of the Carolina coastal estuaries attracted vast numbers of ducks and geese each fall. Wealthy northerners bought many former plantations in South Carolina and converted them into hunting clubs (Kovacik, 1996). Waterfowl populations are now considerably smaller in the Carolinas, and much prime habitat has been lost, degraded, or closed to hunting. Although duck hunting is still enjoyed by many Carolinians, it has taken a back seat to deer hunting and related activities in all but a few locales.

Campground Towns

Family camping is very popular in the Carolinas, a region renowned for its many "resort-style" camping facilities. There are clusters of large campgrounds and RV parks in the western North Carolina mountains, the Outer Banks, the Santee-Cooper region, and especially the Grand Strand (Janiskee, 1990). Many tourists are surprised to learn that Myrtle Beach is the undisputed "Seaside Camping Capital of the World" (Photo 8.10). Myrtle Beach State Park has offered beachfront camping since the 1930s. During 1959-1971, a time when oceanfront land was still cheap and could be developed with few restrictions, lo-

Photo 8.10. Commercial campground at Myrtle Beach, "Seaside Camping Capital of the World." (Courtesy South Carolina Parks, Recreation and Tourism)

cal entrepreneurs developed several clusters of unusually big campgrounds along the beach. Today, the more than 9,000 developed sites (including annual lease sites) in the remaining seven commercial campgrounds represent one of the densest concentrations of campsites anywhere. Myrtle Beach's commercial campgrounds, which function primarily as self-contained resorts for vacationers and weekenders, are so big, diverse, and amenities-rich that they deserve to be called "campground towns." The largest, Ocean Lakes, has three-quarters of a mile of sandy beach and is one of the biggest and most lavishly equipped camping facilities in the world. On the 300-acre premises of this ultra-modern hybrid, which has received the National Association of RV Parks and Campgrounds' RV Park of the Year Award, are nearly 900 transient sites, over 2,000 annual lease sites (for park model trailers), a large mobile home court, some apartment villas, a few beach houses, and 29 miles of paved roads. Campers enjoy such niceties as cable TV and modem-friendly phone jacks, an Olympic-size outdoor pool, an indoor pool, a splash park, a video arcade, tennis courts, a mini-golf course, a gas station, a 17,000 square foot recreation center, church services, golf cart rentals, camper storage, and an RV sales and service facility.

Chitlin Struts and Bubbafests

Of the roughly 26,000 community festivals held each year in America, at least 900 are produced in the Carolinas (Janiskee, 1996). The typical community festival takes place on a weekend, uses public venues such as streets or parks, and draws about 14,000 people. Nearly all employ the "something for everybody" approach to entertainment, offering crowd pleasers such as parades, food booths, music, arts and crafts, children's games, and cookoffs. Communities love to produce festivals. They make money and attract visitors, and even a nonprofit event can help to put a community on the map, provide a showcase for local talent, serve educational purposes, promote cleanups and fixups, get people working together, boost community pride, and give residents something to do.

The prime festival season in the Carolinas, which peaks in April, July (Independence Day),

Photo 8.11. Okra Strut Festival in Irmo, SC.

and October, is anchored by a group of larger scale, more widely publicized events. The region has two mega-festivals, each with an attendance of over 300,000. The Sun Fun Festival kicks off Myrtle Beach's summer season and the Bele Chere Festival celebrates Asheville's vibrant downtown. Among the other major festivals of the Carolinas are events as diverse as Spoleto Festival USA (Charleston's world-class performing arts extravaganza), the Grandfather Mountain Highland Games & Gathering of Scottish Clans (the second-largest event of its kind in the world), and the Freedom Weekend Aloft in Simpsonville, South Carolina (one of America's largest hot air balloon events).

Perhaps, the most interesting festivals in the Carolinas are the dozens of "place celebrations" that spotlight the cultural traits, economic activities, and other things that give the region its distinctive character and make each of its communities unique. Coastal communities like Myrtle Beach (Sun Fun Festival), Wrightsville Beach (Surf, Sun, and Sand Celebration), and Beaufort, SC (Beaufort County Water Festival) celebrate beach resort appeals and the bounty of the sea. Numerous mountain communities celebrate the reality and myth of Appalachian culture with events like Cullowhee's Mountain Heritage Day (*Asheville Citizen-Times*, 1998) and Mountain Rest's Hillbilly Days. At various places and times, other festivals celebrate regional trademarks such as azaleas, dogwoods, textiles, coon dogs, hardwood furniture, catfish, shrimp, chitlins, pork barbecue, watermelon, collards, cotton, tobacco, grits, boiled peanuts, poke sallet, okra, and sweet potatoes (Photo 8.11).

Carolina Thunder

Stock car racing has an international following, but nowhere is it more popular than in the Carolinas and nearby states (Pillsbury, 1996). Southern race fans proudly recall the era of mountain moonshiners with souped-up cars and the early days of organized racing on the beach at Daytona. With ample justification, Carolinians consider their particular part of the South to be the very core of the stock car industry. Although the industry's main promotion and sanctioning organization, NASCAR, was born at Daytona Beach (in 1947), the first superspeedway was built at Darlington, South Carolina. In NASCAR's early years, when stock car racing was a strongly regionalized sport, the Carolinas accounted for a disproportionate share of the races, drivers, and fans. There are three major tracks in the Carolinas—at Darlington, Charlotte, and Rockingham (N.C.), and many communities in the region still have dirt or short-paved tracks. North Carolina plays an especially vital role in building racecars, training drivers, and promoting NASCAR interests. The Charlotte area in particular functions as the "mecca of motorsports," boasting not only the renowned Lowe's Motor Speedway, but also the racing shops of dozens of Nextel Cup and Grand National racing teams (at Lakeside Industrial Park in Mooresville), as well as truck racing shops and several racing museums and related attractions.

The same socio-economic changes that sent old-style dirt track racing into decline in the rural South created huge urban markets for NASCAR's flashy, corporate-sponsored Nextel Cup product. Carolina NASCAR fans fill the stands and luxury suites for the races, swamp the race car exhibits and autograph sessions at the malls, load up on racing souvenirs, and remain fiercely loyal to products endorsed by multimillionaire drivers. Nowhere is this enthusiasm more evident than at the Lowe's Motor Speedway (LMS), a sprawling, ultramodern complex that hosts three major NASCAR races (two Nextel Cup races and the Nextel All-Star Challenge) and various other stock car, truck, kart, and dirt bike races. Each 160 mph-plus Nextel Cup race on the 1.5-mile, 24-degree banked oval draws over 167,000 spectators, one of the largest live audiences for any sports event in America. Of course, rapt attention to the race itself is not mandatory at a NASCAR event. Whether they are at the LMS, the Nextel Cup speedways at Darlington and Rockingham, or "minor league" facilities like the Greenville-Pickens Speedway, fans seem to be not so much watching a race as immersing themselves in the cacophonous sounds, sights, and smells of the speedway milieu and enjoying a great big party.

Although stock car racing will continue to be a very important spectator sport and economic activity in the Carolinas, the region is slated to play a less salient role in the sport's evolution. NASCAR is strongly committed to high-profile, capital-intensive growth and change that projects the sport into major new markets in America and abroad. Given that Nextel Cup teams can compete in only about three-dozen races a year, races ideally should be allocated to the largest and most lucrative market areas. The Carolina market is comparatively small, so it came as little surprise when the long-standing Winston Cup race at North Wilkesboro, North Carolina was moved to the Dallas-Ft. Worth market in 1977, and the venerable Darlington track lost one of its two major race dates to a California track.

The Pigskin Cult

In their landmark study of American sports geography, Rooney and Pillsbury (1992) placed North Carolina in the Eastern Cradle region (baseball-basketball-football) and most of South Carolina in a football-crazy region they dubbed "Pigskin Cult." The latter designation is an apt one, since Carolinians are extremely fond of high school and college football. The prime reason is that there have been so few other choices for fall weekend entertainment. Pro football arrived in the Carolinas only recently, and while soccer has caught on with youngsters, it draws few adult spectators. Athletic programs in Carolina high schools and colleges traditionally have allocated most of their money and better athletes to football and basketball. For these and related reasons, fall weekends in the Carolinas have come to mean Friday night lights and tailgating Saturdays.

High school football mania reaches its zenith in west Texas, but Carolinians also support the game as few others do. In Carolina small towns and suburbs (and throughout the South), a winning football team makes a huge contribution to a community's sense of well-being. Some fans even seem to believe that winning the state championship—or just beating the rival town's team—is the very wellspring of pride and contentment. The hallmark feature of a Pigskin Cult high school is a capacity crowd in a 10,000-seat stadium on Friday night. What really sets these programs apart, however, is their whatever-it-takes commitment to winning football. Youngsters with the right stuff are steered into football and not "wasted" on minor sports. In high school these budding athletes receive first-rate coaching and conditioning, plus special treatment to pump their egos and their grades. Taxpayers and booster clubs dig down deep to fund stadiums, weight rooms, and other facilities that rival those of many colleges. Head coaches—typically the best-paid employees of their schools—keep their jobs only by using these resources to win consistently.

For all their love of the high school game, Carolinians are even fonder of the collegiate variety. A big reason is tailgating. Many newcomers are surprised to learn how important college football is to the Carolina culture and economy, and truly amazed to see how much time, effort, and money goes into tailgating in this part of the country. On football Saturday mornings, many supermarkets, fast fooderies, and liquor stores do a land office business as fans get ready to party. Roads converging on the stadiums feed torrents of cars, conversion vans, and motor homes into the parking lots where tailgate partying kicks into high gear and continues for hours on end. In the Pigskin Cult region, tailgating with fellow football fans is at least half the fun of the football experience. Some stadiums even have nearby condominium parking areas with party facilities—the most novel example of which is the Cockaboose Railroad, a collection of several dozen elaborately-outfitted cabooses on a railroad spur next to the University of South Carolina's Williams-Brice Stadium (Photo 8.12).

The NCAA Division I-A teams in the Carolinas get the great majority of the fans, money, and media attention. In Pigskin Cult territory, that means extraordinary numbers of fans, mind-boggling amounts of money, and excruciatingly detailed media coverage. Atop the heap are the five teams in the Atlantic Coast Conference (ACC), the region's lone Southeastern Conference (SEC) affiliate (the University of South Carolina), and

Photo 8.12. Cockaboose Railroad tailgating stadium-side at the University of South Carolina in Columbia.

several other nationally recognized programs like East Carolina and the Citadel. Although the quality of play is sometimes deficient, support for college football remains phenomenally strong. The numbers speak volumes, especially in South Carolina. Although the Palmettto state has fewer than five million people, it supports two major football programs with attendance and financial support that rivals any program in America. Clemson University has an 81,000-seat stadium that is usually jampacked, plus a large athletic booster club (IPTAY) that raises $10 million or more a year. Lusting for a repeat of its 1981 national title performance, CU shelled out nearly $2 million in one 10-year period just to get rid of coaches it did not want. Nothing illustrates the Pigskin Cult obsession better than fan support for the University of South Carolina football Gamecocks. Despite occasional flashes of excellence, including a best-ever 10-2 season, a Heisman trophy winner, and a pair of victories over Ohio State University in the Outback Bowl, the Gamecock worksheet reflects almost perfect mediocrity (475-482-44 after 108 seasons) and only four bowl wins had been logged by 2006. A low was reached in the late 1990s with a string of 21 consecutive defeats. Nevertheless, the football Gamecocks play in an 80,250-seat gridiron palace with extensive Executive Club seating, tailgating facilities second to none, and average attendance that perennially ranks in the top 15 nationally.

Hoops Mania

As the end of the football season nears in the Carolinas, excitement builds toward the start of the basketball season. The college basketball fans of the Carolinas are abundant, devoted, and truly blessed. Few areas of similar size in the country, if any, have so many teams with such admirable credentials. Clemson, South Carolina, and other teams have their good years, but the most conspicuously successful programs are those of UNC Chapel Hill, Duke, Wake Forest, and North Carolina State. College basketball is king in North Carolina, the state that anchors the basketball-crazy Atlantic Coast Conference. Duke, which in

1999 went 16-0 in the AAC and won the national championship, would be the gold standard program in just about any other part of the country. But the University of North Carolina is the king of Tobacco Road. Like Kentucky and Duke, UNC is expected to make a solid run at the national championship every year. When the Tar Heels' legendary coach Dean Smith retired in 1997 after 36 seasons at the helm, he had notched 879 victories and was the winningest coach in the sport. His program had yielded 27 consecutive seasons with 20 or more wins, 28 top 15 finishes, 13 ACC championships, 11 trips to the Final Four, and two national titles—not to mention many former assistant coaches who are now successful head coaches and many players (including Michael Jordan and James Worthy) who have excelled in the NBA.

The Hockey Invasion

New leisure trends and patterns are evolving at a rapid rate in the Carolinas. These days, many more Carolinians are educated, affluent, urban, and urbane, and there are many more northerners, Hispanics, and others whose lifestyles and values do not fit the traditional Carolina mold. It is interesting to note the recreational consequences of these variables operating in combination with technological innovation. More Carolinians are enjoying off-road vehicle riding, windsurfing, and recreational shopping, but fewer are interested in quail hunting, cane pole fishing, and cockfighting. Dozens of other leisure activities are waxing, waning, arriving on the scene, or taking new shape. Nothing illustrates this dynamic change better than the invasion of ice hockey.

Professional ice hockey was virtually nonexistent in the South until the late 1980s, but by 1998 the southern states had acquired 40 franchises in seven leagues. In the Carolinas alone, the 1998 complement included six minor league hockey teams in three different leagues, plus the Carolina Hurricanes, an NHL franchise in Raleigh (Photo 8.13). There are many reasons for this dramatic expansion, which is still underway. American interest in hockey zoomed when

Photo 8.13. Carolina Hurricanes game. (Courtesy Carolina Hurricanes)

Wayne Gretzky, probably the greatest player of all time, was traded to Los Angeles from Edmonton in 1988. Minor league hockey's subsequent surge in the Carolinas and throughout the South reflects influences such as an abundant supply of transplanted Northern fans, the growth of inline skating, the aggressive marketing of franchises in smaller cities, the appeal of affordable prices and skillful promotions, and disgruntlement with the runaway prices and perceived arrogance of the major league sports. Whatever the reasons, and even though the NHL Hurricanes have struggled to thrive,[viii] minor league hockey seems here to stay. As cities acquired franchises and built new arenas, many more Carolinians became avid fans. The region has even given the sport its own hockey-as-theater slant with things like dasherboards advertising "3 Little Pigs BBQ."

Endnotes

i Those who need detailed, practical information about tourism and recreation opportunities in the Carolinas can easily get it. State and regional travel guides are available at book stores and libraries. The state tourism agencies of North Carolina and South Carolina both supply comprehensive state visitor guides that are available free of charge at places such as the welcome centers on interstate highways. Other good sources of information include chamber of commerce brochures, the travel and sports sections of major regional newspapers such as *The Charlotte Observer* and *The State* [Columbia, SC], and Internet web sites maintained by tourism promotion agencies or private firms.

ii Tourism and recreation fall under the general heading of leisure behavior. "Leisure" stems from the Latin root *licere*, which means "to choose." It is appropriate to think of leisure as "choosing time," since it implies freedom to use time for purposes other than performing work or discharging obligations. Swimming, fishing, skiing, and other leisure activities that produce immediate gratification associated with pleasurable physical, emotional, esthetic, social, or educational experiences are lumped under the general heading of recreation. Tourism has been defined many different ways, but is commonly considered to be any leisure activity that entails travel to other communities or distant places.

iii The South Carolina Department of Parks, Recreation and Tourism publishes annual reports on the state's tourism industry. When calculating benefits, tourism is deemed worthy of a large economic multiplier. In 1997, for example, the $6.6 billion that visitors spent in the state was assumed to have yielded a total economic impact of $14.3 billion (SCPRT, personal communication).

iv. "American Travel Survey" statistics, U.S. Department of Transportation Bureau of Transportation Statistics (www.bts.gov/programs/ats/).

v. Many of the interesting things to see and do in the Carolinas are lesser known or out of the way attractions that may not be mentioned in popular state and regional guides. Among the publications that provide more detailed information about these less frequently visited places and events are the following: McLean, N. 1996. *Explore South Carolina: A Guide for Families, Teachers, and Youth Groups*. Columbia, S.C.: MCL Publications; Pitzer, S. 1996. *North Carolina Off the Beaten Path: A Guide to Unique Places*. 3rd ed. Old Saybrook, CT: Globe Peqout Press; Price, W. F. 1996. *South Carolina Off the Beaten Path: A Guide to Unique Places*. Old Saybrook, CT: Globe Peqout Press; and Todd, C. W. and S. Wait. 1997. *South Carolina, A Day at a Time*. Orangeburg, S.C.: Sandlapper Publishing Company.

vi Beach music, the realm of performers like The Drifters and The Four Tops, is peppy rhythm and blues that invokes feelings of being young, carefree, and having fun at the beach with good friends. The shag, a rather complex "slow jitterbug" that is danced to beach music, was born on Ocean Drive in the 1940s and reached its heyday there in the 1950s and 1960s. Shagging competitions and the low-budget 1989 movie "Shag" have attracted a good deal of national attention to this phenomenon. Every year for more than a decade, the Myrtle Beach C&VB has produced shagging competitions and parties to boost visitor industry. About 12,000 mostly middle-aged shaggers now gather each April and September at the Ocean Drive clubs in North Myrtle Beach.

vii. There are over 390 units in America's national park system, but only about 60 are formally called "National Park" (the rest being named historic site, monument, parkway, etc.). Thus, while the Blue Ridge Parkway's 17 million visits a year makes it the most visited unit in the park system, the roughly nine million visits a year to Great Smoky Mountains National Park make it the busiest National Park in America and the world. The park is so popular because it charges no entrance fee and is big and pretty, ideally suited for "windshield touring," linked to the heavily-traveled Blue Ridge Parkway, located within easy driving distance of half the country's population, and easily accessible from the interstate highways that carry northern vacationers to and from the Florida vacationlands.

viii. The franchise, then known as the Hartford Whalers, was a financial failure in Connecticut when it was moved to Raleigh in 1997 and renamed the Carolina Hurricanes. The Hurricane franchise struggled to build an adequate fan base in the face of difficulties like being based in Greensboro for several seasons pending the completion of a $158 million arena in Raleigh, having ticket prices higher than those of any other major league sport, playing many games on weeknights, and serving a market area that has comparatively few hockey-savvy people and hard core NHL fans. The Hurricanes sold only 5,800 season tickets for the 2000-2001 season, but averaged over 12,000 attendance per game by 2004.

References

Asheville Citizen-Times. 1998. Festival Honors Mountain Heritage. In *Mountain Travel Guide: The Complete Destination Guide of Western North Carolina*, ed. C. Currie and L. Carrington, 23-24. Asheville, NC: Asheville Citizen-Times.

Bower, E. L. 1996. Mystery and Treasure: The Charleston Complex. In *Snapshots of the Carolinas: Landscapes and Cultures*, ed. D. G. Bennett, 17-21. Washington, D.C.: Association of American Geographers.

Boyle, J. 1998. Feast for the Eyes: Foliage Makes WNC a Colorful Autumn Paradise. In *Mountain Travel Guide: The Complete Destination Guide of Western North Carolina*, ed. C. Currie and L. Carrington, 3-4. Asheville, NC: Asheville Citizen-Times.

Brent, M. T. 1997. *Coastal Resort Morphology as a Response to Transportation Technology*. Unpublished Doctoral dissertation, University of Waterloo.

Butler, R. 1980. The Concept of a Tourist Area Cycle of Evolution: Implications for Management of Resources. *Canadian Geographer* 24 (1), 5-12.

Enscore, S. 1996. Old Salem's Pennsylvania Heritage. In *Snapshots of the Carolinas: Landscapes and Cultures*, ed. D. G. Bennett, 29-32. Washington, D.C.: Association of American Geographers.

Fodor's Travel Publications, Inc. 1998. *The Carolinas and Georgia: Fodor's '98*. New York: Fodor's Travel Publications, Inc.

Gares, P. A. 1996. Shoreline Changes along the Cape Hatteras National Seashore. In *Snapshots of the Carolinas: Landscapes and Cultures*, ed. D. G. Bennett, 73-76. Washington, D.C.: Association of American Geographers.

Hines, E. and Smith, M. 1996. Gold Mining in North Carolina. In *Snapshots of the Carolinas: Landscapes and Cultures*, edited by D. Gordon Bennett, pp. 43-47. Washington, DC: Association of American Geographers.

Janiskee, R. L. 1990. Resort Camping in America. *Annals of Tourism Research* 17 (3), 385-407.

_____. 1994. Recreational Billfishing in South Carolina: Toward Restraint and Replenishment. In *Proceedings, 1993 Southeastern Recreation Research Conference*, ed. H. A. Clonts, 45-55. General Technical Report SE-90. Asheville, N.C.: Southeastern Forest Experiment Station.

_____. 1996. Community Festivals in the Carolinas. In *Snapshots of the Carolinas: Landscapes and Cultures*, ed. D. G. Bennett, 57-61. Washington, D.C.: Association of American Geographers.

_____ and P. Chirico. 1997. Managing for Low-Impact Recreation and Ecotourism in South Carolina's ACE Basin. In *Proceedings of the 1996 Northeastern Recreation Research Symposium*, ed. W. F. Kuentzel, 293-295. USDA Forest Service General Technical Report NE-232. Radnor, PA: U.S.D.A. Forest Service, Northeastern Forest Experiment Station.

_____, L. S. Mitchell, and J. H. Maguire. 1996. Myrtle Beach: Crowded Mecca by the Sea. In *Snapshots of the Carolinas: Landscapes and Cultures*, ed. D. G. Bennett, 217-220. Washington, D.C.: Association of American Geographers.

_____ and P. L. Drews. 1998. Rural Festivals and Community Reimaging. In *Tourism and Recreation in Rural Areas*, ed. R. W. Butler, C. M. Hall, and J. Jenkins, 157-175. Chichester, UK: John Wiley.

Jones, M. S. 1996. The Cherokee Reservation. In *Snapshots of the Carolinas: Landscapes and Cultures*, ed. D. G. Bennett, 33-36. Washington, D.C.: Association of American Geographers.

Kovacik, C. F. 1996. Plantations and the Low Country Landscape. In *Snapshots of the Carolinas: Landscapes and Cultures*, ed. D. G. Bennett, 3-6. Washington, D.C.: Association of American Geographers.

_____ and J. J. Winberry. 1989. *South Carolina: The Making of a Landscape*. Columbia: University of South Carolina Press.

Leib, J. I. 1996. Resort Development, Tourism and Cultural Survival in the Gullah Sea Islands. In *Snapshots of the Carolinas: Landscapes and Cultures*, ed. D. G. Bennett, 225-229. Washington, D.C.: Association of American Geographers.

Lewis, C. H. 1998. *Horry County, South Carolina 1730-1993*. Columbia: University of South Carolina Press.

Marks, S. A. 1991. *Southern Hunting in Black and White: Nature, History, and Ritual in a Carolina Community*. Princeton, N.J.: Princeton University Press.

Mason, D. S. 1996. The Buck Stops Here. In *Snapshots of the Carolinas: Landscapes and Cultures*, ed. D. G. Bennett, 49-52. Washington, D.C.: Association of American Geographers.

Mayfield, M. W. and J. DeHart. 1996. Carolina Whitewater. In *Snapshots of the Carolinas: Landscapes and Cultures*, ed. D. G. Bennett, 91-94. Washington, D.C.: Association of American Geographers.

McLean, N. 1996. *Explore South Carolina: A Guide for Families, Teachers, and Youth Groups*. Columbia, S.C.: MCL Publications.

Mitchell, L. S. 1996. Is Hilton Head Island a Hedonists's Haunt or Hermitage or Hospice? In *Snapshots of the Carolinas: Landscapes and Cultures*, ed. D. G. Bennett, 57-61. Washington, D.C.: Association of American Geographers.

Moore, T. G., and Ross, T. E. 1996. Native Americans in the Carolinas. In *Snapshots of the Carolinas: Landscapes and Cultures*, ed. D. G. Bennett, 125-129. Washington, D.C.: Association of American Geographers.

Morris, G. 1993. *North Carolina Beaches: A Guide to Coastal Access*. Chapel Hill: University of North Carolina Press.

Mullen, D. and G. Green. 1997. *The Insiders' Guide to Myrtle Beach and the Grand Strand*. 4th ed. Manteo, N.C.: Insiders' Publishing, Inc.

Pillsbury, R. 1996. Carolina Thunder: The Changing Scene. In *Snapshots of the Carolinas: Landscapes and Cultures*, ed. D. G. Bennett, 53-56. Washington, D.C.: Association of American Geographers.

Pitzer, S. 1996. *North Carolina Off the Beaten Path: A Guide to Unique Places*. 3rd ed. Old Saybrook, CT: Globe Peqout Press.

Price, W. F. 1996. *South Carolina Off the Beaten Path: A Guide to Unique Places*. Old Saybrook, CT: Globe Peqout Press.

Richards, C. 1998. Winter Wonderland: WNC Slopes Cater to Winter Sports Enthusiasts. In *Mountain Travel Guide: The Complete Destination Guide of Western North Carolina,* ed. C. Currie and L. Carrington, 35-38. Asheville, N.C.: Asheville Citizen-Times.

Rooney, J. F., Jr. and R. Pillsbury. 1992. Sports Regions of America. *American Demographics,* November, 31-39.

_____. 1992. *Atlas of American Sport.* New York: Macmillan Publishing Company.

Soule, P. T. 1996. North Carolina's Climate. In *Snapshots of the Carolinas: Landscapes and Cultures,* ed. D. G. Bennett, 65-68. Washington, D.C.: Association of American Geographers.

South Carolina Department of Parks, Recreation and Tourism. 1997. *The Tourism Industry in South Carolina, 1996 Annual Report.* Columbia, S.C.: South Carolina Department of Parks, Recreation and Tourism.

Stuart, A. W. 1996. The Charlotte Urban Region. In *Snapshots of the Carolinas: Landscapes and Cultures,* ed. D. G. Bennett, 109-113. Washington, D.C.: Association of American Geographers.

Summerlin, C. and Summerlin, V. 1997. *Traveling the Southern Highlands: A Complete Guide to the Mountains of Western North Carolina, East Tennessee, Northeast Georgia, and Southwest Virginia.* Nashville, TN: Rutledge Hill Press.

Todd, C. W. and S. Wait. 1997. *South Carolina, A Day at a Time.* Orangeburg, S.C.: Sandlapper Publishing Company.

Waccamaw Regional Planning and Development Council. 1997. *Waccamaw Region Population and Economic Study, 1997.* Georgetown, S.C.: Waccamaw Regional Planning and Development Council.

_____. 1998. *1998 Grand Strand Region Economic Forecast.* Georgetown, S.C.: Waccamaw Regional Planning and Development Council and Charleston Southern University.

Winberry, J. J. 1996. Gullah: The People, Language, and Culture of the Sea Islands of South Carolina. In *Snapshots of the Carolinas: Landscapes and Cultures,* ed. D. G. Bennett, 11-15. Washington, D.C.: Association of American Geographers.

Chapter 9

HIGHER EDUCATION IN THE CAROLINAS

Roger Winsor
Appalachian State University

HISTORY

Four-Year Colleges and Universities

This chapter will trace the evolution, spatial diffusion and characteristics of higher education in the Carolinas. It also will discuss higher education's role in producing a more highly skilled labor force and the development of research parks.

The College of Charleston, South Carolina was established in 1770 and was the first college or university in the Carolinas. Two years later, Salem College followed in Winston-Salem, North Carolina. The University of North Carolina at Chapel Hill, established in 1789, was the nation's first public state-supported university. The University of South Carolina at Columbia, founded in 1801, was the third public state-supported university established. Figure 9.1 and Appendix 9.1 illustrates the diffusion of four-year colleges across the Carolinas. With the exception of the College of Charleston, all of the colleges established between 1770 and 1820 were located in the Piedmont. Likewise, most of those established between 1820 and 1870 were also in the Piedmont, with a few established on the Coastal Plain at Charleston, S.C. and in the Mountains at Mars Hill and Brevard in North Carolina. There was also a strong Piedmont focus to colleges established between 1870 and 1975 in the Carolinas.

Many of the first colleges and universities in the Carolinas were established by religious denominations. Salem College was established in 1772 by the Moravians to educate women. Baptists established Furman University in Greenville, South Carolina, in 1826; Wake Forest University (1834), initially located in Wake Forest, North Carolina but moved to Winston-Salem in the 1950s; Limestone College, Gaffney, South Carolina in 1845; and Chowan College, Murfreesboro, North Carolina in 1848. The Lutherans established a Lutheran Theological Seminary in 1830 at Columbia, South Carolina. Presbyterians established both Davidson College (1837) at Davidson, North Carolina and Erskine College (1837) at Due West, South Carolina; Guilford College in Greensboro, North Carolina was established in 1837 by the Society of Friends (Quakers); Catawba College, Salisbury, North Carolina was established in 1851 by the Reformed Church. Methodists established three institutions of higher education, two in 1838, Greensboro College, in Greensboro, North Carolina; and Trinity College, which began in rural northwestern Randolph County, North Carolina. In 1892, Trinity College was moved to Durham and renamed Duke University. Methodists also established Brevard College, Brevard, North Carolina in 1854 (Drake, 1964, 118-150).

The Medical University of South Carolina, established in 1824, is the Southeast's oldest medical school. It was started in Charleston because its population was much larger than Columbia's and Charleston had more physicians, a large medical society library, and established clini-

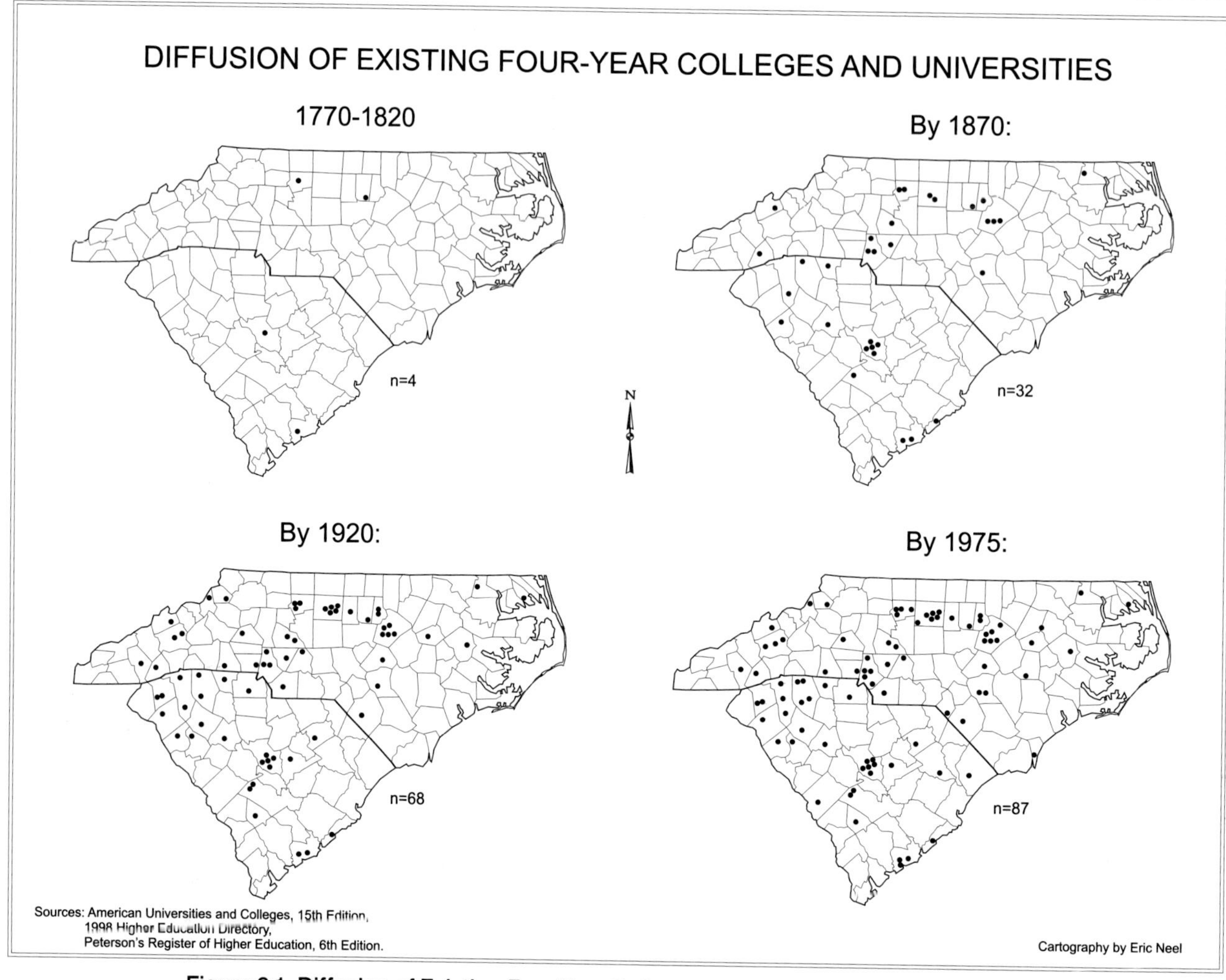

Figure 9.1. Diffusion of Existing Four-Year Colleges and Universities in the Carolinas
Sources: American Council on Education, 1997; Rodenhouse, 1998; *Peterson's Register*, 1993; selected college catalogues.

cal facilities (Edwards, 1986, 13). The University of North Carolina at Chapel Hill began training physicians in 1890 with a two year program (students took the last two years elsewhere). It became a four year program in 1954. The medical program at Wake Forest University began in 1902, Duke University's started in 1930, and East Carolina University in Greenville, North Carolina and the University of South Carolina at Columbia both began in 1977.

The University of North Carolina at Chapel Hill, whose law school was established in 1843, is the oldest in the two Carolinas. The law school at University of South Carolina at Columbia was founded in 1866. Wake Forest University's program began in 1894, Duke University's in 1930, North Carolina Central University's in Durham in 1939, and Campbell University's, at Buie's Creek, N.C. in 1975.

The pharmacy program at The University of North Carolina at Chapel Hill is the oldest in the Carolinas, being established in 1880. The Medical University of South Carolina started its program in 1881, the program at University of South Carolina at Columbia began in 1924, and Campbell University's followed in 1986.

The University of North Carolina at Chapel Hill established the Carolinas' first dental school in 1949. The program at the Medical University of South Carolina was established in 1964. The Veterinary Medicine program at North Carolina State University in Raleigh, which was established in 1979, is the only such program in the two Carolinas.

Four land-grant colleges with corresponding agricultural experiment stations and extension services, were established in the Carolinas. Initially, they primarily focused on agriculture and engineering: North Carolina State University in Raleigh in 1887, Clemson University in Clemson, South Carolina in 1889, North Carolina Agricultural and Technical University in Greensboro, in 1891; and South Carolina State University in Orangeburg, in 1898. Today, North Carolina State University (1887), North Carolina Agricultural and Technical University (1891), Clemson University (1893), University of South Carolina at Columbia (1908), Duke University (1910), and the University of North Carolina at Charlotte (1979) all offer engineering degrees.

A Dual System?

After the Civil War, as a part of a de jure segregated higher education system, historically black colleges were established in both Carolinas to educate former slaves. In North Carolina, this began in 1865 with the founding of Shaw University in Raleigh, which was established by Baptists. In 1867, four additional institutions of higher education were established: Barber-Scotia College in Concord and Johnson C. Smith University in Charlotte were both established by Presbyterians; Saint Augustine's College in Raleigh was founded by Episcopalians; and Fayetteville State was a state-supported institution (Figure 9.2 and Appendix 9.2). Initially, Barber-Scotia only educated women, but became coeducational in 1954. In 1873, Bennett College was established by the Methodist Church to meet the needs of young black women. North Carolina

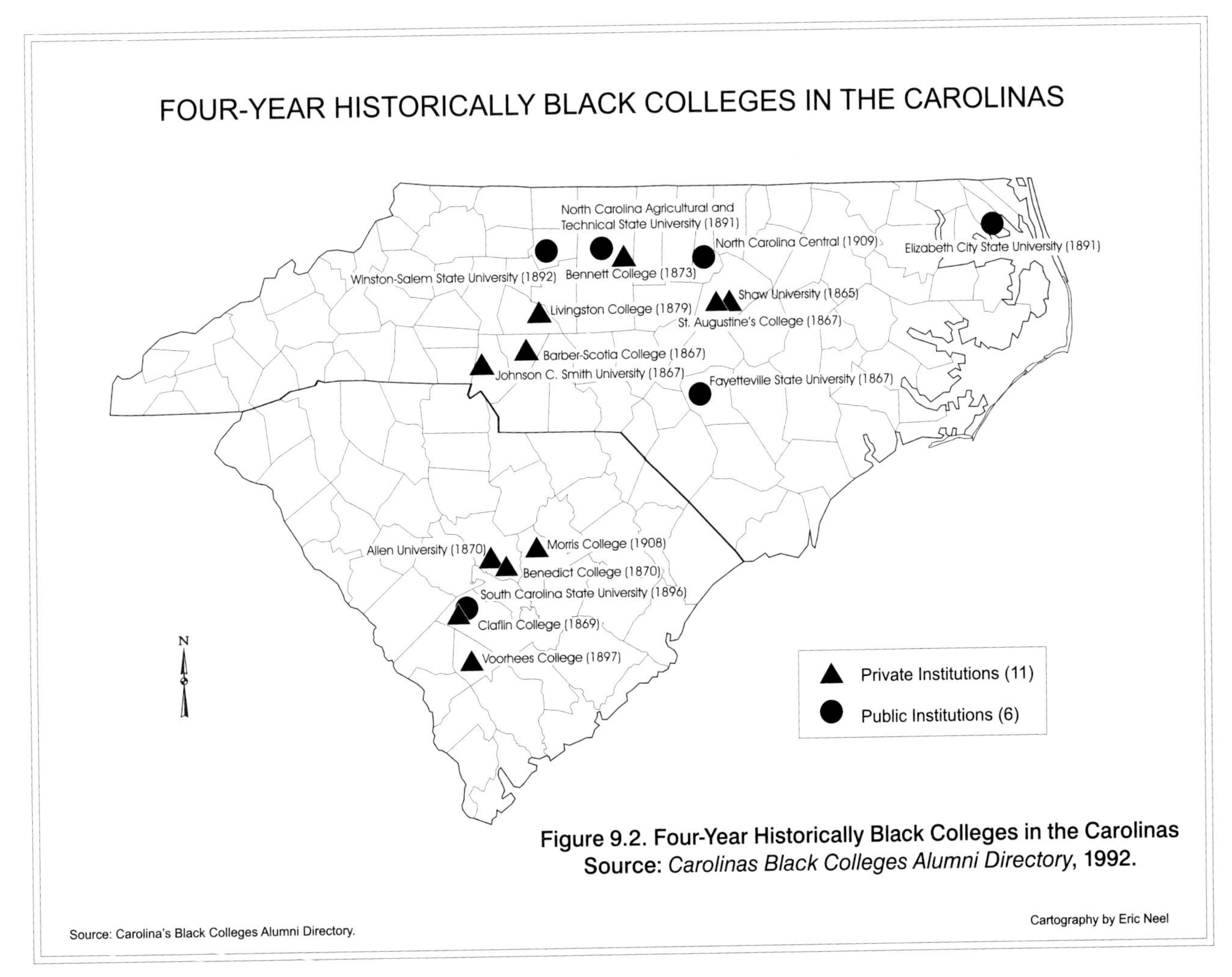

Figure 9.2. Four-Year Historically Black Colleges in the Carolinas
Source: *Carolinas Black Colleges Alumni Directory*, **1992.**

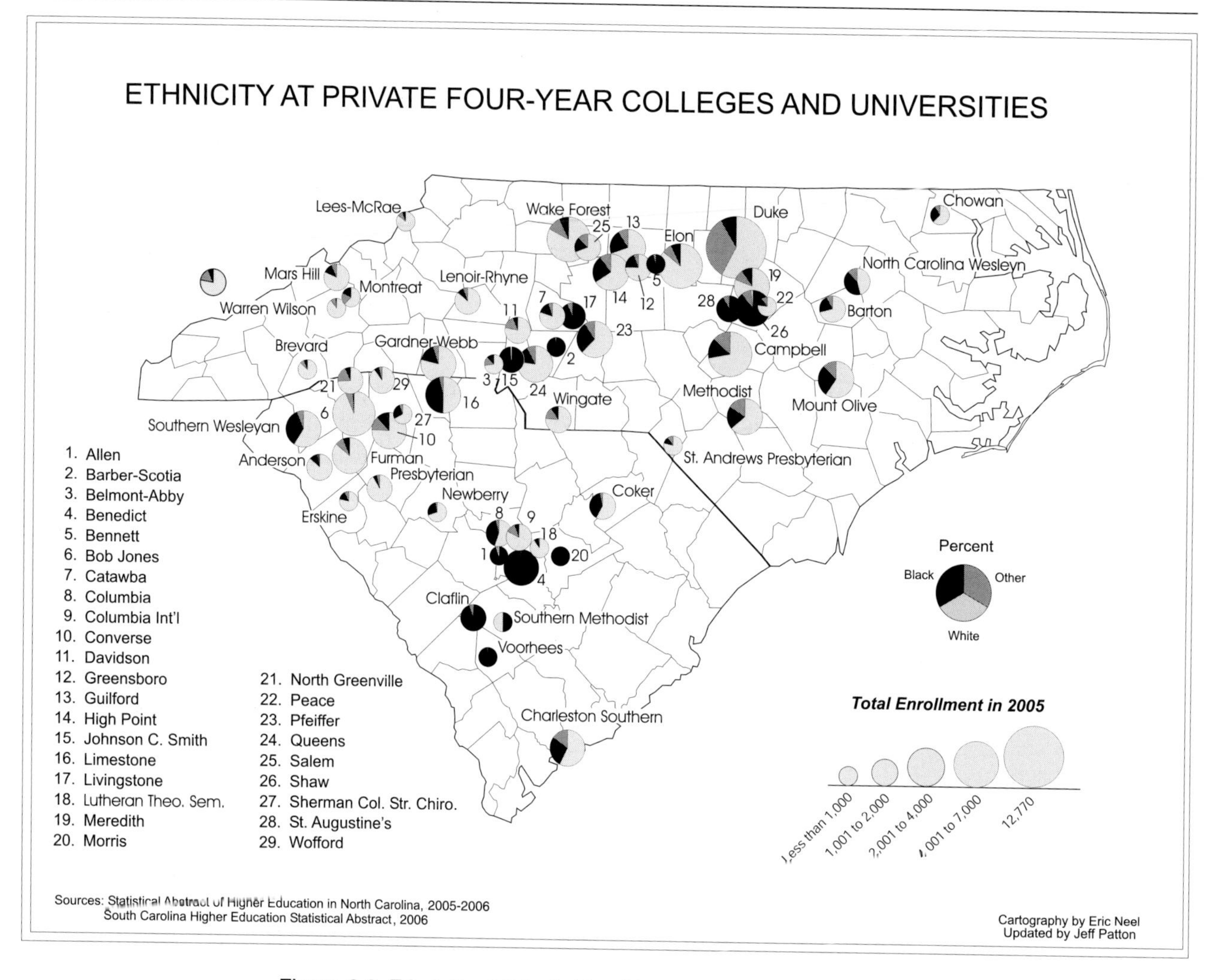

Figure 9.3. Ethnicity at Private Four-Year Colleges and Universities
Sources: University of North Carolina, 2000 <www.northcarolina.edu/content.php/assessment/reports/previousabs.htm> and South Carolina Commission on Higher Education, 2000 <www.che400.state.sc.us/web/stats.htm>

has more publicly-supported historically black universities than any other state: Elizabeth City State University, Fayetteville State University, North Carolina Agricultural and Technical State University in Greensboro, North Carolina Central University in Durham, and Winston-Salem State University (Healy, p. A30.) Except for Elizabeth City State, all are located in the Piedmont from Fayetteville in the east to Winston-Salem in the west. In 1971, North Carolina incorporated these five public historically black universities into the unified University of North Carolina system.

South Carolina's first historically black college was Claflin College in Orangeburg founded in 1869 by the Methodist Church. The African Methodist Episcopal and Baptist Churches established Allen University and Benedict College, both located in Columbia, the following year. In 1872, South Carolina established a state-supported Agricultural College and Mechanical Institute for Colored Students as a part of private Claflin College in Orangeburg. In 1896, this became South Carolina State College, was renamed South Carolina State University in 1992, and was consolidated into South Carolina's state-supported higher education system (McMillian, 1952, 167-200; Carolinas Black College Alumni Directory, 1992, 73-78).

As is illustrated by Figures 9.3 and 9.4, ethnically, the system of four-year private and public higher education in the Carolinas continues to be very bifurcated with the historically black private and public colleges and universities continuing to attract predominantly black student populations. In 2000, the North Carolina historically white four-year colleges and universities which attracted the largest proportion of black students were North Carolina Wesleyan College (42%), Chowan College (29%), Pfeiffer University (28%), Mount Olive College (28%), High Point University (24%), University of North Carolina at Pembroke (23%), Guilford College (21%), and Methodist College (20%). The South Carolina historically white schools which attracted the largest proportions of black students were Southern Methodist College (50%), Limestone College (46%), Columbia College (40%), Francis Marion University (40%), Coker College (39%), Southern Wesleyan University (35%), Charleston Southern University (27%), Newberry College (26%), Sherman College of Straight Chiropractic (26%), The University of South Carolina Upstate (26%), Charleston Southern University (22.4%), Southern Wesleyan University (21.3%), The University of South Carolina at Aiken (25%), Winthrop University (25%), and Lander University (25%). (University of North Carolina, 2000 <www.northcarolina.edu/content.php/assessment/reports/ previousabs.htm> and South Carolina Commission on Higher Education, 2000 <www.che400.state.sc.us/web/stats.htm>.) Most of these schools had increased their percent African American enrollment by five to ten percentage points.

ETHNICITY at PUBLIC FOUR-YEAR COLLEGES and UNIVERSITIES

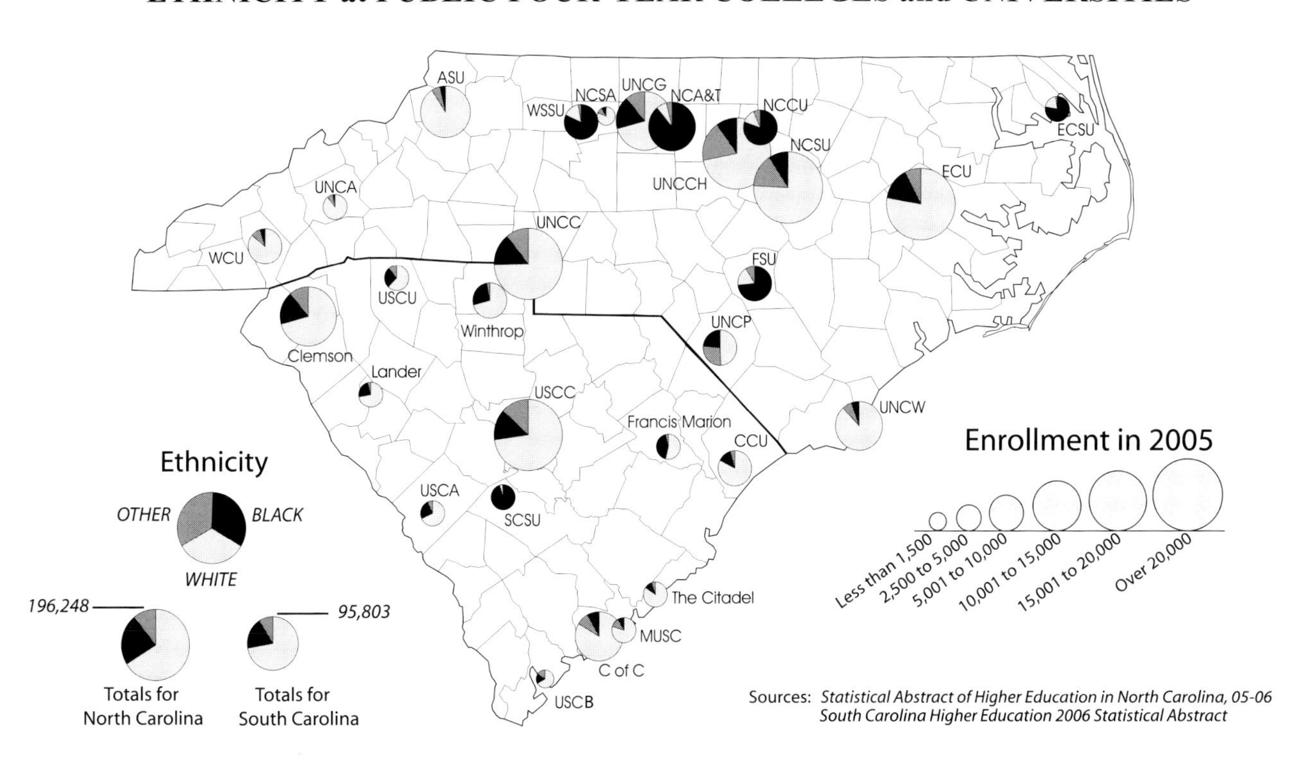

Figure 9.4. Ethnicity at Public Four-Year Colleges and Universities
Sources: University of North Carolina, 2000 <www.northcarolina.edu/content.php/assessment/reports/previousabs.htm> and South Carolina Commission on Higher Education, 2000 <www.che400.state.sc.us/web/stats.htm>
(Cartography by Leslie Meadows. Updated by Jeff Patton)

Patterns of Four-Year Colleges and Universities

Figure 9.5 shows that most of the four-year colleges in the Carolinas are located in the Piedmont Crescent, which extends in an arc from Pembroke, Fayetteville, Raleigh, Durham, Chapel Hill, Greensboro, Winston-Salem, Charlotte (N.C.), Rock Hill, Spartanburg, Greenville, to Clemson (S.C.). Several outliers include Elizabeth City, Greenville, Wilmington, Boone, Asheville and Cullowhee (N.C.) and Charleston and Columbia (S.C.).

With the exception of the University of North Carolina at Chapel Hill and the University of South Carolina at Columbia, private institutions were the Carolinas' higher education innovators, delivering higher education prior to the development of a system of state-supported colleges and universities and initially dominating higher education in the Carolinas. Figures 9.3 and 9.4 and Tables 9.1 and 9.2 show that today public institutions dominate college and university enrollments in the Carolinas. At the aggregate level, Table 9.1 notes that more than twice as many students in the Carolinas attend state-supported colleges and universities (272,422) than privately supported institutions (112,800), and that North Carolina (252,149) has almost twice as many students enrolled in four-year colleges and universities as South Carolina (133,073). Table 9.2 notes that at the individual level the state-supported colleges and universities have much larger enrollments than privately supported institutions, with The University of South Carolina at Columbia (27,065), North Carolina State University (26,496), and The University of North Carolina at Chapel Hill (25,042), having the largest enrollments. Duke University (13,789) is the Carolinas'

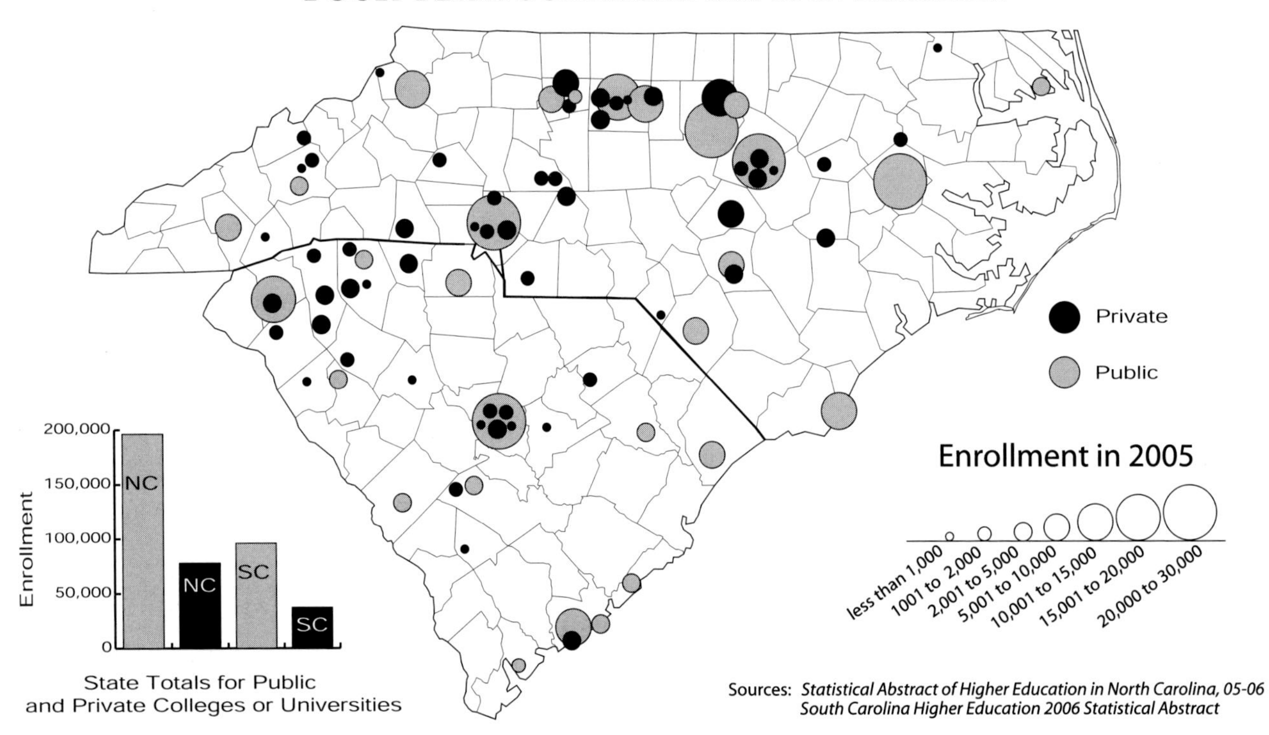

Figure 9.5. Distribution and Resident Credit Headcount Enrollment at Four-Year Colleges in the Carolinas. Sources: University of North Carolina, 2006 <www.northcarolina.edu/content.php/assessment/reports/previousabs.htm> and South Carolina Commission on Higher Education, 2006 <www.che400.state.sc.us/web/stats.htm> (Cartography by Leslie Meadows. Updated by Jeff Patton)

biggest private university, ranking ninth overall, and Wake Forest University (6,622), another private institution, ranks fifteenth in student enrollment (University of North Carolina, 2006. <www.northcarolina.edu/content.php/assessment/reports/previousabs.htm> and South Carolina Commission on Higher Education, 2000. <www.che400.state. sc.us/web/stats.htm>).

The Carolina's first two-year college was Louisburg College in Louisburg, North Carolina established in 1787. Saint Mary's College in Raleigh, North Carolina was established in 1842 as an Episcopal School for women, and Mitchell College in Statesville, North Carolina was established in 1852 by Presbyterians to educate women.

Table 9.1. Resident Credit Headcount Enrollments at Four-Year Colleges and Universities in the Carolinas, Fall 2005

State	Private Four-Year Colleges	Public Four-Year Colleges	TOTALS
North Carolina	75,530	176,619	252,149
South Carolina	37,270	95,803	133,073
TOTALS	***112,800***	***272,422***	***385,222***

Sources: University of North Carolina, 2006 <www.northcarolina.edu/content.php/assessment/reports/previousabs.htm> and South Carolina Commission on Higher Education, 2006 <www.che400.state.sc.us/web/stats.htm>

Table 9.2. Largest Resident Credit Headcount Enrollments at Four-Year Colleges and Universities, Fall 2005

College or University	Credit Headcount Enrollment
University of South Carolina at Columbia	27,065
North Carolina State University	26,496
University of North Carolina at Chapel Hill	25,042
East Carolina University	20,934
University of North Carolina at Charlotte	18,077
Clemson University	17,165
University of North Carolina at Greensboro	14,324
Appalachian State University	13,811
Duke University	13,789
College of Charleston	11,332
University of North Carolina at Wilmington	11,021
North Carolina Agricultural and Technical	10,381
Western Carolina University	7,648
North Carolina Central University	7,205
Wake Forest University	6,622
Winthrop University	6,480

Sources: University of North Carolina, 2006 <www.northcarolina.edu/content.php/assessment/reports/previousabs.htm> and South Commission on Higher Education, 2006 <www.che400.state.sc.us/web/stats.htm>

Two-Year Colleges

Buncombe County, North Carolina, established the Carolinas' first tuition-free public junior college in 1927. In 1930, the North Carolina State Supreme Court upheld the right of Buncombe County to finance post-secondary education with local taxes. This program offered post-secondary education, liberal arts, technical and vocational programs such as pre-nursing and pre-aviation (Wiggs, 1989, 1).

Since the 1950s, North Carolina governors, like Luther Hodges, have noted that the state needed a trained work force in order to attract more industry to the state (Wiggs, 1989, 5). The idea of a statewide public community college system received more momentum in 1962 after the release of a report written by C. Horace Hamilton, a professor of Rural Sociology at North Carolina State University. The report noted that among the 50 states North Carolina ranked 47th in the number of college-age people attending college and that community colleges, unlike four-year colleges, could provide statewide post-secondary education within commuting distances of citizens' residences without the expense of dormitories (Brooks et al, 1997, 389; Wiggs, 1989, 7-8).

North Carolina's Community College System began in 1963, under the Governorship of Terry Sanford, when the state merged its junior colleges and public industrial education systems, both of which had been publicly funded since 1957, into one unified system reporting both to the State Board of Higher Education and to the State Board of Education (Wiggs, 1989, 7). From 1963 until 1980, North Carolina's Community College system was a part of the State Board of Education, after which it became the State Board of Community Colleges (Wiggs, 1989, vii-viii).

South Carolina's Technical College System began in 1964, under the Governorship of Ernest Hollings, as a mechanism to attract industry to the State. In the 1960s, South Carolina's economy, like that of North Carolina, was dominated by agriculture and textiles and its workers earned less than the national average income (Duffy, 1997, 431).

In the mid-1960s, South Carolina's elected leaders realized that, while the state experienced industrial growth over the previous ten years, this trend could not be maintained if the state failed to provide "a continuous flow of skilled trained broadly educated workers to man the increasingly technical requirements of our highly automated society" (South Carolina General Assembly, 1966, 4). The legislature also was concerned about the fact that South Carolina had the nation's smallest proportion of its college-age population enrolled in college (South Carolina General Assembly Interim Report, 1966, 5).

Both states presently rely on publicly supported systems of community and technical colleges as a factor in industrial recruitment. Over the last decade, North Carolina has been among the top states in attracting new manufacturing plants, in part because of the job training provided at the state's community colleges (Bleakley, 1996, A1). In 1958, with some of the nation's lowest high school graduation rates, North Carolina dealt with a reputation of having a poorly trained and unskilled workforce and being the first state in the nation to offer new companies "free, customized training for their employees" (Schmidt, 1997, A29-30). South Carolina also offered this to in-coming firms (Mallory, 1993, 102). Retraining a workforce, which previously was employed in agriculture, furniture or textiles to work in a new industry is seen as an investment, rather than a "giveaway" because it prepares residents to work in a new industry (Schmidt, 1997, A30). This training "pays for itself" because the state gains new tax revenue from the new firms locating in the state and has a more highly paid and more highly trained employees (Schmidt, 1997, A30). "The community college with its flexible and accessible short courses, certificate programs, associate degree programs, and innovative programs directed to new and expanding industries was ideally suited to workforce development" (Brooks et al, 1997, 391).

Appendix 9.3 notes that North Carolina started constructing a system of state-supported two-year colleges and universities before South Carolina. North Carolina has a series of 58 publicly-supported community colleges compared to the 21 campuses in South Carolina's publicly-supported technical colleges and the two-year campuses of the University of South Carolina system.

Appendix 9.3 and Figure 9.6 show that most of the two-year colleges were established in a sixteen-year interval between 1957 and 1972. While most of the initial two-year colleges were private, the states started to establish state-supported two-year colleges throughout their respective states in the late 1950s and early 1960s, so that today most of the two-year colleges are state supported.

As Appendix 9.3 illustrates, North Carolina began its state-supported two-year college system at Wayne Community College at Goldsboro in 1957, and subsequently established 25 two-year colleges between 1957 and 1963. In the same time interval, South Carolina established seven two-year colleges.

There was a strong Piedmont focus to the first colleges, with eleven of the eighteen initial campuses on the Piedmont, three in the Mountains, and four in the Coastal Plain (Wiggs, 1989, 6). As Appendix 9.3 and Figure 9.6 note, the pattern of the diffusion of state-supported two-year colleges in North Carolina is a hierarchical one, down the urban hierarchy with many of the first state-supported two-year schools which started in 1958 and 1959 established in, or in proximity to, some of the state's larger communities: Asheville, Burlington, Durham, Goldsboro, Greensboro, High Point, Raleigh, and Wilmington.

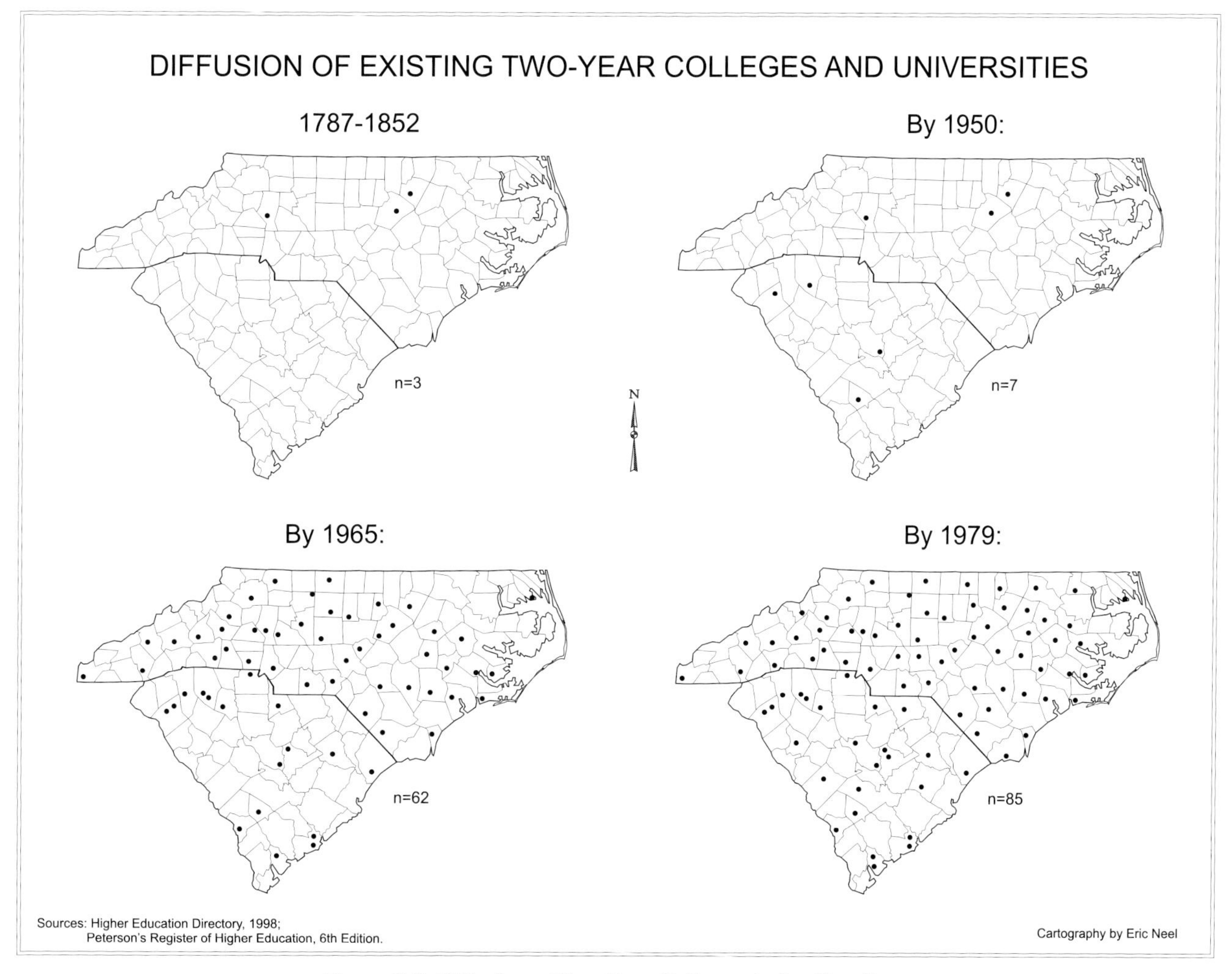

Figure 9.6. Diffusion of Two-Year Colleges in the Carolinas
Sources: Rodenhouse, 1998; Peterson's Register, 1993; selected college catalogs.

More than half of the earliest campuses established in the South Carolina system were located in more rural settings: Beaufort, Lancaster, Sumter, and Pendleton. In the same interval, the state of South Carolina established two-year colleges in Spartanburg, Greenville and Rock Hill.

South Carolina had two historically-black two-year colleges: Clinton Junior College in Rock Hill was established in 1894 by the African Methodist Episcopal Church to educate freed slaves, and Denmark Technical College in Denmark is a state-supported institution established in 1948 as a part of South Carolina Trade School System (*Carolinas Black Colleges Directory*; Mac Millian, pp. 41-44, 59-77) In 1950, several years after North Carolina established junior colleges in Asheville, Charlotte and Wilmington (which later would become campuses of the University of North Carolina system, e.g., the University of North Carolina-Asheville), North Carolina established a publicly supported black junior college in Charlotte, Carver College (Wiggs, 1989, 2).

Table 9.3 notes that the total headcount enrollment at North Carolina's 58 public community colleges (198,339) is more than double that of South Carolina's 16 public technical colleges and four two-year campuses of the University of South Carolina (78,883). (See University of North Carolina, 2006. <www.northcarolina.edu/content.php/assessment/reports/previousabs.htm> and South Carolina Commission on Higher Education, 2000. <www.che400.state. sc.us/web/stats.htm>.)

Table 9.3. Total Headcount Enrollment at Public Community, Technical and Two-Year Colleges, Fall 2005

College or University	Total Headcount Enrolment
North Carolina	198,339
South Carolina	78,883
TOTALS	**277,222**

Sources: University of North Carolina, 2006 <www.northcarolina.edu/content.php/assessment/reports/previousabs.htm> and South Carolina Commission on Higher Education, 2006 <www.che400.state.sc.us/web/stats.htm>.

By 2005, the community/technical colleges and other two-year schools in the Carolinas had much greater proportions of African Americans than did the four-year institutions. Only a fifth of North Carolina's historically white four-year colleges and universities were 21 to 28 percent African American, in addition to one at 42 percent, and over two-fifths of South Carolina's were 20 to 50 percent African American. But for the two-year schools, nearly three-fifths of those in North Carolina and 95 percent of the historically white ones in South Carolina were 20 to 70 percent African American. As Table 9.4 and Figure 9.7 note, public community and technical colleges with the largest enrollments are located in the Carolina's most populous metropolitan areas. Central Piedmont Community College in Charlotte, North Carolina is the Carolina's largest public community or technical college with 16,440 students. Greenville Technical College in Greenville, South Carolina and Wake Technical Community College in Raleigh, North Carolina have enrollments of 13,357 and 12,236, respectively. Trident Technical Community College in Charleston, South Carolina enrolls 11,407, Midlands Technical College in Columbia, South Carolina enrolls 10,779. (See University of North Carolina, 2006. <www.northcarolina.edu/content.php/assessment/reports/previousabs.htm> and South Carolina Commission on Higher Education, 2000. <www.che400.state.sc.us/web/stats.htm>.)

Table 9.4 lists the 15 public community/ technical and two-year colleges in the two Carolinas enrolling more than 5,000 students. At the other end of the spectrum, North Carolina has 27 and South Carolina has eight public community, technical, and two-year colleges with total headcount enrollments of 2,000 to 4,999. North Carolina has 18 South Carolina has eight public community, technical, and two-year colleges with total headcount enrollments with total headcount enrollments of less than 2,000. These campuses tend to be located in the more rural, less populated parts of the two states. (See University of North Carolina, 2006. <www.northcarolina. edu/content.php/assessment/reports/previousabs.htm> and South Carolina Commission on Higher Education, 2000. <www.che400.state.sc.us/web/stats.htm>.)

Table 9.4. Headcount at Public Community, Technical and Two-Year Colleges with Enrollments Exceeding 5,000, Fall 2005

Community College	City	Headcount Enrollment
Central Piedmont Community College	Charlotte, N.C.	16,440
Greenville Technical College	Greenville, S.C.	13,357
Wake Technical Community College	Raleigh, N.C.	12,236
Trident Technical College	Charleston, S.C.	11,407
Midlands Technical College	Columbia, S.C.	10,779
Guilford Technical Community College	Jamestown, N.C.	9,814
Fayetteville Technical Community College	Fayetteville, N.C.	8,408
Cape Fear Community College	Wilmington, N.C.	7,463
Forsyth Technical Community College	Winston-Salem, N.C.	6,996
Asheville-Buncombe Tech. Community College	Asheville, N.C.	6,259
Pitt Community College	Greenville, N.C.	6,085
Durham Technical Community College	Durham, N.C.	5,495
Horry-Georgetown Technical College	Conway, S.C.	5,362
Rowan-Cabarrus Community College	Salisbury, N.C.	5,220
Gaston College	Gastonia, N.C.	5,094

Sources: University of North Carolina, 2006 <www.northcarolina.edu/content.php/assessment/reports/previousabs.htm> and South Carolina Commission on Higher Education, 2006 <www.che400.state.sc.us/web/stats.htm>.

TOTAL UNDERGRADUATE ENROLLMENT AT PUBLIC COMMUNITY, TECHNICAL AND TWO-YEAR COLLEGES, FALL 2005

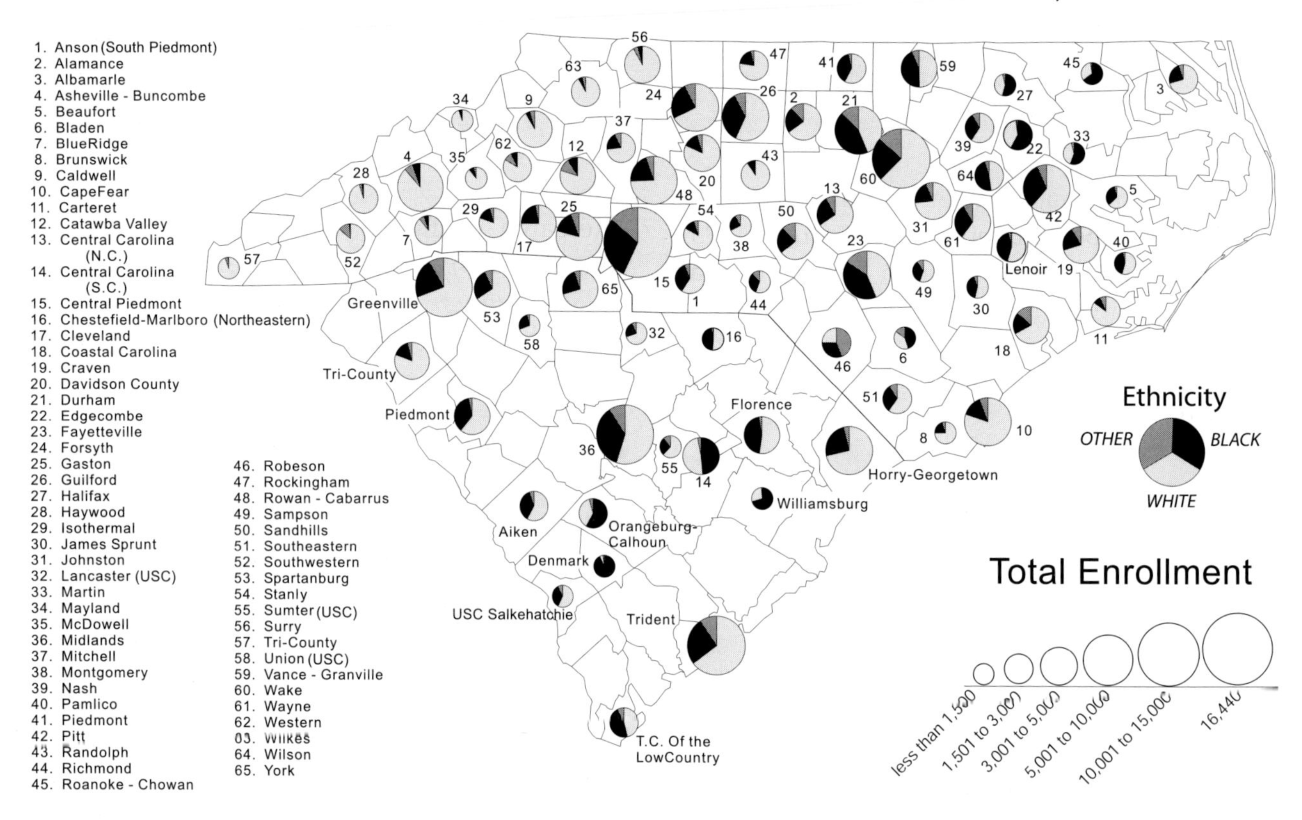

Sources: *Statistical Abstract of Higher Education in North Carolina, 05-06*
South Carolina Higher Education 2006 Statistical Abstract

Cartography by Jim Young and Breece Robertson, updated by Jeff Patton.

Figure 9.7. Total Full Time and Grand Total Enrollment at Public Community, Technical and Two-Year Colleges, Fall 1996

Source: United States Department of Education, 1997, <http:www.nces.ed.gov/ipeds>.

Table 9.4 and Figure 9.7 highlight the fact that at most public community, technical and two-year colleges located throughout the small-towns and more rural areas, the ratio of grand total enrollment to full-time enrollment is 2:1. This 2:1 ratio is exceeded at the metropolitan campuses, such as Midland's Technical College, Trident Technical College, Greenville Technical College, Wake Technical Community College, Fayetteville Technical Community College, Guilford Technical Community College and Forsyth Technical Community College. At Charlotte's Central Piedmont Community College the ratio between full-time enrollment and grand total enrollment exceeds 3:1. Perhaps, this is a function of employers in the larger metropolitan areas being in endeavors which require their labor force must continually up-date their skills by taking classes at community or technical colleges.

Education as an Engine of Economic Growth: Carolina Research Parks

Research Triangle Park

In the 1930s, 75 percent of North Carolina's manufacturing production and 80 percent of the manufacturing wage earners came from textiles, furniture, lumber and tobacco (C. K. Brown, 1931, 139). Although North Carolina had more than 120 colleges and universities, after World War II, new technology-based and scientific industries were not locating within the state (Hamley, 1982, 60). In the 1950s, North Carolina had the nation's second-lowest per-capita income and suffered from a "brain drain" (Heinlein, 1995, 75). Although the state was touted as having some of the South's finest colleges and universities, two-thirds of the science students trained at Duke University, North Carolina State University and the University of North Carolina at Chapel Hill relocated upon graduation because they were unable to find suitable employment in North Carolina (Hodges, 1960, 20; Alexander, 1977, 150).

As early as 1952, Professor Howard W. Odum, who founded the University of North Carolina at Chapel Hill's Institute for Research in Social Science, discussed multi-university cooperation focused mainly on research in the social sciences (Wilson, 1967, 2-7; Larrabee, 1991, 63-64). The idea of a science and technology-based North Carolina Research Park began with Romeo H. Guest of Greensboro. Guest planned and developed industrial sites and was a director of the North Carolina State University's Textile Foundation, as well as its Engineering Foundation (Larrabee, 1991, 59). By the early 1950s, he observed a decline in the southward shift of industry. In his travels, he observed the convergence of industry along Massachusetts Route 128 and in proximity to Harvard University and the Massachusetts Institute of Technology (of which he was a graduate). This led him to think about encouraging research laboratories in the vicinity of Duke University, the University of North Carolina at Chapel Hill, and North Carolina State University. "Guest thought was that research would create new products and breed new, rather than borrowed, industry in the state" (Hamilton, 1966, 255). By June, 1954, Guest had coined the name "Research Triangle of North Carolina" (Hamilton, 1966, 255).

By August, 1954, Guest realized that state leadership was needed and sought help from Robert M. Hanes, president of Wachovia Bank in Winston-Salem and chair of the Committee on Industry of the State Board of Conservation and Development (Hamilton, 1966, 256). Guest, Hanes, and others met with Governor Luther H. Hodges on December 1, 1954. Following this meeting Malcolm E Campbell, Dean of the School of Textiles at North Carolina State University, and Professor William A. Newell, his research chief, prepared "A Proposal for the Development of an Industrial Research Center in North Carolina" (Hamilton, 1966, 256). "On the premise that research is a big business and attracts industry, they recommended co-ordinating the existing resources for the state to build a great center for industrial research" (Hamilton, 1966, 256).

The Research Triangle Park was a "conscious attempt" to build a governmental and industrial research complex upon and around a strong academic community (Hodges, 1960, 19). The Research Triangle Park (RTP) is located within a few miles of Duke University, North Carolina State University, and the University of North Carolina at Chapel Hill (Figure 9.8). RTP is a non-profit subsidiary of the three universities, with any profits produced going to them (Research Park Thrives, 1966, 185; Hamilton, 1966, 267). RTP scientists often hold adjunct professorships at the adjacent universities (Research Park Thrives, 1966, 185). In addition to the use of university faculty as consultants and access to university library resources, such locations produce their own synergies and "intellectual ferment"

(Hodges, 1960, 19). Firms locating in the RTP, not only have proximity to skilled and trainable labor but also provide opportunities for their workers to upgrade their skills by earning a Master's degree or a Ph.D. (Carter, 1965, 869).

The Research Triangle Committee was chartered on October 1, 1956 (Wilson, 1967, 12). To retain the area's Campus-like appearance in the 6,900-acre park, only 15 percent of each firm's land could be built upon (Hamilton, 1966, 272). Companies locating in the Research Triangle Park agree to devote at least nine percent of site operations to research, but most exceed this (Toronto Star).

Some of the first firms to locate in the RTP were the American Association of Textile Chemists and Colorists (technical and scientific textile society), ASTRA (nuclear reactor design), Beaunit Fibers (synthetic fibers), Chemstrand Corporation a subsidiary of Monsanto Chemicals (synthetic fibers and polymers), International Business Machines Corporation (computer terminals), National Environmental Health Science Center (environmental factors on human health), Technitrol (specialized computers), and the United States Department of Agriculture Forest Service Research Station (forest physiology and pathology) (Hodges, 1960, 22; Wilson, 1967, 33-42). By 2001, the largest of the Research Triangle Park's 136 employers were International Business Machines (14,000—information technology, having developed the bar code scanner here), Nortel Networks (8,500—telecommunications), Glaxo-Wellcome (4,885—pharmaceuticals, having developed the AZT anti-AIDS drug here), Ericsson (2,000—telecommunications),

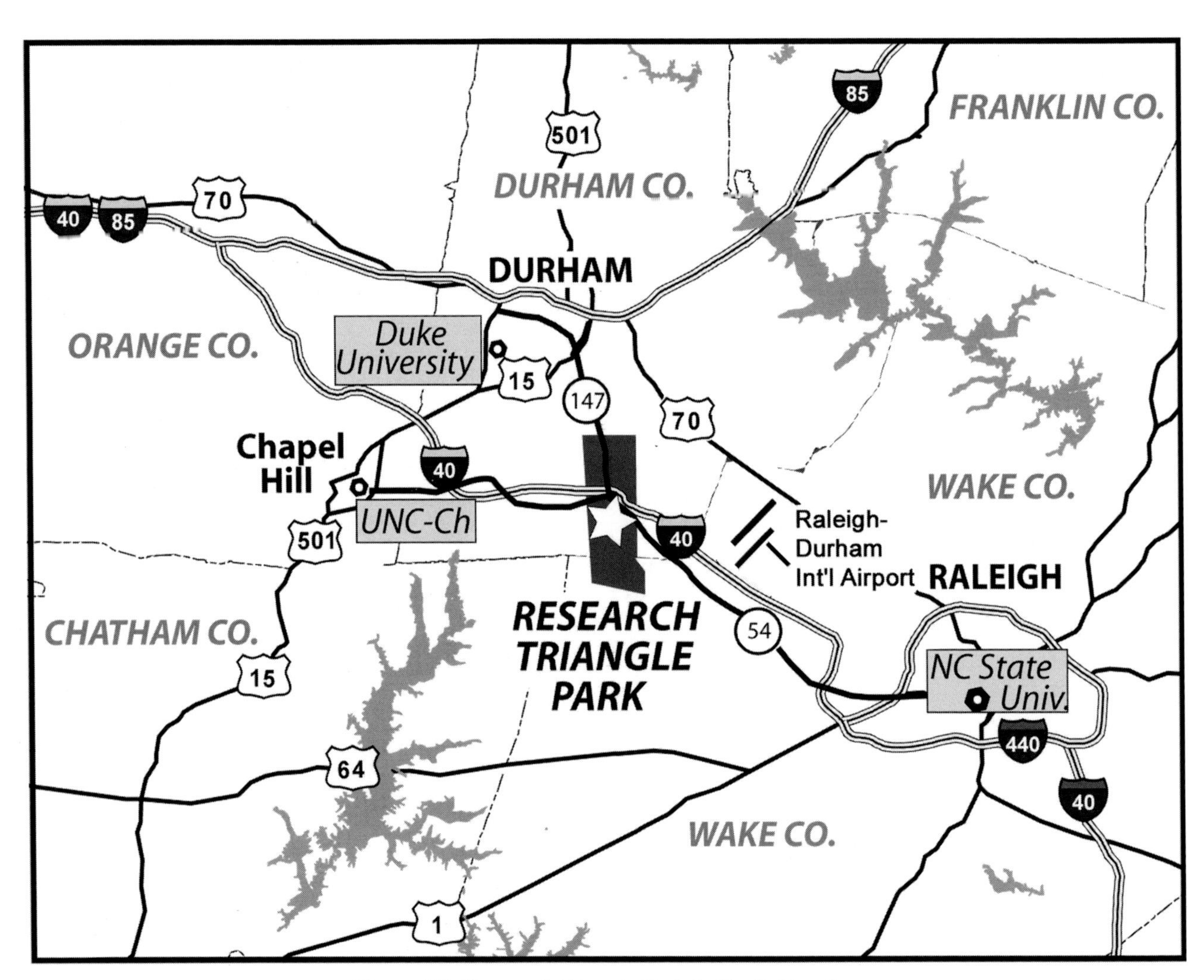

Figure 9.8. Research Triangle Park in North Carolina

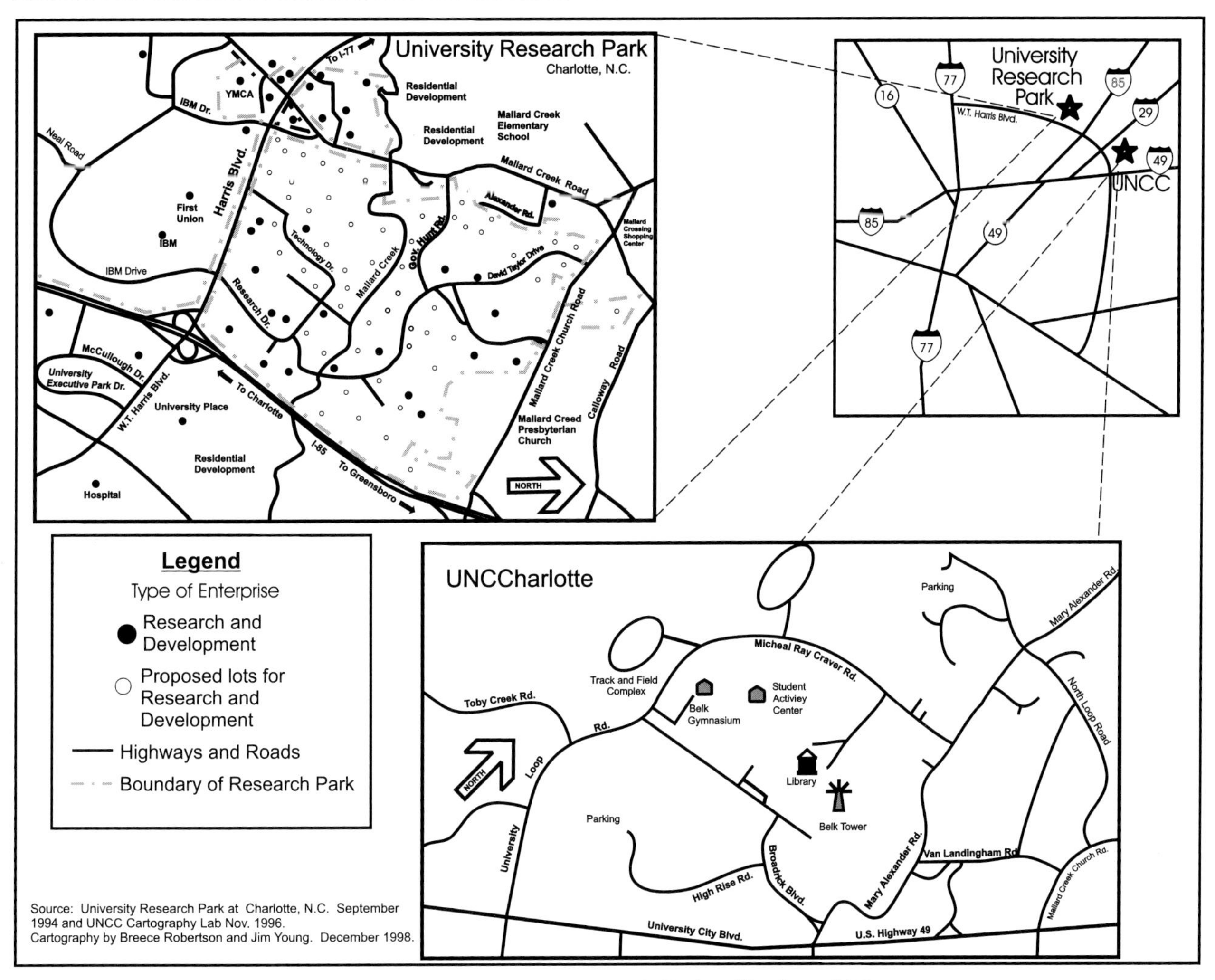

Figure 9.9. University Research Park—Charlotte, N.C.

Cisco Systems, (2,000—networking technologies), Research Triangle Institute (1,750—contract research for government and industry), the United States Environmental Protection Agency (1,734—human and ecological effects of environmental pollutants), and the National Institute of Environmental Health Sciences (1,000—biomedical research). See (www.rtp.org/rtpfacts). Recently, the Research Triangle Park was identified as one of 46 international high-tech hubs (Hillner, 2000, 258-271).

The Research Triangle Park has become a "must-see" stop for international tourists from China, Ghana, Great Britain, Finland, Haiti, India, Morocco, Nepal, Nigeria, Singapore and the United Arab Emirates. RTP's Jamie Nathanson notes "Most are coming because they're trying to start their own research parks ... They're looking at the area ... what we did, what makes (RTP) so successful" (Strow, 1997, 1-2).

University Research Park

In 1966, when W.T. "Bill" Harris, a Charlotte supermarket mogul, became president of the Charlotte Chamber of Commerce, he appointed J. Paul Lucus to head a committee to develop the Chamber's 1966 program. One of the items of focus was development of a research park between the University of North Carolina at Charlotte (UNCC) and Davidson College (Colvard et al, 1988, 3).

Dean W. Colvard, who ten years earlier was a member of Governor Luther Hodges Research Triangle Park Committee, became UNCC's first chancellor in 1966. In the 1960s, many of Charlotte's workers were employed in low-wage distribution jobs (University Research Park, 1994, 1). From the beginning, UNCC's leadership believed that Charlotte, which is North Carolina's largest city, and its Piedmont region needed more high-tech industry and that education would provide an attractive environment for high wage firms (Colvard et al, 1988, 5-6). University Research Park was chartered on November 28, 1966 (Figure 9.9).

While the RTP was "created to reap the benefits of an existing concentration of research talent ... University Research Park was born out of the vision of business leaders in the state's largest city and commercial center" (Colvard et al, 1988, 5). The foci of the two research parks are different, which reflects the differences between the two locales. While the RTP with proximity to the several research universities and its large concentration of scientists has North Carolina's "brain trust," University Research Park "lacked an essential ingredient, a research university." Therefore, Charlotte focused on its strengths, its large banks (Speizer, 1996, F1). In contrast to the RTP, Charlotte "has largely ignored science in favor of commerce," so the University Research Park "is closer to a suburban office and data processing center than a research park," attracting functions such as corporate offices and light manufacturing (Speizer, 1996, F1). Gary Shope, Vice- President of the Research Triangle Foundation, views the University Research Park as complementary, not competition. "We really never bump into each other" (Speizer, 1996, F1).

Some of the first firms to locate in University Research Park were Collins and Aikman (textiles), Reeves Brothers (textiles), Allstate (insurance) and International Business Machines

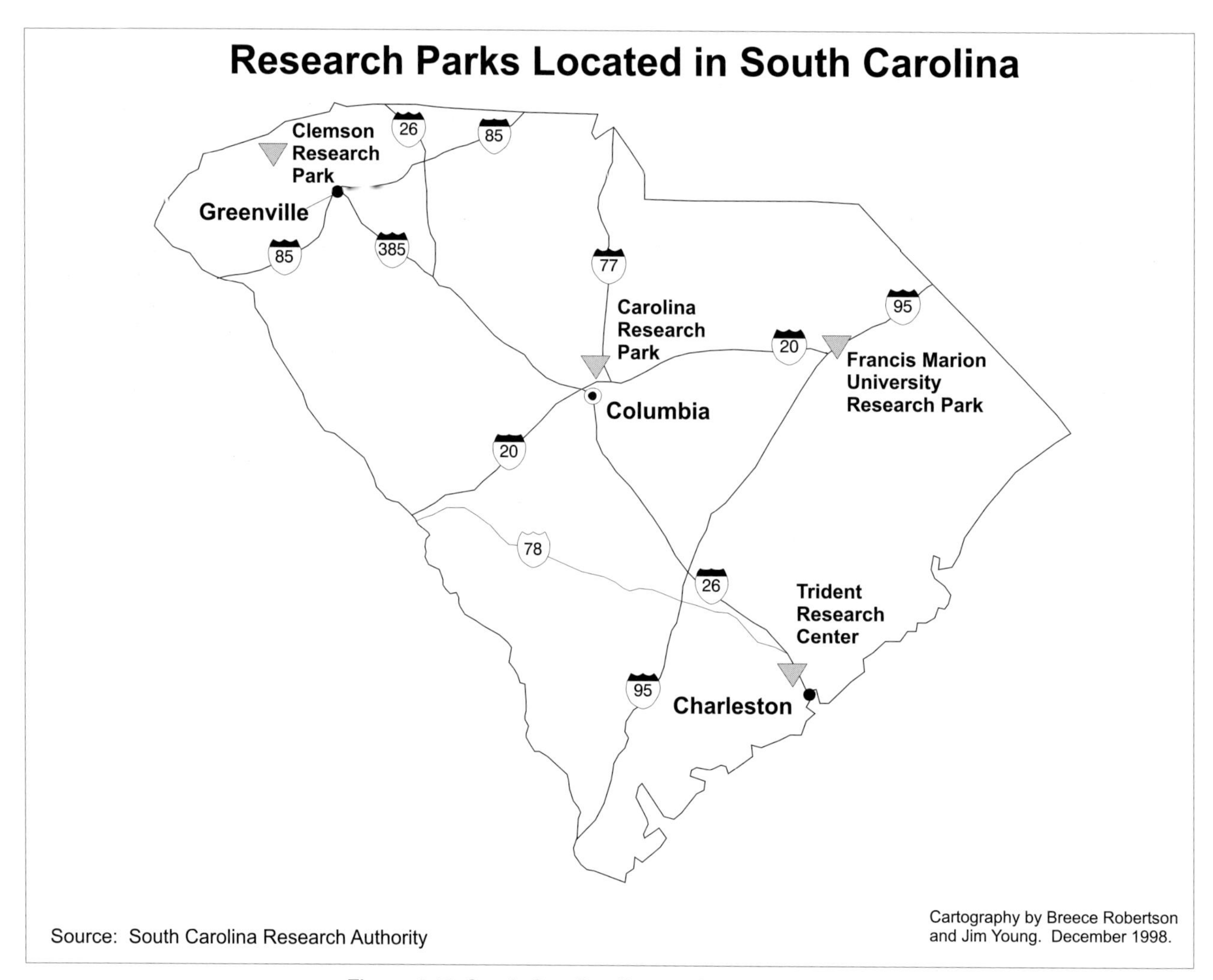

Figure 9.10. South Carolina Research Authority Parks

(computers) (Colvard et al, 1988, 7, 17-18, 20-21, 26, 35-36). Some of the newer firms locating in University Research Park are Dow Jones (printing a regional edition of the *Wall Street Journal*), Electric Power Research Institute (power plant inspection and maintenance systems), First Data Corporation (hospital information management systems and customer service for national credit card companies), First Union Bank (check processing, bank credit card and smart card operations) and Verbatim (computer data storage) (Burgess, 1996, 5C; Speizer, 1996, F1).

South Carolina Research Authority

The South Carolina Research Authority (SCRA), with headquarters in Columbia, was established in 1983 to develop research parks and high-technology programs within the state. SCRA was the nation's first coordinated statewide system of research parks (Long, 1987, 1). Currently, the SCRA manages Carolina Research Park in Columbia close to the University of South Carolina; Charleston Research Park with proximity to the College of Charleston, the Medical University of South Carolina, the Citadel, and Charleston Southern University; Clemson Research Park near Clemson University; and Francis Marion University Research Park associated with the state-supported university of the same name and the city of Florence (South Carolina Research Authority <www.scra.org>). (See Figure 9.10.)

Some of the firms at Carolina Research Park are Dana (CV joint design & fabrication), and SCT Utility Systems (software design). Firms at Charleston Research Park are Battelle (Rapid Acquisition of Manufactured Parts—RAMP—on demand manufacturing of non-standard parts for the military and other clients), Arthur D. Little (RAMP), and Northrop Grumman (RAMP). Firms at Clemson Research Park are Environmental Systems Engineering (environmental research), Mikron (plastic injection molding) and the Tile Council of America (ceramic tile research). The Francis Marion University Research Park was established in 1997 and currently houses the Pee Dee Education Center, a consortium of Francis Marion University and 19 school districts (www.scra.org, www.pdec.state.sc.us). South Carolina's research parks have more of a manufacturing oriention than the biotechnology-microelectronics focused RTP or the office function-focused University Research Park.

Acknowledgments

The author has benefited from the assistance of Leslie Meadows, Eric W. Neel, Deborah "Breece" Robertson, John D. Boyd, Neal G. Lineback, Jim Young and Bobby H. Sharp, all of Appalachian State University; Jeff Patton of UNC Greensboro; Linda Balfour and Gary Barnes at the General Administration of the University of North Carolina and Camille Brown of the South Carolina Commission on Higher Education.

Thanks!

Appendix 9.1.
Chronological Establishment of Colleges and Universities

Name	Place	Date Established
College of Charleston	Charleston, S.C	1770
Salem College	Winston-Salem, N.C.	1772
University of North Carolina at Chapel Hill	Chapel Hill, N.C.	1789
University of South Carolina at Columbia	Columbia, S.C.	1801
Medical University of South Carolina	Charleston, S.C.	1824
Furman University	Greenville, S.C.	1826
Lutheran Theological Seminary	Columbia,S.C.	1830
Wake Forest University	Winston-Salem, N.C.	1834
Davidson College	Davidson, N.C.	1837
Erskine College	Due West, S.C.	1837
Guilford College	Greensboro, N.C.	1837
Duke University	Durham, N.C.	1838
Greensboro College	Greensboro, N.C.	1838
Citadel	Charleston, S.C.	1842
Limestone College	Gaffney, S.C.	1845
Chowan College	Murfreesboro, N.C.	1848
Catawba College	Salisbury, N.C.	1851
Brevard College	Brevard, N.C.	1853
Columbia College	Columbia, S.C.	1854
Wofford College	Spartanburg, S.C.	1854
Mars Hill College	Mars Hill, N.C.	1856
Newberry College	Newberry, S.C.	1856
Peace College	Raleigh, N.C.	1857
Queens College	Charlotte, N.C.	1857
Shaw University	Raleigh, N.C.	1865
Barber-Scotia College	Concord, N.C.	1867
Fayetteville State University	Fayetteville, N.C.	1867
Johnson C. Smith University	Charlotte, N.C.	1867
Saint Augustine's College	Raleigh, N.C.	1867
Chaflin College	Orangeburg, S.C.	1869
Allen University	Columbia, S.C.	1870
Benedict College	Columbia S.C.	1870
Lander University	Greenwood, S.C.	1872
Bennett College	Greensboro, N.C.	1873
Belmont Abby College	Belmont, N.C.	1876
Livingstone College	Salisbury, N.C.	1879
Presbyterian College	Clinton, S.C.	1880
Pfeiffer University	Misenheimer, N.C.	1885
Winthrop University	Rock Hill, S.C.	1886
Campbell University	Buie's Creek, N.C.	1887
North Carolina State University	Raleigh, N.C.	1887
Univ. of North Carolina-Pembroke	Pembroke, N.C.	1887
Clemson University	Clemson, S.C.	1889

Converse College	Spartanburg, S.C.	1889
Elon University	Elon, N.C.	1889
Western Carolina University	Cullowhee, N.C.	1889
Elizabeth City State University	Elizabeth City, N.C.	1891
Lenoir-Rhyne College	Hickory, N.C.	1891
Meredith College	Raleigh, N.C.	1891
North Carolina A & T University	Greensboro, N.C.	1891
University of North Carolina at Greensboro	Greensboro, N.C.	1891
North Greenville College	Tigerville, S.C.	1892
Winston-Salem State University	Winston-Salem, N.C.	1892
Warren Wilson College	Asheville, N.C.	1894
Wingate University	Wingate, N.C.	1896
South Carolina State University	Orangeburg, S.C.	1896
Voorhees College	Denmark, S.C.	1897
Appalachian State University	Boone, N.C.	1899
Lees-McRea College	Banner Elk, N.C.	1900
Barton College	Wilson, N.C.	1902
Gardner-Webb University	Boiling Springs,N.C.	1905
Southern Wesleyan University	Central, S.C.	1906
East Carolina University	Greenville. N.C.	1907
Coker College	Hartsville, S.C.	1908
Morris College	Sumter, S.C.	1908
North Carolina Central University	Durham, N.C.	1910
Anderson College	Anderson, S.C.	1911
Montreat College	Montreat, N.C.	1916
High Point University	High Point, N.C.	1924
University of North Carolina at Asheville	Asheville, N.C.	1927
Columbia International University	Columbia, S.C.	1923
Bob Jones University	Greenville. S.C.	1927
University of North Carolina at Charlotte	Charlotte, N.C.	1946
University of North Carolina at Wilmington	Wilmington, N.C.	1947
Southeastern Baptist Theol. Sem.	Wake Forest, N.C.	1950
Mount Olive College	Mount Olive, N.C.	1951
Coastal Carolina University	Conway, S.C.	1954
Methodist College	Fayetteville, N.C.	1956
North Carolina Wesleyan College	Rocky Mount, N.C.	1956
Saint Andrews College	Laurinburg, S.C.	1958
University of South Carolina at Aiken	Aiken, S.C.	1961
North Carolina School of the Arts	Winston-Salem, N.C.	1963
Charleston Southern University	Charleston, S.C.	1964
University of South Carolina at Spartanburg	Spartanburg, S.C.	1967
Francis Marion University	Florence, N.C.	1970
Sherman College of Chiropratic	Spartanburg, S.C.	1973
East Coast Bible College	Charlotte, S.C.	1975

Sources: American Council on Education, 1997; Rodenhouse, 1998; *Peterson's Register*, 1993); selected college catalogs.

Appendix 9.2.
Chronological Establishment of Four-Year Historically-Black Colleges and Universities in the Carolinas

Name and Type	Place	Date Established
Shaw University (private)	Raleigh, NC	1865
Barber-Scotia College (private)	Concord, NC	1867
Fayetteville State Univ. (public)	Fayetteville, NC.	1867
Johnson C. Smith Univ. (private)	Charlotte, NC	1867
St. Augustine's College (private)	Raleigh, NC	1867
Claflin College (private)	Orangeburg, SC	1869
Allen University (private)	Columbia, SC	1870
Benedict College (private)	Columbia, SC	1870
Bennett College (private)	Greensboro, NC	1873
Livingston College (private)	Salisbury, NC	1879
Elizabeth City State Univ. (public)	Elizabeth City, NC	1891
N.C. Agri. & Tech. St. Univ. (pub.)	Greensboro, NC	1891
Winston-Salem State Univ. (public)	Winston-Salem, NC	1892
South Carolina State Univ. (public)	Orangeburg, SC	1896
Voorhees College (private)	Denmark, SC	1897
Morris College(private)	Sumner, SC	1908
North Carolina Central Univ. (pub.)	Durham, NC	1909

Source: *Carolina's Black College Alumni Directory*; American Council on Education, 1997; selected college catalogs.

Appendix 9.3.
Chronological Establishment of Two-Year Colleges

Name	Place	Date Established
Louisburg College	Louisburg, N.C.	1787
Saint Mary's College	Raleigh, N.C.	1842
Mitchell Community College	Statesville, N.C.	1852
Spartanburg Methodist College	Spartanburg, S.C.	1911
Columbia Junior College	Columbia, S.C.	1935
Forrest Junior College	Anderson, S.C.	1946
Denmark Technical College	Denmark, S.C.	1948
Wayne Community College	Goldsboro, N.C.	1957
Alamance Community College	Graham, N.C.	1958
Central Carolina Community College	Sanford, N.C.	1958
Davidson County Community College	Lexington, N.C.	1958
Durham Technical Community College	Durham, N.C.	1958
Guilford Technical Com. College	Jamestown, N.C.	1958
Lenoir Community College	Kinston, N.C.	1958
Wake Technical Community College	Raleigh, N.C.	1958
Wilson Tech. Community College	Wilson, N.C.	1958
Asheville-Buncombe Tech. C.C.	Asheville, N.C.	1959
Cape Fear Community College	Wilmington, N.C.	1959
Univ. of South Carolina-Beaufort	Beaufort, S.C.	1959
Univ. of South Carolina-Lancaster	Lancaster, S.C.	1959
Catawba Valley Community College	Hickory, N.C.	1960
College of the Abermarle	Elizabeth City, N.C.	1960
Fayetteville Tech. Comm. College	Fayetteville, N.C.	1961
Pitt Community College	Greenville, N.C.	1961
Rowan-Cabarrus Community College	Salisbury, N.C.	1961
Spartanburg Technical College	Spartanburg, S.C.	1961
Anson Community College	Polkton, N.C	1962
Central Carolina Techn. College	Sumter, S.C.	1962
Greenville, Technical College	Greenville, S.C.	1962
Pamlico Community College	Grantsboro, N.C.	1962
Randolph Community College	Asheboro, N.C.	1962
Tri County Technical College	Pendelton, S.C	1962
York Technical College	Rock Hill, S.C.	1962
Carteret Community College	Morehead City, N.C.	1963
Central Piedmont Comm. College	Charlotte, N.C.	1963
Coastal Carolina Comm. College	Jacksonville, N.C.	1963
Gaston College	Dallas, N.C.	1963
Rockingham Community College	Wentworth, N.C.	1963
Sandhills Community College	Pinehurst, N.C.	1963
Caldwell Comm. C. & Tech. Insti.	Hudson, N.C.	1964
Florence Darlington Tech. College	Florence, S.C.	1964
Forsyth Technical Com. College	Winston-Salem, N.C.	1964
Isothermal Community College	Spindale, N.C.	1964

James Sprunt Community College	Kenansville, N.C.	1964
McDowell Technical Comm. College	Marion, N.C.	1964
Richmond Community College	Hamlet, N.C.	1964
Southeastern Community College	Whiteville, N.C.	1964
Southwestern Community College	Sylva, N.C.	1964
Surry Community College	Dobson, N.C.	1964
Tri County Community College	Murphy, N.C.	1964
Trident Technical College	Charleston, S.C.	1964
Western Piedmont Comm. College	Morganton, N.C.	1964
Cleveland Community College	Shelby, N.C.	1965
Craven Community College	New Bern, N.C.	1965
Haywood Community College	Clyde, N.C.	1965
Horry Georgetown Tech. College	Conway,S.C.	1965
Nielson Electronics Institute	Charleston, S.C.	1965
Robeson Community College	Lumberton, N.C.	1965
Sampson Community College	Clinton, N.C.	1965
Univ. of South Caro.-Salhahatchie	Allendale, S.C.	1965
Univ. of South Carolina-Union	Union, S.C.	1965
Wilkes Community College	Wilkesboro, N.C.	1965
Organgeburg-Calhoun Tech. College	Orangeburg, S.C.	1966
Piedmont Technical College	Greenwood, S.C.	1966
Univ. of South Carolina-Sumter	Sumter, S.C.	1966
Beaufort Community C	Washington, N.C.	1967
Bladen Community College	Dublin, N.C.	1967
Edgecombe Community College	Tarboro, N.C.	1967
Halifax Community College	Weldon, N.C.	1967
Martin Community College	Williamston, N.C.	1967
Montgomery Community College	Troy, N.C.	1967
Nash Community College	Carrigae, N.C.	1967
Roanoke-Chowan Community College	Ahoskie, N.C.	1967
Blue Ridge Community College	Flat Rock, N.C.	1969
Chesterfield-Marlboro Tech. C.	Cheraw, S.C.	1969
Johnston Community College	Smithfield, N.C.	1969
Vance-Granville Community College	Henderson, N.C.	1969
Williamsburg Technical College	Kingstree, S.C.	1969
Piedmont Community College	Roxboro, N.C.	1970
Mayland Community College	Spruce Pine, N.C.	1971
Stanley Community College	Albermarle, N.C.	1971
Aiken Technical College	Aiken, S.C.	1972
Tech. College of the Low Country	Beaufort, S.C.	1972
Midlands Technical College	Columbia, S.C.	1974
Brunswick Community College	Supply, N.C.	1979

Sources: Rodenhouse, 1998; *Peterson's Register*, 1993; selected college catalogs.

References

Adams, F. 1998. Buckle on the Sunbelt. *Change* 10, 13-15.

Alexander, T. 1977. A Park That Reversed the Brain Drain. *Fortune* 95, 148-153.

American Council on Education. 1997. American Universities and Colleges 15th ed. New York: Walter de Gruyter, 1997.

Augenblick, Van de Water and Associates. 1986. *Higher Education in South Carolina, Agenda for the Future: Report to the South Carolina Commission on Higher Education.* Denver: Augenblick, Van de Water and Associates.

Balfour, L. 1998. *Statistical Abstract of Higher Education in North Carolina,* 1997-1998. Chapel Hill: University of North Carolina.

Bennett, L. The Five and Ten Bastille. 1980. *Ebony* 35 110-122.

Bleakley, F. R. 1996. Ready To Work: To Bolster Economies, Some States Rely More on Two-Year Colleges. *Wall Street Journal,* A1, A12. New York.

Branch, T. 1988. *Parting the Waters: America in the King Years 1954-1963.* New York: Simon and Schuster.

Brooks, J. B., K. L. Joss, and B. Newsome. 1997. North Carolina's Community Colleges: The Connection to the Workforce. *Community College Journal of Research and Practice* 21, 387-396.

Brown, C. K. 1931. Industrial Development in North Carolina. *Annals of American Academy of Political and Social Science* 153, 133-140.

Brown, M., ed. 1998. *South Carolina Higher Education, 1998 Statistical Abstract.* Columbia: Commission on Higher Education.

Burgess, H. 1996. Credit Service to Bring 400 Jobs to Charlotte's University Research Park. *The Herald.* Rock Hill, S.C., 5C.

Carolinas Black Colleges Alumni Directory. 1992. Durham: Carolinas Black Colleges Alumni Directory. Carpenter, C. C. 1970.

Carpenter, C. 1970. *The Story of Medicine at Wake Forest University.* Chapel Hill: University of North Carolina Press.

Carter, L. J. 1965. Research Triangle Seeks High-Technology Industry. *Science* 150, 867-871.

_____. 1978. Research Triangle Park Succedes Beyond Its Promoter Expectations. *Science* 200, 1469-1470.

Chafe, W. H. 1980. *Civilities and Civil Rights: Greensboro, North Carolina and the Black Struggle for Freedom.* New York: Oxford University Press.

Chaffin, N. C. 1950. *Trinity College, 1839-1892; The Beginnings of Duke University.* Durham: Duke University Press.

Christensen, R. 1998. What if the RTP Hadn't Been Built. *News and Observer,* A20. Raleigh.

Colvard, D. W., D. M. Orr, Jr., and M. D. Bailey. 1988.*University Research Park: The First Twenty Years.* Charlotte, North Carolina: Urban Institute at the University of North Carolina at Charlotte.

Cresap, McCormick and Paget. 1962. *Higher Education in South Carolina.* New York: Cresap, McCormick and Paget.

Drake, W. E. 1964. *Higher Education in North Carolina Before 1860.* New York: Carlton Press.

Duffy, E. South Carolina Technical College System. 1997. *Community College Journal of Research and Practice.* 21, 431-443

Easterby, J. H. 1935. *A History of the College of Charleston.* Charleston: College of Charleston.

Edmonds, R. W. 1929. The Workshop of the Carolinas.*Review of Reviews* 80, 71-74.

Edwards, J. B. 1986. *The Southeast's Oldest Medical School.* New York: Newcomer Society of the U.S.

Eliott, S. L. 1988. *Origins, Elements, and Implications of Cutting Edge Initiatives for Research and Academic Excellence in Higher Education.* Columbia: South Carolina Commission in Higher Education.

50 Who Changed America. 1995. *Ebony* 51, 108-130.

Foust, D. and M. Mallory. 1993. The Boom Belt. *Business Week*, 98-104.

Gibb, J. M., ed. 1985. *Science Parks and Innovation Centres: Their Economic and Social Impact.* Amsterdam: Elsevier.

Gifford, J. F. 1972. *The Evolution of a Medical Center: A History of Medicine at Duke University to 1941.*Durham, North Carolina: Duke University Press.

Graham, F. P. 1944. The First University of the People. *School and Society* 59, 17-20.

Green, E. 1916. *A History of the University of South Carolina:* Columbia: The State Company.

Halberstam, D. 1998. *The Children.* New York: Random House.

Hamilton, W. B. 1966. The Research Triangle of North Carolina: A Study in Leadership for the Common Weal. *South Atlantic Quarterly* 65, 254-278.

Hamley, W. 1982. Research Triangle Park: North Carolina. *Geography* 67, 59-62.

Healy, P. 1996. A Myriad of Problems for Public Black Colleges. *Chronicle of Higher Education* 42, A30-A31, A35.

Heinlein, S. 1995. Brainpower Keeps Research Flowering in Triangle Park. *Business Journal Serving San Jose and Silicon Valley* 21, 75.

Hillner, J. 2000. Venture Capitals. *Wired* 8, 258-271.

Hodges, L. 1960. The Research Triangle of North Carolina: Its Purposes, Its Program and Its Relationship to Industrialization in the South. *State Government* 33, 17-22.

Hoffman, C., T. Snyder, and B. Sonnenberg. 1996. *Historically Black Colleges and Universities, 1976-1994.* Washington, DC: U.S. Department of Education, Office of Educational Research and Improvement, National Center for Education Statistics.

Hollis, D. W. 1951. *University of South Carolina.*Volume I. *South Carolina College.* Columbia University of South Carolina Press.

_____. 1956. *University of South Carolina.*Volume II. *College to University.* Columbia: University of South Carolina Press.

King, A. K. 1987. *The Multicampus University of North Carolina Comes of Age, 1956-1886.* Chapel Hill: University of North Carolina.

Kolcum, E. 1983. Research Triangle Park Spurred Growth. *Aviation Week and Space Technology* 118, 54-57.

Larrabee, C. X. 1991. *Many Missions: Research Triangle*

Institute's First 31 Years. Research Triangle Park, N. C.: Research Triangle Institute.

Levitt, R. 1986. *Research Parks and Other Ventures: The University/Real Estate Connection.* Washington, D.C.: Urban Land Institute.

Lindsey, Q. 1990. Organizing Research in an Information Society. *Society* 27, 63-65.

Lockmiller, D. A. 1939. *History of the North Carolina State College of Agriculture and Engineering.* Raleigh: Edwards and Broughon.

Long, L. 1987. Clemson Research Park A Marketing Analysis and Development Plan. Unpublished Masters thesis. Clemson University.

Luger, M. I. 1984. Does North Carolina's High-Tech Development Work? *American Planning Assocation Journal* 50, 280-289.

McMillian, L. K. 1952. *Negro Higher Education in the State of South Carolina.* Orangeburg, South Carolina: Unknown.

Mallory, M. 1993. Workers Trained to Order—At State Expense. *Business Week* September 27, 102.

Morris, A. D. 1984. *The Origins of the Civil Rights Movement: Black Communities Organizing for Change* (New York: Free Press.

Morton, H. and J. Collins, eds. 1987. *The University of North Carolina at Chapel Hill: The First 200 Years.* Raleigh: Capitol Broadcasting.

North Carolina Board of Higher Education. 1986. *State Supported Traditionally Negro Colleges in North Carolina* .Raleigh: North Carolina Board of Higher Education.

North Carolina Community College System. *A Matter of Facts:The North Carolina Community College System Fact Book.* Raleigh: http://www.ncccs.cc.nc.us.

North Carolina Community College System. 1997. *Annual Statistical Report, 1996-1997.* Raleigh: http://www.ncccs.cc.nc.us.

Officials of UNC Debate Policy on Desegregation. 1994. *Wilmington Star-News*, July 9.

Peterson's Register of Higher Education, 6th ed. 1993. Princeton, New Jersey: Peterson's Guide.

PhDs Among the Possums. 1979. *Economist* 270, 58.

Powell, W. S. 1992. *The First State University: A Pictorial History of the University of North Carolina*, 3rd ed. Chapel Hill: University of North Carolina Press.

Public Higher Education in South Carolina: A Survey. 1946. Nashville, Tennessee: George Peabody College for Teachers.

Research Alive and Well in N.C. 1977. *Time* 110, 46.

Research Park Thrives in Academic Neighborhood. 1966. *Business Week*, December 10, 184-88.

Research Triangle Park Has Job Lessons To Teach Toronto. 1998. *Toronto Star*, April 6 (Lexis-Nexis).

Rodenhouse, M., ed. 1998. *1998 Higher Education Directory*. Falls Church, Virginia: Higher Education Publications.

Schmidt, P. 1997. States Turn to Community Colleges to Fuel Economic Growth. *Chronicle of Higher Education* 43, A29-A30.

Schmidtt, H. The List: RTP Employers. 1997. *Triangle Business Journal* 12, 62.

Singleton, E. M. 1971. A History of the Regional Campus System of the University of South Carolina. Unpublished Ph.D. dissertation. Columbia: University of South Carolina.

South Carolina Commission on Higher Education. 1993. *Achieving the Vision in Difficult Times: A Call to Action for 1993*. Columbia: South Carolina Commission on Higher Education.

South Carolina Commission on Higher Education. 1991. *Choosing South Carolina's Future: A Plan for Higher Education*. Columbia: South Carolina Commission on Higher Education.

South Carolina Commission on Higher Education. 1995. *Connections for Cooperation: Initiatives for 1995*. Columbia: South Carolina Commission on Higher Education.

South Carolina Commission on Higher Education. 1994. *Quality and Service: Initiatives for 1994*. Columbia: South Carolina Commission on Higher Education. South Carolina Commission on Higher Education. 2006. <www.che400.state.sc.us/web/stats.htm>.

South Carolina Commission on Higher Education. 2006. *2006 South Carolina Higher Education Statistical Abstract*. <http://www.che.sc.gov/Finance/Abstract/Abstract-2006web.pdf>. South Carolina Commission on Higher Education. 1968.

South Carolina Commission on Higher Education. 1968. *Survey and Principles for Implementation: South Carolina Two-year Post-High School Education*. Columbia: South Carolina Commission on Higher Education.

South Carolina Commission on Higher Education. 1971. *Two-Year Post-Secondary Education in South Carolina: A Joint Report*. Columbia: South Carolina Commission on Higher Education.

South Carolina General Assembly. 1996. *Interim Report of a Committee Created to Study the Feasibility of Establishing a State Supported System of Junior Colleges*. Columbia: The Committee.

South Carolina General Assembly, Legislative Audit Council. 1978. *Management and Operational Review of the South Carolina Commission on Higher Education*. Columbia: South Carolina General Assembly.

South Carolina Research Authority. 1998. *1997 Annual Report*. Columbia: South Carolina Research Authority.

South Carolina Research Authority. www.scra.org.

Southern Regional Education Board. 1970. *The Black Community and the Community College, Action Programs*. Atlanta: Southern Regional Education Board.

Southern Regional Education Board. 1970. *New Challenges to the Junior Colleges: Their Role in Expanding Opportunities for Negroes, A Progress Report*. Atlanta: Southern Regional Education Board.

Speizer, I. 1996. URP No RTP Copycat. *News and Observer,* April 28, F1. Raleigh.

State of South Carolina. 1991. Recommendations of the Study Committee: *A Report on South Carolina's Need for a Planned System of Public Education Beyond the High School.* Columbia: State of South Carolina.

Strow, D. 1997. RTP's a Must See For International Visitors. *Triangle Business Journal* 13, 1-2.

United States Department of Education, National Center for Education Statistics, Integrated Postsecondary Education Data System (IPEDS). *Fall Enrollment Survey, 1996-1997.* http:www.nces.ed.gov/ipeds.

University of North Carolina. *Statistical Abstract of Higher Education 05-06.* http://www.northcarolina.edu/content.php/assessment/reports/abstract-current. htm.

University Research Park, Charlotte, North Carolina. 1994. *News and Observer,* Special Advertising Section, September. Raleigh.

Wiggs, J. L. 1989. *The Community College System of North Carolina: A Silver Anniversary History, 1963-1988.*Raleigh: North Carolina State Board of Community Colleges.

Wilson, L. R. 1967. *The Research Triangle of North Carolina.* Chapel Hill: Colonial Press.

Windley, J. 1998. The List: RTP Employers. *Triangle Business Journal* 13, 15.

Chapter 10

PLANNING FOR THE FUTURE

D. Gordon Bennett

University of North Carolina at Greensboro

Would not the indigenous natives and the explorers and early settlers of "Carolana" be awestruck at the changes which have transpired through the centuries on this land each happened upon in his life's journey? Who could have imagined the great migrations of so many diverse peoples to this region and the consequential population growth and economic development, along with the resulting ecological impact, that have so transformed the Carolinas?

Envisioning the future is extremely difficult because so many unknown, and even unborn, individuals can trigger unanticipated events and discoveries which can radically alter the direction of recent trends and greatly impact either seemingly overwhelming forces or impenetrable obstacles to change. Indeed, it is not what is expected or sought that always leads to a brighter future, but rather unanticipated ideas and revelations which can catapult society toward a much different future than could have been foreseen. Consider how different the history and geography of the Carolinas and the South would have been had not Eli Whitney chanced to vacation in a cotton-growing area near Savannah in the late eighteenth century. Without his cotton gin, would slavery have survived and led to Civil War? And how different might the Carolinas and the Nation be if their social structure today had not a handful of African American students not mustered the courage to sit at the lunch counter at Woolworth's Department Store in Greensboro, N.C. some 40 years ago. Indeed, what a different area the Raleigh-Durham-Chapel Hill MSA would be had not some four decades ago Romeo Guest from Greensboro and North Carolina Governor Luther Hodges not conceived of a high-tech industrial park near The University of North Carolina at Chapel Hill, North Carolina State University and Duke University.

Yet, planning for the future in light of present knowledge and recent trends is important. Even if every future invention and event can not be foretold, evaluating the best direction in which to move given present and anticipated circumstances can lead to better decision making and help prevent many mistakes from being made. Ideals of the kind of society in which Carolinians wish to live and ideas about how best to move toward current goals should result in a better future.

The Carolinas are continuing to grow in population and economic development, but some parts are increasing much faster than others. Even with considerable efforts being expended to encourage population and economic growth in some stagnating sections of the Carolinas, most economic expansion has continued to be focused largely on the Piedmont Urban Crescent from the Raleigh, N. C. area (the fifth fastest-growing metropolitan area in the United States in the 1990s) to the Greenville, S. C. area, plus a few other metropolitan areas, such as Columbia and Charleston in South Carolina and Wilmington, Fayetteville, and Asheville in North Carolina. In addition, several high-amenity, mainly rural counties—primarily in the mountains and along the coast—have attracted considerable retirement- and resort-related development.

Photo 10.1. Converting the rural fringe of an urban area to commercial uses (Courtesy of U.S. Department of Agriculture)

Most of the population increase and business and manufacturing gains have occurred around the periphery of the largest cities, spilling into the surrounding rural environs and filling the open spaces between the major nodes (Photo 10.1). Building more and better roads and highways between and around cities have allowed people and businesses to move farther out from the older cores. Such developments, called "edge cities," frequently occur around major interchanges of a loop road and radial highways and are composed of a combination of office parks, shopping centers, hotels, restaurants, and condos/apartments. This pattern has resulted in the destruction and enveloping of farmlands, forests, and other open land, thus increasing pressure on the natural resources of these areas, rather than development focused on the rehabilitation of many older sections of the already built-up urban centers. This pattern has been particularly notable and contentious around

Raleigh and Charlotte. Only recently has the development of older industrial parts of cities, often referred to as "brownfields," been seriously considered. But such efforts, as impressive as they might be in individual instances, pale in comparison to the overwhelming combination of development projects surrounding the larger cities.

The Carolinas, especially North Carolina, have been slower than northern and western sections of the United States in undergoing metropolitan sprawl and the concomitant independent incorporation of small communities near or adjacent to the larger cities. As the older cities become encircled by separate governing entities housing relatively more affluent residents who use the streets and other infrastructure and services of the central cities without contributing to their maintenance and rehabilitation through taxes, the residents left behind have to shoulder an increasing fiscal burden. Paradoxically, towns which incorporate to escape city taxes often face major water and sewer and other infrastructure costs themselves as the increasing density of the population outstrips original well and septic systems and road networks. Indeed, development of these outlying residential areas and their associated commercial centers greatly increased traffic on existing narrow rural roads, thus necessitating an enormous expenditure of state, county, and even town funds to improve and expand roads and other infrastructure, whose costs are only minimally contributed to by either the suburban residents or the developers of these areas.

Planning for a more efficient use of the land for residential and business development is essential if the people of the Carolinas are to escape the plight of the northern metropolitan areas from whence many recent newcomers have moved. Metropolitan planning must become a truly metropolitan governing structure with all parts of each metropolitan area participating and working together for the benefit of the whole. Each unit within and outside the metropolitan areas must also utilize citizen-involved land use planning in order to ensure an acceptable balance between the highest and best use of the land and the conserving of the critical natural resources which guarantee the continuation of economic and population growth while maintaining a high quality of life in each community.

There continues to be a major change in the economic structure of the Carolinas, as is true for the nation as a whole. During the last half of the twentieth century, the traditional low-wage, resource-based industries of textiles, tobacco, furniture, and foods were supplemented by electrical, electronic, automotive, rubber, and petroleum-related manufacturing. High-technology industries have become especially noteworthy in the Raleigh/Durham/Chapel Hill (Triangle) region, and to a lesser extent, in the Charlotte and Columbia areas.

As nearly 200,000 workers lost their jobs in textile factories between 1990 and 2004, thousands of high-skilled positions were created in the services sector. North Carolina actually gained 17,000 manufacturing workers between 1980 and 1994, but during the next 10 years, it lost over 185,000 manufacturing jobs. South Carolina lost 30,000 during the first period and an additional nearly 100,000 during the following one. Since 1980, most new jobs were actually the result of expansions of existing businesses and openings of new firms rather than plants moving to the Carolinas from other locations. Although a few large plant openings, such as BMW near Greer, South Carolina and Dell computers in the Triad of North Carolina, have greatly benefited the economy of this region, most economic growth has been by small businesses and industries.

Officials of both these states have become increasingly aware of the need to educate and train workers at a much higher level than in the past, if high-paying, high-technology jobs are to be attracted to this region. Both colleges/universities and community colleges/technical institutes are going to be instrumental in educating the work force adequately to meet the demands of business and industry in the twenty-first century. Indeed, special technical training/education packages have already been included in industrial promotion packages by recruiters in both states when attempting to secure the location of major manufacturing firms. Some officials are already suggesting free education for the first two years in community colleges and technical institutes.

But higher levels of education can not be achieved unless there is a dramatic improvement in the achievement of students in the public schools. In the late twentieth century, the public was not always willing to accept the tax burden and the family role of backing necessary improvements in the public schools or of demanding action by appropriate public officials. Until elected officials are convinced that not only parents, but also taxpayers, demand programs and facilities in public schools that will continue to attract high-level jobs will needed improvements be forthcoming. There are different responsibilities on the part of state and county governments in funding programs and facilities for public education.

The economic restructuring that has occurred in the Carolinas has simultaneously resulted in considerable individual financial hardships for many and great economic opportunities for others. This change will undoubtedly continue as lower-wage countries siphon off more and more low-level manufacturing jobs. Indeed, even though as recently as the early 1990s the Carolinas continued to lead the Nation in the percentage of all workers being employed in manufacturing (almost one-fourth compared to less than a sixth for the U. S. as a whole), North Carolina dropped to tenth and South Carolina to eleventh place in 2004. Increasing shares of workers in this region and in the nation have been being accounted for by the service sector (Figure 10.1). By 2002, more than half the 1990 jobs in textiles and nearly two-thirds of those in apparel had been lost to technological innovation and to China and other countries. Moreover, nearly two-thirds of the jobs in leather products, half those in printing and publishing, two-fifths of those in industrial machinery, a third of the ones in primary metals, and a fourth of the ones in tobacco products, lumber products and electronic equipment (including computers) had disappeared. Bright spots in manufacturing were limited to transportation equipment and fabricated metals, whose jobs increased by over 80 percent and 40 percent, respectively.

Between 1980 and 1994, the proportion of workers in manufacturing declined by 13 percentage points in the Carolinas and by six points in the United States, while the share of persons in services rose by 13 points in this region and by eight points in the Nation. By the mid-1990s, workers in manufacturing still outnumbered those in all services in North Carolina, but employees in services had become more numerous than those in industry in South Carolina (Figure 10.2). For the United States, the number of persons working in services was almost double those in manufacturing. By 2004, the proportion in manufacturing had declined to just 15 percent in the Carolinas and to 11 percent in the U.S. as a whole. Between 1990 and 2005, the number of workers in manufacturing had decreased by 412,000, or by a third, while employment in the service sector had more than doubled, having added more than 1,123,000 jobs. Health services and business services had more than doubled and tripled, respectively, by 2004. The finance, insurance and real estate (FIRE) category had grown by 60 percent. However, jobs in the retail sector had dropped by more than a sixth and those in wholesale trade had leveled off.

An additional aspect of economic restructuring is that many of the declining types of manufacturing have been greater generators of hazardous waste than most of the newer industries replacing them. Thus, even though new businesses might increase total waste disposal, they do not do so to the same extent as older ones did. Nevertheless, the amount of waste, especially hazardous waste, is so great that it poses a serious threat to the environment, particularly the air and water. This hazard emanates not just from manufacturing concerns, but also from agriculture, notably hog farms, as well as from lumber processing, golf courses, and cities, many of which often find themselves not in compliance with water treatment standards. This issue must be addressed before irreversible pollution of the groundwater supply occurs and future population growth and economic development are curtailed. The great difficulty in solving this problem is not just developing the technological means to protect the water, but determining the financial plan that will pay for the necessary abatement procedures. During the late 1990s, the

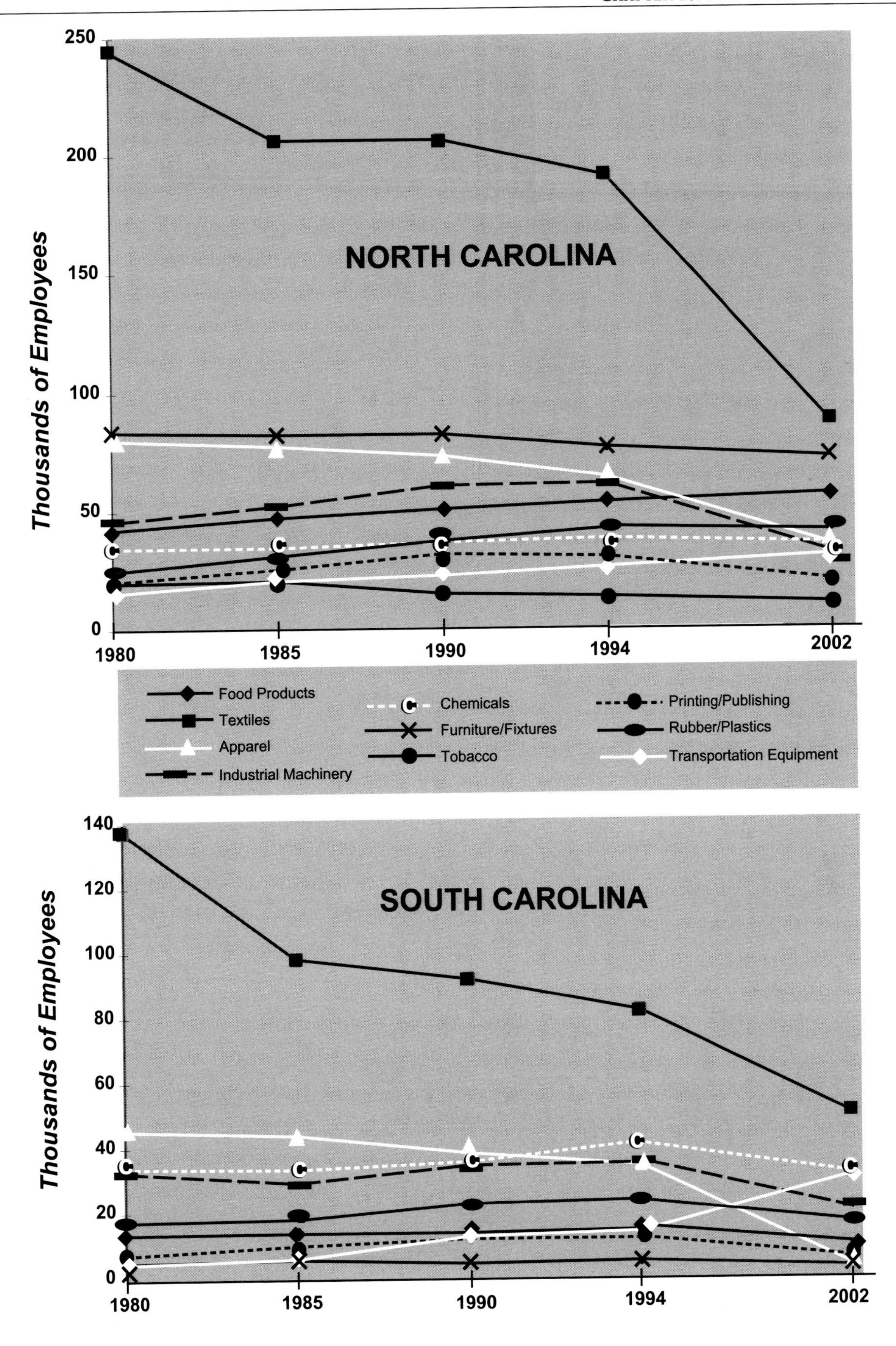

Figure 10.1. Employment in major industries in the Carolinas, 1980-2004.

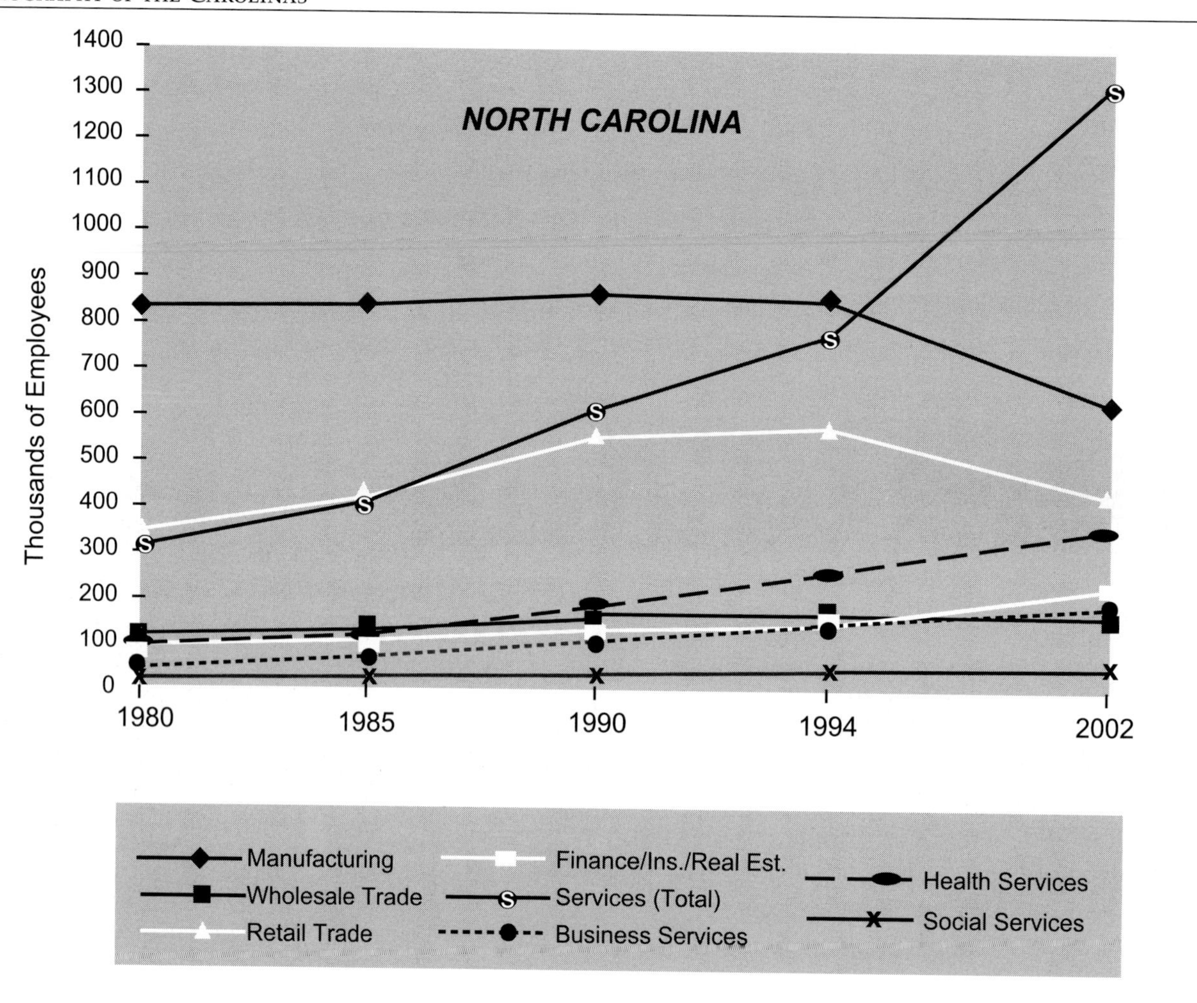

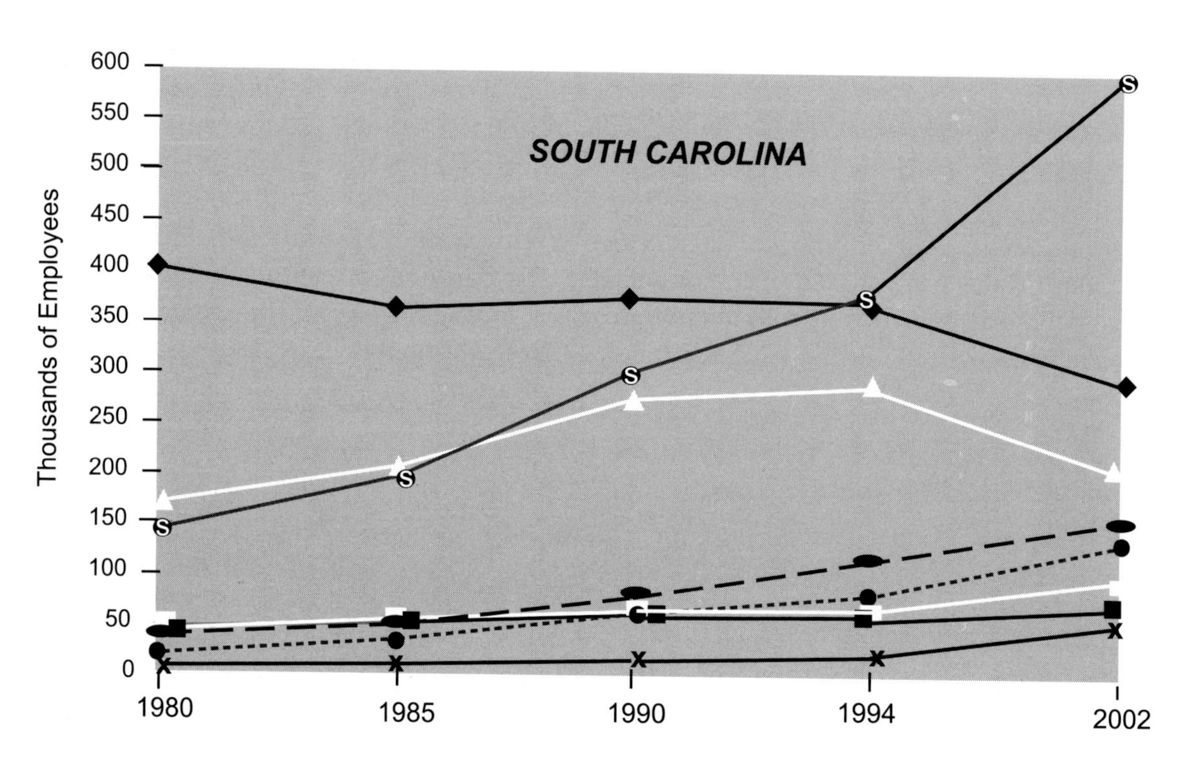

Figure 10.2. Employment in major sectors of the economy in the Carolinas, 1980-2004.

Photo 10.2. Part of the redeveloped waterfront district in Wilmington, NC—one urban area in the Carolinas to which many persons are retiring.

issue of hog waste disposal and the related environmental threat occupied much of the time of the North Carolina legislature.

One of the growth sectors of the economy of the Carolinas for the last quarter of the twentieth century has been retirement development (Photo 10.2). Both states have shared in the financial benefits of thousands of retirees moving into this region. In fact, North Carolina as a retirement destination rose in rank from 27th in the 1960s to 17th in the 1970s to 7th in the 1980s to 5th during the 1990s. Although the high-amenity coastal and mountain areas have gained most of the attention in this regard, several cities and rural parts of the Piedmont have also attracted large numbers of retirees. Studies of the mountain and coastal counties have shown that retirees moving into the Carolinas have disproportionately high incomes and are major consumers, thus having a very positive impact on the economy of the receiving areas. This trend is likely to continue and even to accelerate during the first quarter of the twenty-first century as the baby boomers explode into the retirement years. Even here, though, careful planning for the future is imperative. Given that the surge of persons of retirement age is known and that many parts of the Carolinas are attractive retiree destinations, planning for the environmental, infrastructural and other public impacts of residential and associated economic development projects will prevent many negative consequences which might otherwise take place.

Therefore, population growth and economic development related to retirees will require thoughtful planning not only for roads, water and sewer, and environmental safeguards, but also for education and health needs. Retirees often seek locations that offer cultural and educational opportunities, such as community college or college/university courses on a wide variety of subjects. This group is likely to account for an increasing share of the enrollment in community colleges, colleges and universities in retirement destination areas. Although any expanding population requires greater health care, retirees have disproportionate and special medical requirements. They should provide the necessary additional market in many rural areas to attract doctors and other medical services and facilities that might well have otherwise avoided these more sparsely populated remote locations. About two-thirds of the elderly die from cancer, heart disease, strokes, and cirrhosis of the liver. Another special health need of retirees is long-term health care, and as the baby boomers age through the retirement years, this need will increase dramatically.

Whether the consideration is general population growth and economic development or a particular aspect, such as retiree in-migration, planning for the future must take into account the long-term impact of continuing to develop residentially in such a low-density manner that the cost of building and maintaining the infrastructure could become prohibitive. Already, inadequate funds are being allocated to maintain and rehabilitate older roads and water and sewer systems, even as new ones are being demanded by an expanding population. Some experts believe that computers and telecommunications will decrease commuting substantially in the future. However, even though this might be true for a few occupations, thus far, commuting seems to be continuing to increase. Most new roads and highways are at or beyond capacity before they can be built and opened to traffic.

Some community leaders have been calling for better interurban and intraurban transit, but most of the public still prefers relatively cheap use of private vehicles and the roads necessary on which to operate them with limited congestion. Will some intolerable degree of traffic congestion or air pollution be the catalyst for more acceptance of public transport or will less available and much higher cost gasoline some time in the future be the only factor that will necessitate an increase in the use of public transit?

Unfortunately, the longer land use is developed in a low-density sprawl across the landscape, the more difficult it will be to design an economically efficient mass transit system. Already, most sections of the Carolinas, even the largest cities, are so sparsely populated that in recent decades most cities have experienced a decline or only slight improvement in public transit service. In an era of plentiful and inexpensive gasoline, making a successful argument for planning for mass transit has been a "hard sell," but rising gas prices could bring a reappraisal.

Unforeseen technological breakthroughs in energy and waste disposal might make many of these concerns of planning for the future unnecessary. However, a sudden unexpected crisis resulting from a culmination of one or more economic, infrastructural or environmental disasters might force the public into an untenable situation which might require a prolonged period of serious inconvenience or suffering and societal life-style restructuring before the problem can be accommodated. In the meantime, personal and regional economic hardships might become widespread.

In order to minimize the potential negative consequences and dislocations of such possible calamities, several steps can be taken in planning for the future:

1. Residential and overall densities should be increased within settled areas so that shorter trips to work and to shop are needed and that these journeys could be more easily accommodated by public transit, if necessary.
2. Residential areas should include more types of land use, instead of being used for only housing. Grocery and drug stores and other frequently used establishments should be integrated into new and, where possible, old neighborhoods, so as to minimize the distance needed to be traveled in order to obtain items which are purchased frequently.
3. Planning for the likely use of mass transit at some future time should be included in any new development scheme, whether residential, office, retail, or industrial—similar to neo-traditional town planning areas. Such planning should also include the integration of mass transit for older, more densely populated areas and flexible public transit for newer, more sparsely populated sections.

Conclusion

Whether the people of the Carolinas choose to plan adequately for the future and to take the necessary steps to avoid possible calamities or whether they just let future events govern their lives and those of their children is a decision only they can make. Whether they opt for maximum short-term gain and life-style preference or for a more conservative sustainable long-term use of the land and natural resources, only time will tell. There are community and state leaders calling for the kinds of educational programs, health care systems, infrastructural maintenance and construction, land use planning, and environmental safeguards that will better prepare the Carolinas for potential problems in the future. However, others regard near-term lower taxes and unbridled, unconstrained development as a better alternative for moving the region through the twenty-first century. The alternative chosen by Carolinians and their leaders will determine not only the society in which the citizens of the region live for the next 25 years, but also their ability to handle the possible and the unexpected problems that might well face them in the second quarter of the twenty-first century. Will the people of the Carolinas plan wisely for the future or will future events eventually severely limit the options which they have?

INDEX

A

Abbeville 69, 77, 106
Aberdeen 199
Africa 1, 14, 16, 23, 24, 26, 27, 38, 53, 63, 67, 70, 121
African-Americans. *See* blacks
Africans 53, 63, 66, 67, 70, 91
age structure 161, 167, 168, 172, 177, 178, 179, 181, 183
agriculture
  census 143, 145, 151
  Charles Town 61
  hazardous waste 244
  indentured servants 2
  introduction of 54
  labor force 99
  Mountain Region 10
  Piedmont 7
  slaves 72
Aiken 86, 100
Alamance 5, 81
Albemarle 5, 17, 25, 56, 59, 69
Albemarle Sound 4, 60, 61, 63, 65, 162, 178
  Civil War 83
Alcatel-Lucent 125, 133
Algonquian Indians 56, 57
Allendale 106
Allen University 218
Allstate 230
Altria Group 123
American Airlines 118
American Association of Textile Chemists and Colorists 228
American Brands 122
American Drew 124
American LeFrance 126
American Tobacco Company 8, 88, 121
Anderson 77, 88, 206
Anderson, James R. 143
annexation 109, 111, 112
Anson County 151
Appalachian Mountains 9, 14, 16, 26, 35, 37
Appalachian Plateau 27, 31
Appalachian State University 163, 231
apparel industry 98, 100, 101, 103, 120, 127
  decline 95, 102, 109, 115, 244
  outsourcing 120
Arrick Farms 106
Arthur D. Little 231
Asheboro 199
Asheville 220, 223, 224
  Biltmore House 10
  climate 48
  economy 119
  population 89, 96, 162, 164
  railroad 79, 90
  tourism 194, 196, 197, 206
Asheville Airport 99
Asheville Basin
  manufacturing 127
Ashley River 61, 195
Asians 53, 183
ASTRA 228
Atlanta-Sandy Springs-Marietta GA MSA 107
Atlanta and Charlotte Air Line Railroad 90
Atlantic Coast Line Railroad 90
Avondale Mills 101

B

B. A. T.. *See* British American Tobacco
baby boom 169, 181, 247, 248
Back Country 53, 65, 66, 68, 69, 71, 72, 73, 74, 77, 78, 79
  cotton industry 75
  cotton trade 73
  economy 72
  Scotch-Irish settlers 71
balanced growth 98, 99
Baltimore and Ohio Railroad 76
Bank of America 129, 130, 131
Bank of America Stadium 200
Banner Elk 198
Barbados 61
Barber-Scotia College 217
Barden Inlet 17
Barefoot Landing 190, 191
Barnett Bank 131
barrier islands 17, 18, 19, 20, 31, 41, 67, 188, 193, 195
basketball 8, 201, 207, 208
Bassmaster Tournament Trail 202
Bath 5, 63
Battelle 231
Bayer 133
Beaufort
  Civil War 83
  climate 48
  colleges 224
  cotton 87
  employment decline 106
  French settlement 59
  population growth 135
  port 67
  settlement 5
  slave labor 75
  tourism 135, 191, 194

Beaufort County 40, 41, 42, 162
Beaufort County Water Festival 206
Beaufort Inlet 17
Beaunit Fibers 228
Beautancus 154
Benedict College 218
Bennett College 217
Bentonville Battlefield 194
Bermuda High (high pressure cell) 36
Bertie County 40, 44
Bethabara 5
Biltmore House 10, 194, 197
birth rate 163, 173, 174
Blackbeard 2
Black Mountain 27
blacks 182
  AIDS 177
  cigarette industry 182
  freedmen 84
  health care 175
  higher education 219, 224
  labor force 81
  leaders 84
  mortality 177
  out migration 171, 183
  Piedmont 165, 183
  population 74, 89
  poverty 181
  slave labor 75
  slave population 78
  South Carolina 89
  tenantcy system 86
  textile industry 182
  tobacco farming 175
Bladen County 105, 152
Blowing Rock 124, 198
Blue Ridge Mountains 5, 7, 22, 27, 28, 50, 76, 141, 164, 177
  rivers 29
  South Mountain Belt 80
  tobacco industry 155
  vegetation 40
Blue Ridge Parkway 48, 196, 197, 198
Blue Ridge Province 22, 26, 27, 28, 29, 31, 35, 39, 40, 48
BMW 118, 119, 126, 243
Bodie 193
Bogue Inlet 17, 19
boll weevil 6, 87, 145, 153, 154, 157
  eradication 154, 158
Bonnet, Stede 2
Boone 27, 198, 220
Boone Hall Plantation 195
Bowens Fault 26
Branch Banking & Trust (BB&T) 129, 131
Brattonsville 194
Brevard College 215
Brevard Fault 22, 26
Brewer Mine 80
Bridgestone-Firestone 101
British American Tobacco 122
Broad River 56, 75
Broadway at the Beach 189, 190, 191
broilers 143, 148, 149, 151, 158, 166
broilers business 149
Brooke Group 122
Brookgreen Gardens 6
Brown and Williamson 121
brownfields 243
Broyhill Furniture 123, 124
Brunswick County 65, 77
Bull Durham 88
Buncombe County 222
burley tobacco 50, 145
Burlington House 120
Burlington Industries 120
Burlington Worldwide Apparel 120

C

Cabarrus County 25, 100, 123
  gold, discovery of 79
Caldwell County 80, 124, 125
Calhoun, John C. 82
Camden 56, 69, 71, 194
  cotton trade 73
  railroad 77
  river transportation 75
Cameron Mills 97
Campbell, Malcolm E. 227
Campbell University 216
camping 193, 196, 201, 204
Camp Lejeune 112, 128, 163
canal system 75
cancer 146, 174, 176, 178, 179, 183, 248
Cannon Mills 100
Canton 48
Cape Fear
  barrier islands 67
  exploration 57
  indigo 66
  lumber 66
  naval stores 65, 66
  plantation economy 87
  rice 65, 66, 67
  settlement 60, 61, 63, 64
  slaves 66, 71
Cape Fear River 5, 20, 29, 66, 77, 83
Cape Hatteras National Seashore 193
Cape Hatteras
Cape Hatteras Lighthouse 2, 3, 17
Cape Lookout 17, 18
Cape Lookout National Seashore 193
Carlisle Finishing 120
Carl Sandburg Home National Historic Site 194

Carolana 1, 241
Carolina Bays 20, 21
Carolina Bays Parkway 191
Carolina Beach 19, 191
Carolina Belt 80
Carolina Central Railroad 79
Carolina Hurricanes (NHL) 209
Carolina Main Street 95, 98, 103, 106, 131, 136, 137
  airport 118
  air traffic 117
  automotive cluster 126, 127
  balkanization 113
  employment 115, 133, 134, 135
  financial industry 130, 131
  growth rate 107
  influence on growth 109
  knowledge industry 125
  location advantage 116
  manufacturing decline 105
  mass transit 116
  metropolitan areas 107
  Piedmont Triad 120
  population growth 107
  shape 114
  South Carolina 116
  Spersopolis 112
  tobacco industry 121
  urban area 113
Carolina Panthers (NFL) 200
Carolina Research Park 227, 231
Carolinas
  agriculture 91, 153
  banks 131
  industrialization 87
  population 89, 107, 168, 183
  racial composition 165
  urban growth 95, 98
Carolina Slate Belt 25
Carolina Turkeys (Company) 151
Carroll's Foods 150, 151
Cary 109, 112, 133, 164, 183
Catawba/Wateree River 75
Catawba College 215
Catawba Indian Tribes 56, 58
Catawba Path 69
Catawba River 69
census 73, 74, 78, 81, 98, 106, 146, 166, 170, 172, 175, 179, 183
Central Piedmont Community College 224, 226
Centura. *See* RBC Centura Bank
Chanler, J. A. 97
Chapel Hill 119, 182, 220, 227
  first public university 8
  Research Traiangle Park 131
Charleston 178
  airport 119
  antebellum influence 72
  Civil War 6, 81, 83, 84
  colleges 220
  commuters 114
  cotton industry 75
  cotton trade 73
  fishing 202
  growth 95, 169
  medical school 215
  metropolitan area 179
  naval base 128
  population 89, 96, 109, 161, 171
  railroad 76, 79, 90
  rice production 169
  settlement 2, 4
  slaves 179
  Spoleto Festival 206
  tourism 135, 187, 194, 195
  transportation 75, 76
Charleston International Airport 119
Charleston Museum 195
Charleston Navy Yard 128
Charleston Research Park 231
Charleston Southern University 219, 231
Charles Town 2, 5, 61, 63, 65, 66, 67, 68, 69, 70, 72, 169, 195
  indigo export 67
  rice export 67
  settlement 4, 61
Charles Towne Landing 195
Charlotte 199, 206, 217, 224, 226, 229, 230, 243
  airport 117, 118, 119
  banks 130
  bedroom communities 109
  car racing 128
  colleges 220
  cotton mills 97
  financial industry 129, 130, 131
  founding of 5
  gold, discovery of 10
  growth 112
  high technology 243
  highways 116
  manufacturing 8, 98, 127
  mass transit 116
  mint 79
  population 89, 96, 109, 163, 164, 182
  power plant 136
  railroad 77, 79, 90
  real estate development 113
  sports teams 200
  textile industry 88
  tourism 187
  urban growth 107
Charlotte-Gastonia-Rock Hill metropolitan area 181
Charlotte Belt 25
Charlotte Bobcats (NBA) 200
Charlotte Metrolina area 162

Charlotte MSA 107
Charlotte Sting (WNBA) 200
Chatham County 149, 151, 182
Chatooga River 196
chemical industry 8, 28, 29
Chemstrand Corporation 228
Cheraw 45
Cherokee 197
  tourism Indian Tribe 187
Cherokee Indian Reservation 196
Cherokee Indian Tribe 9, 56, 57, 58, 69, 106, 165, 196
Cherry Point Marine Air Station 128
Chesterfield County 45, 80
chickens 6, 101, 148, 149, 166, 175, 181, 182
Chimney Rock Park 196
Chinese 124, 166
Chowan College 215, 219
Chowan County 146
Chowan River 60, 61
cigarette manufacturing machines 98
Cigarette Revolution 145
Cisco 133, 229
city services 243
Civil War 6, 53, 81, 83, 84, 87, 91, 170, 194, 195, 217, 241
  railroads 83
Claflin College 218
clay 16, 21, 22, 26, 43, 44, 46, 47, 152, 198
Clemson 220
Clemson Research Park 231
Clemson University 126, 208, 217, 231
climate 1, 35, 39, 40, 45, 89, 164, 187, 188, 195
  maritime 41
  mountain region 10, 48
  variations 6, 48
climate variations 40, 50
Clingman's Dome 27
Clinton 152
Clinton Junior College 224
Coastal Plain 6, 7, 17, 21, 35, 40, 43, 164
  agriculture 44, 165
  blacks 175, 177
  climate 44
  colleges 215, 223
  cotton 73, 154, 158
  crop diversification 146
  cropland 141
  demographics 181
  diseases 172, 177, 181
  economy 155, 184
  employment 134
  English settlers 72
  flue-cured tobacco 158
  forests 40, 44
  geology 16, 17, 44
  Hispanics 166
  hog farming 105
  Indian Tribes 56
  livestock trade 69
  manufacturing 6
  mortality rate 173, 174
  Outer Plain 20
  population 162
  poverty 181
  railroads 78, 90
  rivers 23, 29
  row crops 158
  sediment 31
  sediment deposits 26
  settlement 54, 69
  settlers 5
  slaves 74
  soils 141
  soybeans 147
  swamps 44
  tobacco industry 66, 77, 87, 88, 98, 145, 146, 157
  trade 79
  transportation 65
  Upper Plain 21
  vegetation 40, 44
Coastal Plain Province 39, 53
Cofitachequi 56, 57, 59
Coker College 219
College of Charleston 215, 231
colleges
  blacks 217, 218
  enrollment 222
colleges, land grant 217
Collins & Aikman Corporation 230
Columbia
  air traffic 119
  bus transportation 116
  Civil War 83
  colleges 215, 216, 218, 220
  commuters 114
  fishing 202
  high technology 243
  hydroelectric power 88
  population 96, 107
  railroad 76, 90
  research facilities 231
  river transportation 75
  selection as capital 72
  settlement 7, 23
Columbia College 219
Colvard, Dean W. 229
Committee on Industry of the State Board of Conservation and Development 227
CommScope 125
Community College System (N. C.) 222
community festivals 201, 205
Concord 88, 112, 123, 137, 200, 217
Cone, Moses H. 194, 198
Cone Denim 120
Cone Family 10

Cone Mills 120
Congaree River 202
conifers 1
Consolidated Deisel 128
Continental Airlines 118
Cooper River 61, 67, 75
Core Based Statistical Areas 106
corn 6, 7, 50, 55, 57, 65, 66, 77, 83, 141, 143, 145, 147, 148, 152, 153, 158, 172
  farm income 146
Corning Cable Systems 125
Cornwallis 5
Corolla 193
cotton 73, 74, 75, 91, 143, 157, 172
  antebellum economy 77
  boll weevil 154
  Charleston 76
  Coastal Plain 154
  export 73, 75
  farm income 146
  farm size 143, 154
  Hispanics 175
  mechanization 145
  mills 81, 88, 97, 98, 100
  North Carolina 77
  Piedmont 78, 96
  plantations 53, 194
  post-bellum production 86
  price 81, 87
  price support 154
  production increase 154
  resurgence 153, 158
  Sea Island 75
  slaves 5, 78, 169, 170, 241
  soil erosion 47
  South Carolina 73
  spindles 102
  tariffs 81
  tenant system 86
  transportation 154
  varieties 73
cotton, black seed 73, 75
cotton, green seed 73, 75
Cotton, Mexican 75
cotton gin 5, 73, 241
cotton industry 72, 98, 153
  decline 153
  Piedmont 155
  wages 99
Croatoan 2, 60
cropland 141, 143, 146, 147, 148, 155
Cross-Island Parkway 193
Cross Creek 65, 71, 72
Crowders Mountain 22
Cuddy Farms 150
Cullowhee 206
Culpepper's Rebellion 5
Cumberland (County) 88, 163
Currituck (County) 107, 135, 165

D

Daimler-Chrysler 128
dairy farms 156
Dana 231
Dan River-Danville Basin 26
Dare 2, 162
Darlington 128, 206, 207
Daufuskie (Island) 191
Davidson (County) 88, 200
Davidson College 215, 229
Davie County Basin 26
death rate 172, 173, 174, 176
de Ayllon, Lucas Vasquez 2, 56, 58
Debidue Island 19
degenerative diseases 174
Deliverance 196
Delta Airlines 118
Delta Woodside 98
demographic transition 172
Denmark Technical College 224
de Soto, Hernan 56
Dillon 106, 181, 200
Discovery Place 199
diseases 152, 169, 174
  Indians 56, 58
Dole Foods 100
Douglas Battery 128
Dow Jones 231
Drake, Sir Francis 60
Drexel-Heritage 124
Drum Inlet 17
Duck 193
Duke, Washington 88
Duke Energy 101, 136
Duke Family 8, 10, 88, 121
Duke Homestead 194
Duke University 8, 129, 132, 162, 208, 215, 216, 217, 220, 227, 241
dunes 20, 42, 43, 59, 193
Duplin County 149, 154
Durham
  black labor force 182
  cigarette industry 98
  colleges 215, 216, 220, 223
  employment loss 122
  health care 183
  Hispanics 182
  Research Triangle Park 131
  textile industry 88
  tobacco industry 88, 101
Durham Basin 26

E

East Carolina University 163, 216

economic development 247, 248
economic restructuring 117, 122, 244
Eden 100
Edenton 5, 63, 64, 65, 66
edge cities 242
Edgecombe county 153
Edgefield County 56
Edisto River 67, 69
Eisai 133
elderly 167, 248
Electric Power Research Institute 231
Elizabeth City 128, 220
Elizabeth City State University 218
employment
  Carolina Main Street 112
  Charleston 128
  Charlotte 109
  commuters 114
  decline 102, 244
  farmers 156
  financial industry 129
  freedmen 84
  globalization 119
  Google 124
  Hispanics 105
  losses 124
  manufacturing industry 103, 105, 129
  Piedmont 8
  service industry 100
  South Carolina 106
  textile industry 103
  tobacco industry 121
England
  exploration 60
  indigo trade 70
  maritime power 65
  ocean trade 67
  settlers 61
  textile industry 70
English
  contact with Indian tribes 56
  struggle for Carolinas 58
  treaty with Cherokees 58
English settlements 3, 60, 61, 194
English settlers 1, 2, 63, 71, 169
entisols 40
Ericsson 228
erosion 10, 13, 16, 17, 22, 30, 31, 39, 43, 47, 50, 147, 193
Erskine College 215
Europe 1, 14, 16, 24, 27, 62, 66, 81, 121
export 64, 66, 67, 70, 72, 98, 133, 147, 169
  indigo 66
export economy 99

F

Fairfield (County) 106
Fall Line 7, 17, 21, 22, 23, 44, 55, 69, 75, 78, 155
fall zone 7
Fayetteville 5, 7, 25, 65, 79, 117, 119, 128, 163, 217, 218, 220
  Civil War 84
  settlement 23
Fayetteville State University 218
Fayetteville Technical Community College 226
FedEx 116, 118
fertility 40, 47, 167, 169, 170, 172, 173, 183
fiber optics 125
Fieldcrest 100
Filipino 166
finance 131
finance industry 8, 120, 130, 244
financial services industry 174
First Data Corporation 231
First Union Bank 130, 231
fishing 3, 5, 25, 58, 193, 196, 197, 201, 202, 209
  Myrtle Beach 190
Flat Rock 195
Flatwoods 141, 157
Fleet Marine Force Base 128
Florence 77, 87, 231
Florence Basin 26
Florida Central and Peninsular Railroad 90
Floyd 38
flue-cured tobacco 145, 146, 157
food industry 243
food processing 105, 106
football 8, 197, 200, 201, 207, 208
forest industry 157
forests 1, 9, 42, 46, 48, 49, 50, 66, 69, 190, 195, 197, 203, 204
forests, maritime 42
Forsyth Technical Community College 226
Fort Bragg 119, 128, 198
Fort Fisher 83, 84, 194
Fort Jackson 128
Fort Mill 98, 121
Fort Moultrie 194, 195
Fort Raleigh National Historic Site 193
Fort Sumter 6, 81, 194, 195
Francis Marion University 219
Francis Marion University Research Park 231
Frederick-Edward 124
Freightliner 128
French Broad River 30, 196
French settlers 2, 57, 59, 69
Fripp Island 191
frost-free season 41, 45, 46
fruits 58
Furman University 215
furniture industry 87, 88, 107, 157
  decline 95, 101, 102, 109, 115, 125
  employment 8
  High Pointmanufacturing 123
  labor 227

Las Vegas 123
Lenoir 124
ownership 98, 124
Piedmont 123
wages 99, 124, 243

G

Gaffney 88, 215
Galey and Lord 120
Gaston County 80, 115, 154
Gastonia 88, 128
geologic plates 1, 13
Georgetown 59, 67, 106, 165, 188
fishing 202
Spanish exploration 2
German settlers 2, 9, 63, 169
Glassy Mountain 22
Glaxo-Wellcome 228
global economy 95, 99, 100, 104, 106, 111, 120, 121, 128, 133, 137
global warming 37, 136
gold 25, 79
Gold Hill 80
Gold Kist 106
Goldsboro 77, 119, 128, 151, 223
Civil War 84
railroad 79
golf 156, 162, 187, 188, 191, 192, 196, 198, 200, 201, 205, 244
Gondwanaland 23, 26
Google 116, 124, 125, 136
Goose Creek 109
Gordillo 1, 57, 58
Grande Dunes Resort 189
Grandfather Mountain 27, 196, 198, 206
Grand Strand 19, 57, 188, 189, 191, 202, 204
tourism 187
Graniteville 100
Graniteville Manufacturing Company 100
Grant, Marshall 154
Granville, Lord Proprietor 63
Granville County 80
Great Philadelphia Wagon Road 169
Great Rift Valley 26
Great Smoky Mountains 27
Great Smoky Mountains National Park 27, 28, 48, 196
Great Valley 2, 5, 7, 71
Greensboro
airport 117
climate 46
colleges 215, 220, 223
commuters 114
FedEx hub 116
Guilford Battlefield 5
manufacturing 7, 128
population 89, 163, 164
textile industry 88, 97
tobacco industry 121, 123
urban expansion 112
Woolworth Sit-In 241
Greensboro-High Point and Winston-Salem metropolitan area 120
Greensboro College 215
Greenville
South Mountain Belt 80
Greenville, N.C.
colleges 216
information industry 162
metro area 163
Greenville, S.C.
automobile industry 126
colleges 215, 220, 224
commuters 114
economy 112
employment 113
Hispanics 182
metropolitan area 181
railroad 77, 90
service industry 166
textile industry 88, 98
Greenville-Pickens Speedway 206
Greenville-Spartanburg-Anderson metropolitan area 165, 181
Greenville-Spartanburg International Airport 118
Greenville-Spartanburg manufacturing 8
Greenville County, S.C. 182
Greenville Technical College 224, 226
Greenwood 88
Greenwood County 134
Greer 126, 243
Gregg, William 100
Griffith, Andy 195
groundwater 152, 244
Grove Park Inn 197
growth centers 99
Guest, Romeo H. 227
Guilford Battlefield 5
Guilford College 215, 219
Guilford County 40, 46, 47
Guilford Courthouse 71, 194
Guilford Mills 120
Guilford Technical Community College 226
Gulf Stream 39, 41, 75, 202

H

Haile mine 25
Halifax 65
Halifax County 97
Hamburg 76
Hanes 120
Hanes, Robert M. 227
Hanesbrand 121
Hanes Family 10

Hanging Rock 9, 22
Harbour Town 192
Hard Rock Park 191
hardwood forests 40, 47, 50, 88, 96, 157
Hariot, Thomas 2, 57
Harrah's Cherokee Casino 197
Harris, W. T. 229
harvested cropland 141, 143, 146, 147, 148, 155
Hatteras 2, 39, 193
Hatteras Inlet 17
Haywood County 40, 48
health services industry 8, 129, 174, 175
heart disease 174, 177, 178, 183, 248
Henredon 124
Hickory 124, 125
Hickory Craft 124
Highland Games 198, 206
High Point 8, 88, 123, 124, 128, 131, 223
High Point University 219
highways 11, 115, 116, 118, 189, 191, 196
Hillbilly Days 206
Hillsborough 69
Hilton, William 60, 61
Hilton Head Island 75, 109
  beaches 162
  fishing 202
  Hispanics 166
  real estate development 179
  retirement community 193
  tourism 135, 187, 191, 192, 193
Hispanics
  English language 166, 175
  labor force 105, 106, 146, 166
  migration 53, 166, 171, 175, 181, 182, 183
  Piedmont 183
  population 166, 173, 179, 182
  Wilkes County 177
Hispanics, Cuban 166
Hispanics, Mexican 166
Hispanics, Puerto Rican 166
HIV/AIDS 176, 177
Hmong 166
hockey 197, 201, 209
Hodges, Gov. Luther H. 227
hog farms 244
hogs 6, 64, 69, 143, 151, 152, 154, 155, 158, 162, 244
Hoke County 163
Holden Beach 188
Hollings, Gov. Ernest 222
Holocene Epoch 54
Holshouser, Gov. Jim 99
Honda 118, 126
Honda ATV 126
horizons 42
Horry County 146
hub and spoke 117
Hugo 38
Huguenots 63, 67, 69, 70, 169
Hunt, Gov. Jim 99
hunting 55, 58, 187, 193, 196, 201, 202, 203, 204, 209
hurricanes 38, 41, 143, 193
Hyde County 162

I

I-40 95, 112, 113
I-77 95, 116
I-85 95, 112, 113, 164
IBM (International Business Machines) 126, 129, 132, 133, 228, 230
Ice Age 54
igneous rock 47
IKEA 116
inceptisols 40
India 101
Indian corn 66, 71
Indian hemp 58
Indian old fields 58
Indians
  diseases 58
  population 57, 165
  slaves 57
Indians, Asian 166
Indian trade 60, 61, 64, 67, 69
Indian tribes 56, 61, 68
indigo 2, 65, 70, 72, 73, 91
indigo plantations 64
industrialization 87
infant mortality 173, 177, 179, 183
infectious diseases 172
influenza 174
Interco 124
International Biosphere Reserve 198
International Center for Automotive Research (Clemson University) 126
International Home Furnishings Market 123
International Textile Group 120
interstate highway system 112
Intracoastal Waterway 191
Iredell County 164
Ireland 101

J

Jacksonville 112, 163
Jamestown 2, 3
Japanese 166
Johnson, F. Ross 122
Johnson, Marvin 149
Johnson C. Smith University 217
Johnston County 84, 164, 183
junior colleges 222, 224

K

Kannapolis 100, 137

Kiawah Indian Tribe 61
Kiawah Island 191
Kiawah Island Resort 195
Kill Devil Hills 193
King's Mountain Belt 25, 26
Kings Mountain 5, 71, 194
Kitty Hawk 193
Koreans 166

L

labor shortage 191
Ladd Furniture 124
lagoons 20, 152
Lake Jocassee 202
Lake Moultrie 75
Lake Waccamaw 21
Lancaster 224
Lancaster County 58, 79, 80
Lander University 219
Lang, Mayor 112
Laurens 106
Laurentia 23, 24, 25, 26
Lea Furniture 124
Lenoir 116, 124, 125, 181
Lexington 69, 107, 124
life expectancy 173, 176, 177, 183
Liggett-Myers 121, 122
Limestone College 215, 219
Lincolnton 81
Lineage (Furniture) 124
Linn Cove Viaduct 196, 198
Linville 198
Little Mountain 22
livestock 67, 69, 86, 88, 143, 147, 152
loblolly pine 21, 44, 46
Lords Proprietors 5, 60, 62
Lorillard Tobacco Company 121, 123
Lost Colony 2, 60, 193
Louisburg College 221
Louis Rich 106
Low Country 6, 66, 67, 69, 72, 73
Lowe's Motor Speedway 200, 206
Lucent. *See* Alcatel-Lucent
Lucus, J. Paul 229
Lumbee Indian Tribe 165
lumber 65, 66, 244
lumber industry 44, 99, 227, 244
Lutheran Theological Seminary 215

M

Maggie Valley 196
magnolia 42
maize 6
Manteo 2, 193
manufacturing
  Charleston 231
  emergency vehicles 126
  metropolitan areas 242
  outsourcing 120
  Piedmont 8, 96, 165
  textile equipment 98
  textiles 81
  tobacco 88, 122
manufacturing industry 87, 99
  automobiles 118, 126, 127
  decline 102, 104, 105, 106, 119, 133, 244
  employment 100, 103, 128, 243, 244
  female workforce 171
  gentrification 122
  globalization 121
  job training 222
  North Carolina 88
  outsourcing 133
  pollution 124, 244
  recruitment 243
  textiles 227
  wages 99
March to the Sea 83
Marconi Commerce Systems 128
Marion 87, 106, 219, 231
maritime 41, 42, 195
marsh 1, 21, 54, 75
Mars Hill College 215
Masco 124
Massachusetts Bay Colony 60
mass transit 116, 184, 248
McCormick County 56, 80
Mecklenburg (County) 77, 80, 97, 112, 114, 116, 181, 182, 183
  commuters 114
  Hispanics 182
  population 182
  population growth 163
median age 166, 169, 174
Medical University of South Carolina 215, 216, 231
megalopolis 113
megalopolitan growth 95
metamorphic belts 24
metamorphic rock 14, 17, 22, 23, 25, 26, 27, 44, 46, 47
Methodist College 219
Metrolina 130, 181
metropolitan area 95, 101, 106, 109, 111, 113, 119, 120, 125, 130, 133, 161, 163, 183, 199, 224, 243
  colleges 226
  economic development 241
  ethnic diversity 161, 166
  population 95, 163, 164, 175, 183
metropolitan counties 171, 175
Metropolitan Statistical Areas (MSAs) 106, 163
Mexican 73, 120, 200
Mexican laborers 183
Michelin 126
Micropolitan Statistical Areas (MiSAs) 106
Microsoft 126

Midlands 165
Midlands Technical College 224
Midway Airlines 118
migration 53, 183
Mile-High Swinging Bridge 198
military bases 107
Mitchell College 221
Moore's Creek 5
Mooresville 107, 128, 206
Moravian settlements 63
Moravian town 194
Moses H. Cone Memorial Park 194, 198
Mountain Heritage Day 206
Mountain Region 6, 164
  agriculture 157
  colleges 223
  mortality rate 174
  out migrationpopulation 170
  settlement 169
Mount Airy 195
Mount Olive 151, 219
Mount Olive College 219
Mt. Airy 28
Mt. Gilead 56
Mt. Holly 128
Mt. Mitchell 9, 27, 195
Mt. Pleasant 109
Municipal Association of South Carolina 109
Murdoch, David 100
Murphy, Wendell H. 151
Murphy Family Farms 151, 154
Myrtle Beach
  beaches 19, 162, 188, 190
  festivals 206
  population growth 107, 164
  tourism 135, 178, 187, 188, 189, 190, 191, 193, 204
  whites 165
Myrtle Beach Airport 119

N

Nabisco. *See* RJR/Nabisco
NAFTA 120
Nags Head 193
Nano-Tex 121
Nantahala National Forest 196
Nantahala River 196
NASCAR 8, 190, 191, 200, 206, 207
NASCAR Museum 128
NASCAR Technical Institute 128
Nash Johnson and Sons 150
National Center for Health Statistics 176
National Environmental Health Science Center 228
National Institute of Environmental Health Sciences 229
NationsBank 130, 131
Native Americans 6, 53, 58, 165
  migration to Americas 53
  population 56
Naval Research Facility, Charleston 132
naval stores 64, 65, 66, 67, 69, 72, 78, 97, 169
Nestle 106
Neuse (River) 5, 17, 29, 63, 65
New Bern 5, 63, 65, 195
  tourism 194
Newberry 106
Newberry College 219
New Economy 101, 104, 121
Newell, William A. 227
New England 5, 38, 61, 64, 66, 71, 73, 81, 97, 98, 100, 199
  settlement 60
New Hanover County 65, 71, 165
New River 30
New River Marine Air Station 128
Newton 124
Nolichucky (River) 196
Nor'easter 39
Nortel 133, 228
Northampton County 153
North Carolina
  agriculture 100
  migration to west 78
  population 71, 78
  railroads 90
  urban growth 89
North Carolina Agricultural and Technical State University 217, 218
North Carolina Central University 162, 216, 218
North Carolina Cooperative Extension Service 146
North Carolina Museum of Art 199
North Carolina Railroad 79, 90
North Carolina State University 132, 162, 208, 216, 217, 220, 222, 227, 241
North Carolina Wesleyan College 219
North Carolina Zoological Park 199
North Charleston 109, 119
Northeastern Railroad 79, 90
North Myrtle Beach 189
Northrop Grumman 231
Novartis/Sandoz (Syngenta) 133
Novo-Nordisk 133

O

Ocean Drive 189
Oconaluftee Indian Village 197
Oconee 106
Ocracoke 3, 193
Ocracoke Inlet 17
Old Salem 194
Olmsted, Frederick Law 198, 199
OMNIMAX theater 199
Onslow County 71, 163
Orangeburg 181, 204, 218
Orangeburg Scarp 21

Oregon Inlet 17, 202
Orton Plantation 6
Outer Banks 2, 3, 6, 17, 18, 19, 63, 65, 77, 162, 193, 204
  Civil War 83
  tourism 187, 191

P

Paleo-Indians 54
Pamlico River 63
Pamlico Sound 56, 59, 63, 83
Pangaea 13, 14, 16
Pangea III 26
Paramount Carowinds 200
Paris Mountain 23
Parris Island 57, 59, 60
Parris Island Marine Corps Training Camp 128
Patriots Point (naval museum) 195
Pea Island National Wildlife Refuge 193
peanuts 6, 143, 146, 147, 153, 157
Pee Dee Education Center 231
Pee Dee Indian Tribe 166
Pee Dee River 5, 19, 29, 76
Pee Dee tobacco region 87, 165
pellagra 172
Pembroke 220
Pender County 165
Pendleton 224
Pepsi 195
Perdue Farms 106
Person County 80
Pfeiffer University 219
PGA Tour Championship 198
Philadelphia Wagon Road 69
Philip Morris Corporation 121, 123
physiography 1
Piedmont
  blacks 165
  Carolina Belt 80
  cattle industry 156
  climate 36, 46
  colleges 215, 218, 220, 223
  corn 78
  cotton industry 73, 75, 77, 91, 98, 155
  demographics 78, 174, 182, 183
  economy 95, 184
  economy, antebellum 72
  economy, colonial 69
  electrification 101
  erosion 31
  farming 156
  fruit processing 100
  furniture industry 123
  geology 16, 22, 24, 25, 44
  gold 80
  Hispanic labor force 106
  Indian Tribes 56
  industry 98, 104
  Inner Belt 25, 26, 27, 28
  manufacturing 96
  migration 174, 175
  mortality rate 174
  mountain remnants 31
  population growth 182, 183
  professional sports 8
  rail roads 79, 89, 90
  retirees 162, 247
  revolutionary war 5
  rivers 23, 29
  sectional differences 72
  settlement 2, 5, 7, 54, 69, 71, 169
  slavery 78
  slaves 74
  soil 40, 46, 47, 141
  textile industry 88, 171
  tobacco industry 66, 77, 87, 88, 123, 145, 155
  transportation system 112
  urbanization 116
  vegetation 40, 46
Piedmont plateau 22, 30
Piedmont Province 27, 31, 39
  settlement 53
Piedmont Region 6, 7, 35, 40, 46, 164, 181
Piedmont Triad 120
  manufacturing 127
  population 162
Piedmont Triad International Airport 116, 117, 118
Piedmont Triangle. *See* Triangle
Piedmont Urban Industrial Crescent 113, 241
  colleges 220
  manufacturing 7
  population 161
  population growth 164
Pilot Mountain 22
pine 9, 40, 45, 46, 48, 49, 50, 66, 157, 200
pine forest 40, 42, 44, 50, 65, 66
pine forests 65
Pinehurst 22, 162, 187, 198
Pisgah National Forest 48
Pitt County 162
plantation economy 65, 67, 69, 87, 169, 181
plantations 6, 53, 72, 74, 75, 81, 86, 89, 91, 165, 169, 192
  labor shortage 84
Plant System Railroad 90
Pleistocene Epoch 53, 54
pneumonia 174
pollution 10, 115, 175, 184, 248
  Charlotte 136
  groundwater 244
Pope Air Force base 128
population
  aging 166
  armed forces 119
  Asians 166
  Barbados 61

Carolina colony 60
Carolina Main Street 107
Carolinas 89, 95, 107, 109, 165, 169, 175
causes of death 174
Charleston 89
Charles Town 67
Charlotte 182
Civil War, 83
Columbia 119
counties 161
death rate 172
demographics 164, 176, 183
density 243
elderly 167
ethnic diversity 177
foreign-born 105
growth 161
Hispanics 166, 173, 182
immigration 171
median age 172
metropolitan areas 107
migration 183
mortality 177, 182
Mountain region 169
MSAs 163
Myrtle Beach 188, 189
Native Americans 165
natural increase 171, 172, 173
nonfarm 112
North Carolina 71, 78, 109
North Carolina colony 63
out migration 78
Piedmont 174
Piedmont Triad 120
Queensborough 68
rural area 89, 172, 175
settlers 63
slaves 179
South Carolina 73, 74, 107, 111
South Carolina colony 61, 67, 68
urban area 63
urban counties 163
whites 170
white settlers 71
Wilmington 65
population growth 107, 241, 243, 247, 248
Carolinas 161
counties 162
metropolitan areas 242
Piedmont 181
Port Royal 2, 57, 58, 59, 61, 192
Civil War 83
settlement 61
potatoes 71
poultry 143, 149, 151, 155
Powhatan 1
Presbyterians 7, 71, 215, 217, 221
Prestage, Bill 150
Prestage Farms 150, 152
PTI. *See* Piedmont Triad International Airport
pulp mills 157
pulpwood 156

Q

Quakers 4, 5, 7, 69, 215
Qualla Boundary 58, 196

R

railroad 76, 77, 78, 79, 83, 88, 89, 90, 171, 207
Raleigh
airport 118
banks 131
colleges 217, 220, 221, 223, 224
commuters 114
economy 112
Hispanics 182
museum 8
NHL franchise 209
population 89, 96, 109, 163
real estate development 243
Research Triangle Park 131
sandhills 44
selection as capital 72
settlement 23
urban growth 111, 183
Raleigh, Sir Walter 2, 169
Raleigh-Cary MSA 107
Raleigh-Durham 7, 116, 117
Raleigh-Durham International Airport (RDU) 118, 119
Raleigh-Durham-Chapel Hill 243
Raleigh-Durham-Chapel Hill-Cary metropolitan area 162
Raleigh-Durham-Chapel Hill metropolitan area 162, 164, 182, 184, 199, 241
Raleigh-Gaston Railroad 78
Raleigh and Augusta Air Line Railroad 90
Raleigh Belt 24
RBC Centura Bank 131
real estate industry 8, 136, 244
real estate taxes 125
Reed, John 25
Reed Gold Mine 25, 194
Reed Gold Mine State Historic Site 194
Reeves Brothers 230
Reidsville 88, 101, 121
Research Triangle. *See* Triangle
Research Triangle Foundation 230
Research Triangle Institute 229
Research Triangle Park (RTP) 109, 114, 132, 227
retail centers 95, 109, 133, 190, 192
retail industry 8, 105, 175, 244
retirees 193

retirement 2, 6, 10, 162, 168, 171, 178, 182, 183, 184, 241, 247, 248
retirement community 193
Revolutionary War 5, 10, 71, 72, 157, 194, 195
Reynolds Building 122
Reynolds Family 8, 10, 121, 122
Reynolds Tobacco Company 8, 122
Ribaut, Jean 59
rice 67
  colonial economy 5
  decline 87
  export 69
  fields 204
  plantations 2, 64, 73, 77, 194
  production 72, 74
  slaves 70, 72, 169, 170
  wild rice 6
rice plantations 78
Richmond 83, 107
Richmond and Danville Railroad 90
Ridge and Valley Province 27
RJR/Nabisco 122, 123
roads 75, 77, 79, 99, 183, 196, 203, 205, 242, 247, 248
Roanoke Island 193
  settlement 2, 60
Roanoke Rapids 7, 23, 88, 97, 154
Roanoke River 5, 29, 65
Roberson (County) 77
Rock Hill 88, 220, 224
Rockingham 128, 206
Rocky Mount 39, 131
Rose Hill 149, 150, 151
Ross, W. L. 120
Rowan County 71, 80
rural area 105, 106
  age structure 167, 168
  Chatham County 182
  colleges 225, 226
  demographics 172, 176
  dirt track racing 206
  economic development 184, 241
  employment 101, 134, 175
  ethnic diversity 161, 166, 183
  fertility rate 169
  furniture industry 98
  health care 248
  hunting 202
  industrialization 99
  labor 98
  Lumbee Indian Tribe 165
  manufacturing 242
  migration 171, 183
  North Carolina 63, 77
  plant closings 109
  population 89, 112
  poverty 179
  roads 243
  Wilkes County 178
Russians 169
Rutherfordton 79
RV 190, 192, 200, 204

S

Safety Components International 121
Salem 5, 7, 63
Salem College 215
Salisbury 65, 71, 79, 107, 215
Saluda County 106, 166
Saluda River 75
Sampson County
Carl Sandburg 149, 152
Sandhills 21, 22, 44, 45, 46, 151, 162, 198
Sandhills Region 35, 40, 134, 141
Sanford, Gov. Terry 222
San Miguel de Gualdape 2, 58
Santa Elena 2, 58, 60
Santee River 5, 19, 29, 59, 67, 70, 75, 202, 204
Sara Lee Corporation 120, 121, 129
Sassafras Mountain 27
Sauratown Mountains 28
Savannah 76
Savannah Railroad 83
Savannah River 5, 29, 55, 68, 69, 76
Savannah River Site 101, 126, 132
sawmills 157
Schenck, Michael 81
School of Textiles (N.C. State University) 227
School of the Arts, Winston-Salem 8
Scotch-Irish settlers 2, 9, 53, 67, 71, 169, 177, 182
Scott, Gov. Robert 99
Scottish Gathering 198
SCT Utility Systems 231
Seaboard Air Line Railroad 90
Sea Island 70, 73, 74
  Civil War 83
  cotton industry 75
  tourism 191
Sea Island cotton 75, 87
sedimentary deposits 23
sedimentary layers 22, 27
sedimentary rock 14, 17, 24, 27, 44
servants, indentured 2
service industry 100, 131, 133, 197, 243, 244, 248
Seymour Johnson Air Force Base 119, 128
sharecroppers 86, 145, 146
Shaw University 217
Sherman, Gen. William T. 83
Sherman College of Straight Chiropractic 219
Siler City 147, 182
Siouan Indian Language 56, 57, 58
Slate Belt 24, 25, 141, 149, 151, 157
slavery 5, 72, 82, 84, 177, 241
slaves 2, 5, 6, 7, 57, 62, 63, 66, 67, 68, 70, 71, 72, 74, 75, 78, 169, 170, 175, 182, 224

Charleston 179
colleges 217
cotton mills 81
population 78, 169, 179
South Carolina 73
tobacco 66
slavesry
tobacco 66
Smalls, Robert 84
Smithfield Foods 105, 152
Society Hill 81
soils 6, 35, 39, 40, 42, 43, 44, 45, 46, 47, 48, 49, 50, 55, 56, 74, 75, 78, 141, 149, 157
cotton 75
South Carolina
migration to west 74
railroads 90
slaves 73
tobacco industry 87
tourism 135
urban growth 89
urbanization 111
South Carolina National Heritage Corridor Program 195
South Carolina Railroad 76, 90
South Carolina Research Authority (SCRA) 231
South Carolina State University 217, 218
Southeastern Automotive Cluster 127
Southeastern Wildlife Exposition 195
Southern Highland Craft Guild's Folk Art Center 197
Southern Methodist College 219
Southern National Bank 131
Southern Pines 187, 198
Southern Railway 90
Southern Wesleyan University 219
South Georgia Basin 26
South Mountain Belt 80
South of the Border 188, 200
Southport 3, 128
soybeans 6, 7, 143, 146, 147, 148, 152, 153, 158
Spartanburg 77, 88, 126, 220, 224
Spersopolis 112, 113, 156, 157
Spoleto Festival 195, 206
Spring Mills Inc 121
Springs Global 121
Springs Industries 98
Statesville 107, 125, 221
stock car racing 128, 201, 206, 207
Stone Mountain 25, 28
stroke 176, 177, 183
Summerlin's Crossroads 154
Summerville 109
Sumter 81, 106, 107, 128, 196, 224
Sumter National Forest 196
Sun City/Hilton Head Development 193
Sun Fun Festival 206
Sunny Point Military Ocean Terminal 128
Sutter's Mill
gold, discovery of 25
swamps 20, 21, 40, 44, 70, 141, 169, 201
Swansboro 17
sweet potatoes 6
Swiss settlers 5, 63, 69, 169

T

Tar Heel 105, 152, 157, 209
Tar River 66
Teach, Edward Teach. *See* Blackbeard
Technical College System (S.C.) 222
Technitrol 228
textile industry 87, 107
decline 95, 102, 125
employment 103
labor 227
mill towns 88
wages 99, 243
textiles 81, 87, 88, 100, 121, 230, 244
Thomas Built Buses 128
Thomasville 107, 124
Thomas Wolfe Memorial 197
Tidewater 6, 20, 178, 179
economy 184
English settlers 71
mortality rate 174
vegetation 40
Tidewater Region 35, 40, 164
tobacco 6, 7, 88, 157, 172, 244
air-cured Burley 145
antebellum economy 77
cultivation 143
English tax 5
export 65, 71, 97, 98
farm income 146, 154
farm size 143, 146
flue-cured 145, 146, 158
history 64
importance to Carolinas 146
mechanization 145
post-bellum production 86
quota 147
settlers 169
slaves 2, 66, 72, 78
soil erosion 47
South Carolina 87
surpassed by poulty 149
tenant system 86
Virginia economy 69
tobacco, bright 87
tobacco industry 72, 87, 88, 121, 174
Coastal Plain 158
Concord 112
decline 101, 102, 109, 115
labor 227

litigation 122
manufacturing 121
Mountains 155
ownership 122
Piedmont 131
pollution 143
Triad 121
wages 243
Tobaccoville 122
Topsail 191
tornadoes 36
tourism 10, 187, 188, 189, 191, 192, 195, 197, 200
South Carolina 135, 190, 191, 193, 204
tourists 6, 187, 193, 195, 196, 198, 200, 229
Trail of Tears 10
transportation system 75, 78
Treaty of Holston 58
Triangle 112, 131, 164, 182, 183, 227, 228, 229
employment 113
high technology 243
manufacturing 127
population 162
tourism 199
Triangle Transit Authority 116
Triassic Age 17, 22, 26
Triassic Basin 26
Trident Technical Community College 224
Trinity College 215
Tryon Palace, New Bern 194
tuberculosis 174
turkeys 6, 7, 58, 143, 149, 150, 151, 158, 204
Tuscarora 1, 56
two-year colleges 223, 224, 225, 226
Tyrrell County 162, 178, 179
Tyson 148, 150

U

U. S. Department of Agriculture Forest Service Research Station 228
U. S. Environmental Protection Agency 99, 229
U.S. Open 198
U.S.S. North Carolina 194
U.S. Senior Open 198
U.S. Women's Open 198
ultisols 40
Unaka Mountains 27
Unemployment 125
Unifi Corporation 120
Union County 115, 151, 164
United Parcel Service 118, 119
university 163, 215, 220, 227, 230, 231, 247
University City 115
University of North Carolina at Asheville 224
University of North Carolina at Chapel Hill 132, 162, 208, 215, 216, 220, 227, 241
University of North Carolina at Charlotte 217, 229
University of North Carolina at Pembroke 219
University of South Carolina (Columbia) 126, 163, 207, 215, 216, 217, 220, 223, 224, 231
University of South Carolina at Aiken 219
University of South Carolina Upstate 219
University Research Park 229, 230, 231
Unto These Hills 10, 197
Up Country 7
Upper Floridian Aquifer 193
Upper Piedmont Manufacturing Region (UPMR) 98
UPS. *See* United Parcel Service
urban-industrial complex 165
urban area 112, 113
commuting 114
ethnic diversity 161
manufacturing decline 105
municipal governance 112
NASCAR 206
population 63, 89, 135, 171, 183
urban centers 175, 188, 242
urban counties 134
urban development 46, 95, 98, 106, 109, 133
urban growth 89, 99, 111
urbanization 111, 175
urban periphery 106
urban renewal 113, 131
urban sprawl 137, 179
US Airways 117, 118, 129
Uwharrie Mountains 25

V

Vanderbilt, George 197
vegetation 6, 21, 26, 35, 39, 40, 42, 43, 44, 45, 47, 48, 50
vegetation patterns 40
Verbatim 231
Verizon Heritage Golf Tournament 192
Verrazzano, Giovanni 1, 57
vertically integrated 148, 149, 150
Vespucci, Juan 58
Vietnamese 166
Virginia 60
Virginia Dare 2
Volvo Trucks, N.A. 128
Vought-Alenia 119

W

Wachovia Bank 130, 227
Wadesboro Basin 26
Wake County 72
Wake Forest University 8, 122, 208, 215, 216, 221
Wake Technical Community College 224, 226
Wal-Mart FLW Tour 202
Wampler-Longacre 150
Warsaw 150, 151
Watauga County 163
Wateree/Catawba River 56
Wateree River 56, 69, 79
water power 88, 98, 171

Wayne County 151, 209, 223
Weldon 97
well-being 111, 161, 207
Western North Carolina Railroad 79
Whitaker Park 122
White, John 2, 57, 60, 169
white-tailed deer 193
White Lake 21
Whiteville 131
Whitewater Falls 195
Whitney, Eli 5, 73, 241
wholesale trade 8, 244
Wilkesboro 147, 207
Wilkes County 149, 151, 177, 178
  blacks 177
  Civil War 177
  population 177
  slaves 177
Williams, David 81
Williamsburg 87, 106
Wilmington
  airport 119
  beaches 19
  Civil War 83
  colleges 220, 223, 224
  development 162
  economy 119
  export of tar 66
  growth 95
  mortality rate 174
  population 89, 96, 107, 161, 164, 171
  railroad 79
  Revolutionary War 5
  seaport 65, 77
  settlement 5, 65
  slave economy 179
  tourism 194
  trade 65, 79
  whites 165
Wilmington and Manchester Railroad 79
Wilmington-Weldon and Raleigh Railroad 78
Wilmington-Weldon Railroad 78, 90
Wilson 101, 131, 146, 227, 228
Winston 7, 88, 89
Winston-Salem
  apparel industry 121
  banks 130, 131
  cigarette industry 98
  colleges 215, 218, 220
  commuters 114
  manufacturing 8, 128
  School of the Arts 8
  textile industry 122
  tobacco industry 101, 121, 123
  tourism 194
  urban growth 122
Winston-Salem State University 218
Winston Cup 128, 207
Winthrop University 219
Winyah Bay 2, 58, 59, 67, 70
Woolworth's Department Store 241
Wrightsville Beach 191, 206

X

xerophytes 45

Y

Yadkin/Pee Dee River 29
Yadkin River 69
York County 101, 107, 115
Yorktown 5, 71